VLSI
TECHNOLOGY

VLSI TECHNOLOGY

Editor-in-Chief
Wai-Kai Chen

CRC Press
Taylor & Francis Group
Boca Raton London New York

CRC Press is an imprint of the
Taylor & Francis Group, an **informa** business

Editor-in-Chief

Wai-Kai Chen is Professor and Head Emeritus of the Department of Electrical Engineering and Computer Science at the University of Illinois at Chicago. He is now serving as Academic Vice President at International Technological University. He received his B.S. and M.S. in electrical engineering at Ohio University, where he was later recognized as a Distinguished Professor. He earned his Ph.D. in electrical engineering at University of Illinois at Urbana/Champaign.

Professor Chen has extensive experience in education and industry and is very active professionally in the fields of circuits and systems. He has served as visiting professor at Purdue University, University of Hawaii at Manoa, and Chuo University in Tokyo, Japan. He was editor of the *IEEE Transactions on Circuits and Systems, Series I and II*, president of the IEEE Circuits and Systems Society, and is the founding editor and editor-in-chief of the *Journal of Circuits, Systems and Computers*. He received the *Lester R. Ford Award* from the Mathematical Association of America, the *Alexander von Humboldt Award* from Germany, the *JSPS Fellowship Award* from Japan Society for the Promotion of Science, the *Ohio University Alumni Medal of Merit for Distinguished Achievement in Engineering Education*, the *Senior University Scholar Award* and the *2000 Faculty Research Award* from the University of Illinois at Chicago, and the *Distinguished Alumnus Award* from the University of Illinois at Urbana/Champaign. He is the recipient of the *Golden Jubilee Medal*, the *Education Award*, and the *Meritorious Service Award* from IEEE Circuits and Systems Society, and the *Third Millennium Medal* from the IEEE. He has also received more than a dozen honorary professorship awards from major institutions in China.

A fellow of the Institute of Electrical and Electronics Engineers and the American Association for the Advancement of Science, Professor Chen is widely known in the profession for his *Applied Graph Theory* (North-Holland), *Theory and Design of Broadband Matching Networks* (Pergamon Press), *Active Network and Feedback Amplifier Theory* (McGraw-Hill), *Linear Networks and Systems* (Brooks/Cole), *Passive and Active Filters: Theory and Implements* (John Wiley & Sons), *Theory of Nets: Flows in Networks* (Wiley-Interscience), and *The Circuits and Filters Handbook* and *The VLSI Handbook* (CRC Press).

Contributors

Victor Boyadzhyan
Jet Propulsion Laboratory
Pasadena, California

Charles E. Chang
Conexant Systems, Inc.
Newbury Park, California

Wai-Kai Chen
University of Illinois
Chicago, Illinois

John Choma, Jr.
University of Southern California
Los Angeles, California

John D. Cressler
Auburn University
Auburn, Alabama

Sorin Cristoloveanu
Institut National Polytechnique de Grenoble
Grenoble, France

Donald B. Estreich
Hewlett-Parkard Company
Santa Rosa, California

Thad Gabara
Lucent Technologies
Murray Hill, New Jersey

Jan V. Grahn
Royal Institute of Technology
Kista-Stockholm, Sweden

Peter J. Hesketh
The Georgia Institute of Technology
Atlanta, Georgia

Karl Hess
University of Illinois
Urbana, Illinois

Kazumi Inoh
Toshiba Corporation
Isogo-ku, Yokohama, Japan

Hidemi Ishiuchi
Toshiba Corporation
Isogo-ku, Yokohama, Japan

Hiroshi Iwai
Toshiba Corporation
Isogo-ku, Yokohama, Japan

Yasuhiro Katsumata
Toshiba Corporation
Isogo-ku, Yokohama, Japan

Pankaj Khandelwal
University of Illinois
Chicago, Illinois

Hideki Kimijima
Toshiba Corporation
Isogo-ku, Yokohama, Japan

Isik C. Kizilyalli
Lucent Bell Laboratories
Orlando, Florida

Stephen I. Long
University of California
Santa Barbara, California

Ashraf Lotfi
Lucent Technologies
Murray Hill, New Jersey

Joseph W. Lyding
University of Illinois
Urbana, Illinois

Samuel S. Martin
Lucent Technologies
Murray Hill, New Jersey

Erik A. McShane
University of Illinois
Chicago, Illinois

Hisayo S. Momose
Toshiba Corporation
Isogo-ku, Yokohama, Japan

Eiji Morifuji
Toshiba Corporation
Isogo-ku, Yokohama, Japan

Toyota Morimoto
Toshiba Corporation
Isogo-ku, Yokohama, Japan

Akio Nakagawa
Toshiba Corporation
Saiwai-ku, Kawasaki, Japan

Philip G. Neudeck
NASA Glenn Research Center
Cleveland, Ohio

Kwok Ng
Lucent Technologies
Murray Hill, New Jersey

Hideaki Nii
Toshiba Corporation
Isogo-ku, Yokohama, Japan

Tatsuya Ohguro
Toshiba Corporation
Isogo-ku, Yokohama, Japan

Mikael Östling
Royal Institute of Technology
Kista-Stockholm, Sweden

Krishna Shenai
University of Illinois
Chicago, Illinois

Meera Venkataraman
Troika Networks, Inc.
Calabasas Hills, California

Kuniyoshi Yoshikawa
Toshiba Corporation
Isogo-ku, Yokohama, Japan

Takashi Yoshitomi
Toshiba Corporation
Isogo-ku, Yokohama, Japan

Contents

1

VLSI Technology: A System Perspective

Krishna Shenai
Erik A. McShane
University of Illinois at Chicago

1.1   Introduction

The development of VLSI systems has historically progressed hand-in-hand with technology innova-tions. Often, fresh achievements in lithography, or semiconductor devices, or metallization have led to the introduction of new products. Conversely, market demand for particular products or specifi-cations has greatly influenced focused research into the technology capabilities necessary to deliver the product. Many conventional VLSI systems as a result have engendered highly specialized technol-ogies for their support.

In contrast, a characteristic of emerging VLSI products is the integration of diverse systems, each of which previously required a unique technology, into a single technology platform. The driving force behind this trend is the demand in consumer and noncommercial sectors for compact, portable, wireless electronics products — the nascent "system-on-a-chip" era.[1-4] Figure 1.1 illustrates some of the system components playing a role in this development.

Most of the achievements in dense systems integration have derived from scaling in silicon VLSI processes.[5] As manufacturing has improved, it has become more cost-effective in many applications to replace a chip set with a monolithic IC: packaging costs are decreased, interconnect paths shrink, and power loss in I/O drivers is reduced. Further scaling to deep submicron dimensions will continue to widen the applications of VLSI system integration, but also will lead to additional complexities in reliability, interconnect, and lithography.[6] This evolution is raising questions over the optimal level of integration: package level or chip level. Each has distinct advantages and some critical deficiencies for cost, reliability, and performance.

Board-level interconnection of chip sets, although a mainstay of low-cost, high-volume manufacturing, cannot provide a suitably dense integration of high-performance, core VLSI systems. Package- and chip-level integration are more practical contenders for VLSI systems implementation because of their compact dimensions and short signal interconnects. They also offer a tradeoff between dense monolithic integra-tion and application-specific technology optimization. It is unclear at this time of the pace in the further

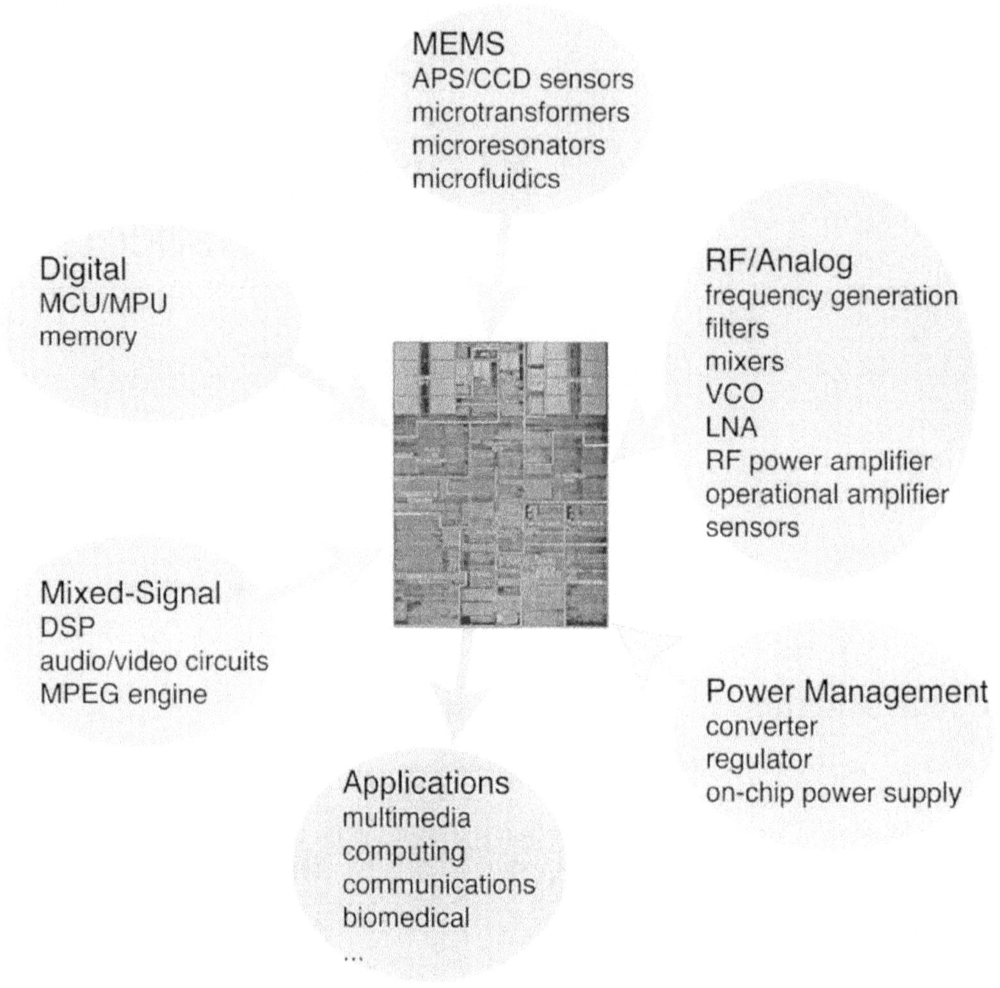

MEMS
APS/CCD sensors
microtransformers
microresonators
microfluidics

Digital
MCU/MPU
memory

RF/Analog
frequency generation
filters
mixers
VCO
LNA
RF power amplifier
operational amplifier
sensors

Mixed-Signal
DSP
audio/video circuits
MPEG engine

Power Management
converter
regulator
on-chip power supply

Applications
multimedia
computing
communications
biomedical
...

FIGURE 1.1   These system components are representative of the essential building blocks in VLSI "systems-on-a-chip."

evolution of VLSI systems, although systems integration will continue to influence and be influenced by technology development.

The remainder of this chapter will trace the inter-relationship of technology and systems to date and then outline emerging and future VLSI systems and their technology requisites. Alternative technologies will also be introduced with a presentation of their potential impact on VLSI systems. Focused discussion of the specific VLSI technologies introduced will follow in later chapters.

Given the level of systems integration afforded by available technology and the diverse signal-processing capabilities and applications supported, in this chapter a "VLSI system" is loosely defined as any complex system, primarily electronic in nature, based on semiconductor manufacturing with an extremely dense integration of minimal processing elements (e.g., transistors) and packaged as a single- or multi-chip module.

1.2   Contemporary VLSI Systems

VLSI systems can be crudely categorized by the nature of the signal processing they perform: analog, digital, or power. Included in analog are high-frequency systems, but they can be distinguished both by design methodology and their sensitivity to frequency-dependent characteristics in biasing and operation. Digital systems consist of logic circuits and memory, although it should be noted that most "digital" systems now also contain significant analog subsystems for data conversion and signal integrity. Power semiconductor devices have previously afforded only very low levels of integration considering their extreme current- and voltage-handling requirements (up to 1000 A and 10 kV) and resulting high temperatures. However, with the advent of hybrid technologies (integrating different materials on a single silicon substrate), partial insulating substrates (with dielectrically isolated regions for power semiconductor devices), and MCM packaging, integrated "smart" power electronics are appearing for medium power (up to 1 kW) applications. A relative newcomer to the VLSI arena is microelectromechanical systems (MEMS). As the name states, MEMS is not purely electronic in nature and is now frequently extended to also label systems that are based on optoelectronics, biochemistry, and electromagnetics.

Digital Systems

Introduction

The digital systems category comprises microprocessors, microcontrollers, specialized digital signal processors, and solid-state memory. As mentioned previously, these systems may also contain analog, power, RF, and MEMS subsystems; but in this section, discussion is restricted to digital electronics.

Beginning with the introduction in 1971 of the first true microprocessor — the Intel 4004 — digital logic ICs have offered increasing functionality afforded by a number of technology factors. Transistor miniaturization from the 10-micron dimensions common 30 years ago to state-of-the-art 0.25-micron lithography has boosted IC device counts to over 10 million transistors. To support subsystem interconnection, multilevel metallization stacks have evolved. And, to reduce static and switching power losses, low-power/low-voltage technologies have become standard. The following discussion of VLSI technology pertains to the key metrics in digital systems: power dissipation, signal delay, signal integrity, and memory integration.

Power Dissipation

The premier technology today for digital systems is CMOS, owing to its inherent low-power attributes and excellent scaling to deep submicron dimensions. Total power dissipation is expressed as

$$
\begin{aligned}
P &= P_{dynamic} + P_{static} \\
&= (P_{switching} + P_{short\text{-}circuit} + P_{leakage}) \\
&= V_{DD}^2 f \sum_n a_n c_n + V_{DD} \sum_n i_{sc_n} + V_{DD} \sum_n (1 - a_n) i_{leak_n}
\end{aligned}
\tag{1.1}
$$

where V_{DD} is the operating supply; f is the clock frequency; per node a_n is the switching activity; c_n is the switching capacitance; i_{scn} is the short-circuit current; and i_{leakn} is the leakage current (subthreshold conduction and junction leakage). From this expression it is apparent that the most significant reduction in power dissipation can be accomplished by scaling the operating supply. However, as V_{DD} is reduced to 1 V, the contribution of leakage current to overall power dissipation increases if transistor V_T is scaled proportionally to V_{DD}. Subthreshold current in bulk CMOS, neglecting junction leakage and body effects, can be expressed as[7]

$$
I_{sub} = \frac{W}{L} I_0 e^{\frac{V_{GS} - V_T}{n\phi_t}} \left(1 - e^{\frac{-V_{DS}}{\phi_t}} \right)
\tag{1.2}
$$

where

$$
I_0 = k'(n - 1)\phi_t^2
\tag{1.3}
$$

$$n = 1 + \frac{\gamma}{2\sqrt{\phi_t}} \tag{1.4}$$

$$k' = \mu C'_{ox} \tag{1.5}$$

$$\gamma = \frac{\sqrt{2q\varepsilon_S N_B}}{C'_{ox}} \tag{1.6}$$

$$C'_{ox} = \frac{\varepsilon_{ox}}{t_{ox}} \tag{1.7}$$

W and L are channel width and length, respectively; ϕ_t is thermal voltage (approximately 0.259 V at 300 K); μ is carrier mobility in the channel; ε_{ox} is gate dielectric permittivity (3.45×10^{-13} F/cm for SiO_2); ε_S is semiconductor permittivity (1.04×10^{-12} F/cm for Si); N_B is bulk doping; and t_{ox} is gate dielectric thickness. This trend is exacerbated if minimal-switching circuit techniques are employed or if sleep modes place the logic into idle states for long periods. Device scaling thus must consider the architecture and performance requirements.

Figures 1.2 and 1.3 show the inverse normalized energy-delay product (EDP) contours for a hypothetical 0.25-micron device.[8] The energy required per operation is

$$E = \frac{P}{f} \tag{1.8}$$

Normalization is performed relative to the best obtained EDP for this technology. Fig. 1.2 shows data for an ideal device and Fig. 1.3 adds non-idealities by considering velocity saturation effects and uncertainty in V_{DD}, V_T, and temperature T. In the ideal device, the dashed lines indicate vectors of normalized constant performance relative to the performance obtained at the optimal EDP point. The switching frequency can be approximated by

$$f = \frac{1}{t_{rise}} = \frac{1}{t_{fall}}$$
$$= \frac{I_{Dsat}}{C V_{DD}} \tag{1.9}$$

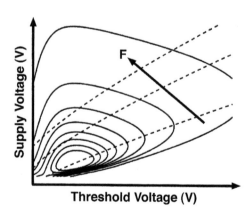

FIGURE 1.2   Inverse normalized EDP contours for an ideal device (after Ref. 8). Dashed lines indicate vectors of constant performance. Arrow *F* shows direction of increasing performance.

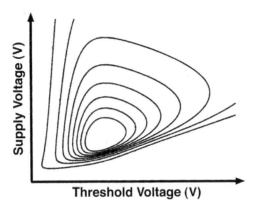

FIGURE 1.3 Inverse normalized EDP contours for a non-ideal device considering velocity saturation and uncertainty in V_{DD}, V_T, and temperature T (after Ref. 8).

and the performance, F, when considered as proportional to f, can be expressed as

$$f = \alpha \frac{(V_{DD} - V_T)^2}{V_{DD}} \tag{1.10}$$

where scaling factor α is applied to normalize performance. These plots illustrate the tradeoffs in optimizing system performance for low-power requirements and highest performance. Frequency can also be scaled to reduce power dissipation, but this is not considered here as it generally also degrades performance.

Considering purely dynamic power losses (CV^2f), scaling the operating supply again yields the most significant reduction; but this scaling also affects the subthreshold leakage since V_T must be scaled similarly to maintain comparable performance levels (see Eq. 1.10). In this respect, fully depleted SOI CMOS offers improved low-voltage, low-power characteristics as it has a steeper subthreshold slope than bulk CMOS. Subthreshold slope, S, is defined as

$$S = \frac{dV_G}{d(\log I_D)} \tag{1.11}$$

This can be expressed (from Ref. 9) for bulk CMOS as

$$S = \frac{\hat{k}T}{q} \ln(10) \left(1 + \frac{C_D}{C_{ox}}\right) \tag{1.12}$$

and for fully depleted SOI CMOS (assuming negligible interface states and buried-oxide capacitance) as

$$S = \frac{\hat{k}T}{q} \ln(10) \tag{1.13}$$

where $\hat{k}$ is Boltzmann's constant (1.38×10^{-23} V·C/K), C_D is depletion capacitance, and C_{ox} is gate dielectric capacitance. Hence, for the same weak inversion gate bias, SOI CMOS can yield a leakage current several orders of magnitude less than in bulk CMOS.

Additional power dissipation occurs in the extrinsic parasitics of the active devices and the interconnect. This contribution can be minimized by salicide (self-aligned silicide) processes that deposit a low sheet resistance layer on the source, drain, and gate surfaces.

Switching Frequency and Signal Integrity

After power dissipation, the signal delay (or maximum switching frequency) of a system is the most important figure-of-merit. This characteristic, as mentioned previously, provides a first-order approximation of system performance. It also affects the short-circuit contribution to power loss since a dc path between the supply rails exists during a switching event. Also, signal delay and slope determine the deviation of a logic signal pulse from an ideal step transition.

Digital systems based on silicon bipolar and BiCMOS technologies still appear for high-speed applications, exploiting the higher small-signal gain and greater current drive of bipolar transistors over MOSFETs; but given the stringent power requirements of portable electronics, non-CMOS implementations are impractical. Emerging technologies such as silicon heterojunction bipolar and field-effect transistors (HBTs and HFETs) hold some promise of fast switching with reduced power dissipation, but the technology is too immature to be evaluated as yet.

The switching rate of a capacitively loaded node in a logic circuit can be approximated by the time required for the capacitor to be fully charged or discharged, assuming that a constant current is available for charge transport (see Eq. 1.9). Neglecting channel-length modulation effects on saturation current, the switching frequency can be written as

$$
\begin{aligned}
f &= \frac{I_{Dsat}}{CV_{DD}} \\
&= \frac{\frac{k'}{2}\frac{W}{L}(V_{GS} - V_T)^2}{CV_{DD}} \\
&\propto \frac{(V_{DD} - V_T)^2}{CV_{DD}}
\end{aligned}
\tag{1.14}
$$

Voltage scaling and its effects on power dissipation have already been discussed. Considering the capacitive contribution, a linear improvement to switching speed can be obtained by scaling node capacitance. Referring to Fig. 1.4 and neglecting interconnect capacitance, the node capacitance of a MOSFET can be expressed as

$$
C_{OUT} = C_{GD} + \kappa(V_{OL}, V_{OH})(C_{db})
\tag{1.15}
$$

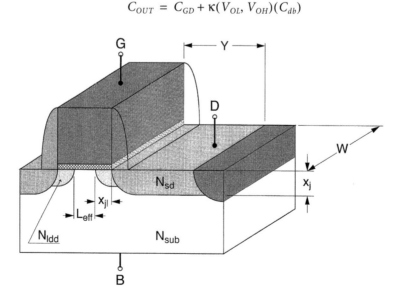

FIGURE 1.4   A MOSFET isometric cross-sectional view with critical dimensions identified.

$$C_{GD} = x_{jl}C_{ox}W + \frac{1}{2}C_{ox}WL_{eff} \tag{1.16}$$

$$C_{db} = C_{j0}WY + 2C_{jsw}(W + Y) \tag{1.17}$$

$$C_{j0} = \sqrt{\frac{q\varepsilon_{Si}}{2\left(\dfrac{1}{N_{sd}} + \dfrac{1}{N_{sub}}\right)\phi}} \tag{1.18}$$

$$C_{jsw} \approx C_{j0}x_j \tag{1.19}$$

The drain-to-body junction capacitance C_{db} is bias dependent, and the scaling factor κ is included to determine an average value of output voltage level. Source/drain diffusion capacitance has two components: the bottom areal capacitance C_{j0} and the sidewall perimeter capacitance C_{jsw}. Although C_{jsw} is a complex function of doping profile and should account for high-concentration channel-stop implants, an approximation is made to equate C_{jsw} and C_{j0}. From Fig. 1.5, it is clear that SOI CMOS has greatly reduced device capacitances compared to bulk CMOS by the elimination junction areal and perimeter capacitances. Another technique in SOI CMOS for improving switching delay involves dynamic threshold voltage control (DTMOS) by taking advantage of the parasitic lateral bipolar transistor inherent in the device structure.[10]

To reduce interconnect resistance, copper interconnect has been introduced to replace traditional aluminum wires.[11] Table 1.1 compares two critical parameters. The higher melting point of copper also reduces long-term interconnect degradation from electromigration, in which energetic carriers dislodge metal atoms creating voids or nonuniformities. Interconnect capacitance relative to the substrate is determined by the dielectric constant, ε_r, and the signal velocity can be defined as

$$v = \frac{c}{\sqrt{\varepsilon_r}} \tag{1.20}$$

Low-ε_r interlevel dielectrics are appearing to reduce this parasitic effect.

TABLE 1.1 Comparison of Interconnect Characteristics for Al and Cu

Material	Specific Resistance ($\mu\Omega$-cm)	Melting Point (°C)
Al	2.66	660
Cu	1.68	1073

Memory Scaling

The two most critical factors determining the commercial viability of RAM products are the total power dissipation and the chip area. For implementations in battery-operated portable electronics, the goal is

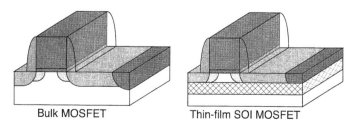

Bulk MOSFET   Thin-film SOI MOSFET

FIGURE 1.5 Cross-sectional views of a MOSFET: bulk and thin-film SOI.

a 0.9-V operating supply — the minimum voltage of a NiCd cell. RAM designs are addressing these objectives architecturally and technologically. SRAMs and DRAMs share many architectural features, including memory array partitioning, reduced voltage internal logic, and dynamic threshold voltage control. DRAM, with its higher memory density, is more attractive for embedded memory applications despite its higher power dissipation.

Figure 1.6 shows a RAM block diagram that identifies the sources of power dissipation. The power equation as given by Itoh et al.[12] is

$$P = I_{DD}V_{DD} \tag{1.21}$$

$$I_{DD} = mi_{act} + m(n-1)i_{hld} + (n+m)C_{DE}V_{INT}f + C_{PT}V_{INT}f + I_{DCP} \tag{1.22}$$

where i_{act} is the effective current in active cells, i_{hld} is the holding current in inactive cells, C_{DE} is the decoder output capacitance, C_{PT} is the peripheral circuit capacitance, V_{INT} is the internal voltage level, I_{DCP} is the static current in the peripheral circuits, and n and m define the memory array dimensions.

In present DRAMs, power loss is dominated by i_{act}, the charging current of an active subarray; but as V_T is scaled along with the operating voltage, the subthreshold current begins to dominate. The trend in DRAM ICs (see Fig. 1.7) shows that the dc current will begin to dominate the total active current at about the 1-Gb range. Limiting this and other short-channel effects is necessary then to improve power efficiency.

Figure 1.8 shows trends in device parameters. A substrate doping of over 10^{18} cm^{-3} is necessary to reduce SCE, but this has the disadvantage of also increasing junction leakage currents. To achieve reduced SCE at lower substrate dopings, shallow junctions (as thin as 15 nm) are formed.[13]

Bit storage capacitors must also be scaled to match device miniaturization but still retain adequate noise tolerance. Alpha-particle irradiation becomes less significant as devices are scaled, due to the reduced depletion region; but leakage currents still place a minimum requirement on bit charge.

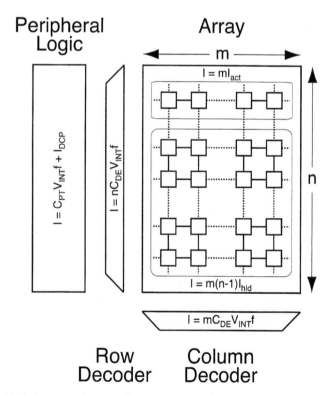

FIGURE 1.6 RAM block diagram indicating effective currents within each subsystem.

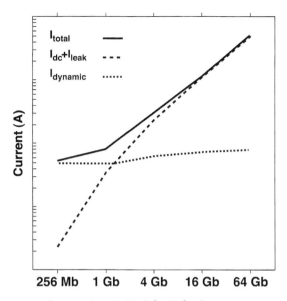

FIGURE 1.7   Contributions to total current in DRAMs (after Ref. 12).

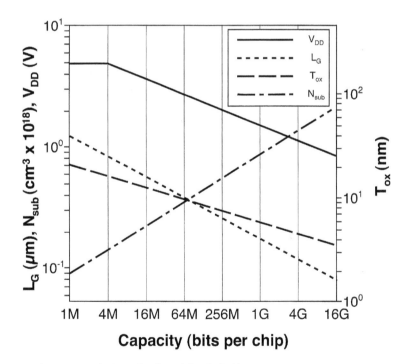

FIGURE 1.8   Trends in DRAM device technology (after Ref. 13).

Figure 1.9 shows that required signal charge, Q_S, has reduced only slightly with increased memory capacity, but cell areas have shrunk considerably. High-permittivity (high-ε_r) dielectrics such as Ta_2O_5 and BST ($Ba_xSr_{1-x}TiO_3$) are required to provide these greater areal capacitances at. reduced dimensions.[14] Table 1.2 lists material properties for some of the common and emerging dielectrics. In addition to scaling the cell area, the capacitor aspect ratio also affects manufacturing: larger aspect ratios result in non-planar interlevel dielectric and large step height variation between memory arrays and peripheral circuitry.

TABLE 1.2 Comparison of High-Permittivity Constant Materials for DRAM Cell Capacitors

Material	Dielectric Constant	Minimum Equivalent Oxide Thickness (nm)
NO	7	3.5 to 4
Ta_2O_5	20–25	2 to 3
BST	200–400	?

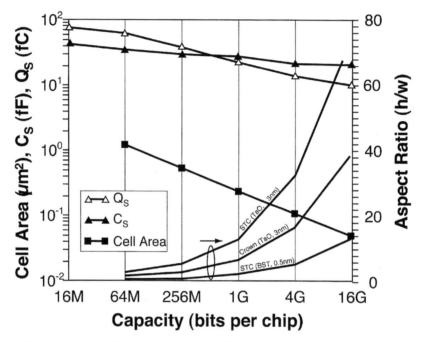

FIGURE 1.9   Trends in DRAM cell characteristics (after Ref. 13).

Analog Systems

Introduction

An analog system is any system that processes a signal's magnitude and phase information by a linearized response of an active device to a small-signal input. Unlike digital signals, which exhibit a large output signal swing, analog systems rely on a sufficiently small signal gain that the linear approximation holds true across the entire spectrum of expected input signal frequencies. Errors in the linear model are introduced by random process variation, intrinsic device noise, ambient noise, and non-idealities in active and passive electronics. Minimizing the cumulative effects of these "noise" contributions is the fundamental objective of analog and RF design.

Reflecting the multitude of permutations in input/output specifications and operating conditions, analog/RF design is supported by numerous VLSI technologies. Key among these are silicon MOST, BJT, and BiCMOS for low-frequency applications; silicon BJT for high-frequency, low-noise applications; and GaAs MESFET for high-frequency, high-efficiency amplifiers. Newcomers to the field include GaAs and SiGe heterojunction bipolar junction transistors (HBTs). The bandgap engineering employed in their fabrication results in devices with significantly higher f_T and f_{max} than in conventional devices, often at lower voltages.[15–17] Finally, MEMS resonators and mechanical switches offer an alternative to active device implementations.

The most familiar application of a high-frequency system is in wireless communications, in which a translation is performed between the high-frequency modulated carrier (RF signal) used for broadcasting and the low-frequency demodulated signal (baseband) suitable for audio or machine interpretation.

Wireless ICs long relied on package-level integration and scaling to deliver compact size and improved efficiency. Also, low-cost commercial IC technologies previously could not deliver the necessary frequency range and noise characteristics. This capability is now changing with several candidate technologies at hand for monolithic IC integration. CMOS has the attractive advantage of being optimal for integration of low-power baseband processing.

Amplifiers

Amplifiers boost the amplitude or power of an analog signal to suppress noise or overcome losses and enable further processing. Typical characteristics include a low noise figure (NF), large (selectable) gain (G), good linearity, and high power-added efficiency (PAE). To accommodate the variety of signal frequencies and performance requirements, several amplifier categories have evolved. These include conventional single-ended, differential, and operational amplifiers at lower frequencies and, at higher frequencies, low-noise and RF power amplifiers.

A challenge in technology scaling is providing a suitable signal-to-noise ratio and adequate biasing at a reduced operating supply. For a fixed gain, reducing the operating supply implies a similar scaling of the input signal level, ultimately approaching the noise floor of the system and leading to greater susceptibility to internal and external noise sources. Large-signal amplifiers (e.g., RF power amplifiers) that exhibit a wide output swing face similar problems with linearity at a lower operating supply.

A low-noise amplifier (LNA) is the first active circuit in a receiver. A common-source configuration of a MOSFET LNA is shown in Fig. 1.10. The input network is typically matched for lowest NF, and the output network is matched for maximum power transfer. Input impedance is matched to the source resistance, R_s, when[18]

$$\omega_0^2(L_1 + L_2)C_{gs} = 1 \tag{1.23}$$

$$\frac{g_m L_1}{C_{gs}} = R_s \tag{1.24}$$

The gain from the input matching network to the transistor gate-source voltage is equal to Q, the quality factor

$$Q = \frac{1}{g_m \omega_0 L_1} \tag{1.25}$$

where ω_0 is the RF frequency. If only the device current noise is considered, then the LNA noise figure can be expressed as

$$NF = 1 + \frac{2}{3}\frac{\omega_0 L_1}{QR_s} \tag{1.26}$$

It is observed that a larger quality factor yields a lower noise figure, but current industry practice selects an LNA Q of 2 to 3 since increasing Q also increases the sensitivity of the LNA gain to tolerances in the passive components. By combining Eqs. 1.24 and 1.25, the device input capacitance C_{gs} can be defined

$$C_{gs} = \frac{1}{R_s Q \omega_0} \tag{1.27}$$

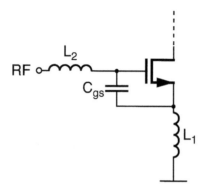

FIGURE 1.10 Common-source LNA circuit schematic.

Assuming that the transistor is in the saturation region and that Miller feedback gain is −1, the contributions to the input capacitance are

$$C = \left(\frac{2}{3}C'_{ox}L_{eff} + C_{gso} + 2C_{gdo}\right)W \tag{1.28}$$

where C_{gso} and C_{gdo} are, respectively, the gate-source and gate-drain overlap capacitances. The bias current (assuming a reasonable value for g_m) can then be obtained from

$$I_{Dsat} = \left(\frac{g_m^2}{2W}\right)\frac{L_{eff}}{k'} \tag{1.29}$$

As device dimensions are reduced, the required biasing current drops. Since cutoff frequency, f_T, also improves with smaller device dimensions, MOSFET performance in RF applications will continue to improve.

Power amplifiers, the last active circuit in a transmitter, have less stringent noise figure requirements than an LNA since the input signal is generated locally in the transmitter chain. Instead, linearity and PAE are more critical, particularly for variable-envelope communications protocols. RF amplifiers typically operate in class AB mode to compromise between efficiency and linearity.[19] Power-added efficiency is defined as

$$PAE = \frac{\eta}{1 - \frac{1}{G}} \tag{1.30}$$

where η is the drain (collector) efficiency (usually about 40 to 75%) and G is the amplifier power gain. This balance is highly sensitive to the precision of matching networks. Technologies such as GaAs, with its high-resistivity substrate, and SOI, with its insulating buried oxide, are best suited for integrated RF power amplifiers since they permit fabrication of low-loss, on-chip matching networks.

Interconnects and Passive Components

Passive components in analog and RF design have the essential role of providing biasing, energy storage, and signal level translation. As device technology has permitted a greater monolithic integration of active devices, a similar trend has appeared in passive components. The quality of on-chip passives, however, has lagged behind that of high-precision discrete components. Two characteristics are required of VLSI interconnects for RFICs: low-loss and integration of high-quality factor passives (capacitors and inductors). As discussed previously, resistive losses increase the overall noise figure, lead to decreased efficiency, and degrade the performance of on-chip passive components. Interconnect and device resistance are minimized by saliciding the gate and source/drain surfaces and appropriately scaling the metallization dimensions. Substrate coupling losses, which also degrade quality factors of integrated passives and can introduce substrate noise, are controlled by selecting a high-resistivity substrate such as GaAs or shielding the substrate with an insulating layer such as in SOI.

In forming capacitors on-chip, two structures are available, using either interconnect layers or the MOS gate capacitance. Metal-insulator-metal (MIM) and dual-poly capacitors both derive a capacitance from a thin interlevel dielectric (ILD) layer deposited between the conducting plates. MIM capacitors offer a higher Q than dual-poly capacitors since, even with silicidation, resistance of poly layers is higher than in metal. Both types can suffer from imprecision caused by non-planarity in the ILD thickness caused by process non-uniformity across the wafer.

MOS gate capacitance is less subject to variation caused by dielectric non-uniformity since the gate oxide formation is tightly controlled and occurs before any back-end processing. MOS capacitors, however, are usually dismissed for high-Q applications out of concern for the highly resistive well forming the bottom plate electrode. Recent work, however, has shown that salicided MOS capacitors biased into strong inversion will achieve a Q of over 100 for applications in the range 900 MHz to 2 GHz.[20]

Inductors are essential elements of RFICs for biasing and matching, and on-chip integration translates to lower system cost and reduced effects from package parasitics. However, inductors also require a large die area and exhibit significant coupling losses with the substrate. In addition to degrading the inductor Q, substrate coupling results in the inductor becoming a source of substrate noise. The Q of an inductor can be defined[21] as

$$Q = \frac{\omega L_s}{R_s} \cdot \text{Substrate loss factor} \cdot \text{Self-resonance factor} \qquad (1.31)$$

where L_s is the nominal inductance and R_s is the series resistance. The substrate loss factor approaches unity as the substrate resistance goes to either zero or infinity. This implies that the Q factor is improved if the substrate is either short- or open-circuited. Suspended inductors achieve an open-circuited substrate by etching the bulk silicon from under the inductor structure.[22] Another approach has been to short-circuit the substrate by inserting grounding planes (ground shields).[21]

Power Systems

Introduction

Power processing systems are those devoted to the conditioning, regulation, conversion, and distribution of electrical power. Voltage and current are considered the inputs, and the system transforms the input characteristics to the form required by the load. The distinguishing feature of these systems is the specialized active device structure (e.g., rectifier, thyristor, power bipolar or MOS transistor, IGBT) required to withstand the electrothermal stresses imposed by the system. The label power integrated circuits (PICs) refers to the monolithic fabrication of a power semiconductor device along with standard VLSI electronics. At the system level, this integration has been made possible by digital control techniques and the development of mixed-signal ICs comprising analog sensing and digital logic. On the technology side, development of the power MOSFET and IGBT led to greatly simplified drive circuits and decreased complexity in the on-chip electronics.

Three types of PIC are identified: smart power, high-voltage ICs (HVICs), and discrete modules. Discrete modules are those in which individual ICs for power devices and control are packaged in a single carrier. Integration is at the package level rather than at the IC. Smart power adds a monolithic integration of analog protection circuitry to a standard power semiconductor device. The level of integration is quite low, but the power semiconductor device ratings are not disturbed by the other electronics.

HVICs are different in that they begin from a standard VLSI process and accommodate the power semiconductor device by manufacturing changes. HVICs are singled out for further discussion as they are the most suitable for VLSI integration. Although the power semiconductor device ratings cannot achieve the levels of a discrete device, HVICs are available for ratings with currents of 50 to 100 A and voltages up to 1000 V.

Two critical technical issues faced in developing HVICs are the electrical isolation of high-power and low-voltage electronics, and the development of high-Q passive components (e.g., capacitors, inductors, transformers). In the following discussion, the characteristics of power semiconductor devices will not be considered, only the issues relating to HVIC integration.

Electrical Isolation

Three types of electrical isolation are available as illustrated in Fig. 1.11.[23] In junction isolation, a p^+ implant is added to form protective diodes with the n^- epitaxial regions. The diodes are reverse-biased by applying a large negative voltage (~ -1000 V) to the substrate. Problems with this isolation include temperature-dependent diode leakage currents and the possibility of a dynamic turn-on of the diode. Additional stress to the isolation regions and interlevel dielectric is introduced when high-voltage interconnect crosses the isolation implants. The applied electric field in this situation can result in premature failure of the device.

A self-isolation technique can be chosen if all the devices are MOSFETs. When all devices are placed in individual wells (a twin-tub process), all channel regions are naturally isolated since current flow is near the oxide–semiconductor interface. The power semiconductor device and signal transistors are fabricated simultaneously in junction and self-isolation, resulting in a compromise in performance.[24] In practice, bulk isolation techniques are a combination of junction and self-isolation since many HVICs exhibit dynamic surges in substrate carriers, corresponding to power device switching, that may result in latchup of low-voltage devices.[25]

Dielectric isolation decouples the fabrication of power and signal devices by reserving the bulk semiconductor for high-voltage transistors and introducing an epitaxial semiconductor layer for low-voltage devices on a dielectric surface. Formation of the buried oxide can be accomplished either by partially etching an SOI wafer to yield an intermittent SOI substrate or by selectively growing oxide on regions intended for low-voltage devices, followed by epitaxial deposition of a silicon film. In addition to providing improved isolation of power semiconductor devices, the buried oxide enhances the performance of signal transistors by reducing parasitic capacitances and chip area.[26]

Interconnects and Passive Components

A key objective in applying VLSI technology to power electronics is the reduction in system mass and volume. Current device technologies adequately provide the monolithic cofabrication of low-voltage and high-voltage electronics, and capacitors and inductors of sufficiently high Q are available to integrate substantial peripheral circuitry. As switching frequencies are increased in switching converters, passive component values are reduced, further improving the integration of power electronics. These integration capabilities have resulted from similar requirements in digital and analog/RF electronics. Conventional VLSI technologies, however, have yet to reproduce the same integration of magnetic materials needed for transformers.

Magnetics integration is hindered by the incompatibility of magnetic materials with standard VLSI processes and concerns over contamination of devices. A transformer cross-section is shown in Fig. 1.12 where the primary and secondary are wound as lateral coils with connections between the upper and lower conductors provided by vias.[27]

1.3 Emerging VLSI Systems

Building on the successes in integrating comprehensive systems from the individual signal processing domains, many next-generation hybrid systems are appearing that integrate systems in a cross-platform manner. The feasibility of these systems depends on the ability of technology to either combine monolithically the unique elements of each system or to introduce a single technology standard that can support broad systems.

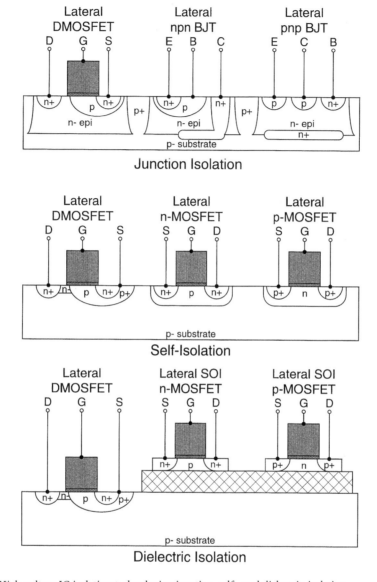

FIGURE 1.11 High-voltage IC isolation technologies: junction, self-, and dielectric-isolation.

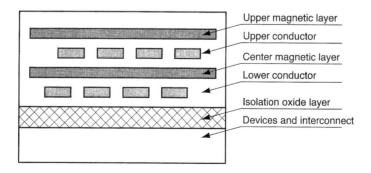

FIGURE 1.12 Cross-sectional view of a VLSI process showing integration of magnetic layers for coil transformers.

Embedded Memory

Computer processors today are bandwidth limited with the memory interface between high-capacity external storage and the processing units unable to meet access rates. Multimedia applications such as 3-D graphics rendering and broadcast rate video will demand bandwidths from 1 to 10 GB/s. A conventional 4-Mb DRAM with 4096 sense amplifiers, a 150-ns cycle time, and a configuration of 1-M × 4-b achieves an internal bandwidth of 3.4 GB/s; however, as this data can be accessed at the I/O pins only in 4-bit segments, external bandwidth is reduced to 0.1% of available bandwidth.

Although complex cache (SRAM) hierarchies have been devised to mask this latency, each additional cache level introduces additional complexity and has an asymptotic performance limit. Since the bandwidth bottleneck is introduced by off-chip (multiplexed) routing of signals, embedded memories are appearing to provide full-bandwidth memory accesses by eliminating I/O multiplexing.

Integration of DRAM and logic is non-trivial as the technology optimization of the former favors minimal bit area, a compact capacitor structure, and low leakage, but in the latter favors device performance. Embedded DRAM therefore permits two implementations: logic fabricated in a DRAM process or DRAM fabricated as a macro in an ASIC process.

Logic in DRAM Process

DRAM technologies do not require multilevel interconnect since their regular array structure translates to very uniform routing. Introducing logic to a DRAM process has been accomplished by merging DRAM front-end fabrication (to capture optimized capacitor and well structures) with an ASIC back-end process (to introduce triple- or quad-level metallization).

Example systems include a 576-Kb DRAM and 8-way hypercube processor[28] and a 1-Mb DRAM and 4-element pixel processor.[29]

DRAM Macro in ASIC Process

DRAM macros currently developed for instantiation in an ASIC process implement a substrate-plate-trench-capacitor (SPT) by modifying deep trench isolation (DTI). Access transistor isolation is accomplished with a triple-well process to extract stray carriers injected as substrate noise.

Fabrication of the stacked-capacitor (STC) structures typical of high-capacity DRAMs cannot be performed in standard ASIC processes without significant modification; this limitation forces additional tradeoffs in the optimal balance of embedded memory capacity and fabrication cost. Example systems include an 8-Mb DRAM and 70-K gate sea-of-gates IC.[30]

Monolithic RFICs

Technologies for monolithic RFICs are proceeding in two directions: all-CMOS and silicon bipolar. All-CMOS has the attractive advantages of ready integration of baseband analog and digital signal processing, compatibility with standard analog/RF CMOS processes, and better characteristics for low-power/low-voltage operation.[31] CMOS is predicted to continue to offer suitable RF performance at operating supplies below 1 V as long as the threshold voltage is scaled accordingly. Silicon bipolar implementations, however, outperform CMOS in several key areas, particularly that of the receiver LNA.

A frequency-hopped spread-spectrum transceiver has been designed in a 1-micron CMOS process with applications to low-power microcell communications.[32,33] A microcell application was chosen since the maximum output power of 20 mW limits the transmission range.

RFICs in bipolar and BiCMOS typically focus on the specific receiver components most improved by non-CMOS implementations. A common RFIC architecture is an LNA front-end followed by a down-mixer. Interstage filtering and matching, if required, are provided off-chip. In CMOS, a 1-GHz RFIC achieved a conversion gain of 20 dB, an NF of 3.2 dB, an IP3 of 8 dBm, with a current drain of 9 mA from a 3-V supply.[34] A similar architecture in BiCMOS at 1 GHz achieved a conversion gain of 16 dB, an NF of 2.2 dB, an IP3 of −10 dBm, with a current drain of 13 mA from a 5-V supply.[35]

Single-Chip Sensors and Detectors

One of the most active areas of system development is in the field of "camera-on-a-chip" in which VLSI-compatible photodetector arrays are monolithically fabricated with digital and analog peripheral circuitry. The older charge-coupled device (CCD) technology is being replaced by active-pixel sensing (APS) technologies in new systems since CCD has a higher per-cell capacitance.[36] APS also offers a monolithic integration of analog-to-digital converters and digital control logic. On-chip digitization of the sensor outputs eliminates the additional power loss and noise contributions incurred in buffering an analog signal for off-chip processing.

Conventional CMOS technology is expected to permit co-fabrication of APS sensors down to feature dimensions as small as 0.25 microns,[37] although leakage currents will become a concern for low-power operation. Other low-power techniques that degrade APS performance such as silicidation and thin-film devices can be controlled by slight process modifications. Silicide blocks eliminate the opaque, low-resistance layer; and thin-film devices, such as those found in fully depleted SOI, can be avoided by opening windows through the SOI-buried oxide to the bulk silicon.

MEMS

Electrothermal properties of MEMS suspended or cantilevered layers are finding applications in a variety of electromechanical systems. For example, suspended layers have been developed for a number of purposes, including microphones,[38] accelerometers, and pressure sensors. By sealing a fluid within a MEMS cavity, pressure sensors can also serve as infrared detectors and temperature sensors. Analog feedback electronics monitor the deflection of a MEMS layer caused by the influence of external stresses. The applied control voltage provides an analog readout of the relative magnitude of the external stresses.

MEMS micropumps have been developed which have potential applications in medical products (e.g., drug delivery) and automotive systems (e.g., fuel injection).[39] Flow rates of about 50 µl/min at 1-Hz cycling have been achieved, but the overall area required is still large (approximately 1 cm^2), presently limiting integration. Beyond macro applications, microfluidics has been proposed as a technique for integrated cooling of high-power and high-temperature ICs in which coolant is circulated within the substrate mass.

1.4 Alternative Technologies

The discussion so far has assumed a VLSI system to be comprised primarily of transistors, with the exception of non-electronic MEMS. Although this implementation has evolved unchallenged for nearly 50 years, several innovative technologies have progressed to the state that they are attracting serious attention as future competitors. Some, such as quantum computing, are still fundamentally electronic in nature. Others, like biological, DNA, and molecular computing, use living cells as elemental functional units.

The chief advantage of these technologies is the extreme power efficiency of a computation. Quantum computing achieves ultra-low-power operation by reducing logic operations to the change in an electron spin or an atomic ionization state. Biological and DNA computing exploit the energy efficiency of living cells, the product of a billion years of evolution. A second attribute is extremely fast computation owing to greatly reduced signal path lengths (to a molecular or atomic scale) and massively parallel (MP) simultaneous operations.

Quantum Computing

Quantum computing (QC) is concerned with the probabilistic nature of quantum states, by which a single atom can be used to "store" and "compare" multiple values simultaneously. QC has two distinct implementations: an electronic one based on the wave-function interaction of fixed adjacent atoms[40] and a biochemical one based on mobile molecular interactions within a fluid medium.[41] In fixed systems, atom placement and stimulation are accomplished with atomic force microscopy (AFM) and nuclear magnetic resonance (NMR), but performance as a system also requires the ability to individually select and operate on an atom.[42] Molecular systems avoid this issue by using a fluid medium as a method of introducing initial conditions and isolating the computation from the measurement. Computational redundancy then statistically removes measurement error and incorrect results generated at the fluid boundaries.

Recent work in algorithms has demonstrated that QC can solve two categories of problems more efficiently than with a classical computer science method by taking advantage of wave function indeterminate states. In search and factorization problems (involving a random search of N items), a classical solution requires $O(N)$ steps, but a QC algorithm requires only $O(\sqrt{N})$; and binary parity computations can be improved from $O(N)$ to $O(N/2)$ in a QC algorithm.[43]

Practical implementation of QC algorithms to very large data sets is currently limited by instrumentation. For example, a biochemical QC system using NMR to change spin polarity has signal frequencies of less than 1 kHz. Despite this low frequency, the available parallelism is expected to factor (with the $O(\sqrt{N})$ algorithm) a 400-digit number in one year: greater than 3×10^{186} MOPS (megaoperations per second).[44]

DNA Computing

In DNA computing, a problem set is encoded into DNA strands which then, by nucleotide matching properties, perform MP search and comparison operations. Most comparison techniques to date rely on conformal mapping, but some reports appearing in the literature indicate that more powerful DNA algorithms are possible with non-conformal mapping and secondary protein interaction.[45] The challenge is in developing a formal language to describe DNA computing compounded by accounting for these secondary and tertiary effects, including protein structure and amino acid chemical properties.[46] Initial work in formal language theory has shown that DNA computers can be made equivalent to a Turing machine.[47]

DNA can also provide extremely dense data storage, requiring about a trillionth of the volume required for an equivalent electronic memory: 10^{12} DNA strands, each 1000 units long, is equal to 1000 T bits.[48] DNA computations employ up to 10^{20} DNA strands (over 11 million TB), well beyond the capacity of any conventional data storage.

Molecular Computing

Molecular computers are a mixed-signal system for performing logic functions (with possible subpicosecond switching) and signal detection with an ability to evolve and adapt to new conditions. Possible implementations include modulation of electron, proton, or photon mobility; electronic-conformation interactions; and tissue membrane interactions.[49] Table 1.3 lists some of the architectures and applications.

"Digital" cell interactions, such as found in quantum or DNA computing, provide the logic implementation. Analog processing is introduced by the non-linear characteristics of the cell interaction with respect to light, electricity, magnetism, chemistry, or other external stimulus. Molecular systems are also thought to emulate a neural network that can be capable of signal enhancement and noise removal.[50]

TABLE 1.3 Summary of Some Architectures and Applications Possible from a Molecular Computing System

Mechanisms and Architectures	Applications
Light-energy transducing proteins	Biosensors
Light-energy transducing proteins (with controlled switching)	Organic memory storage
Optoelectronic transducing	Pattern recognition and processing
Evolutionary structures	Adaptive control

References

1. McShane, E., Trivedi, M., Xu, Y., Khandelwal, P., Mulay, A., and Shenai, K., "Low-Power Systems on a Chip (SOC)," *IEEE Circuits and Devices Magazine*, vol. 14, no. 5, pp. 35-42, June 1998.

2. Laes, E., "Submicron CMOS Technology — The Enabling Tool for System-on-a-Chip Integration," *Alcatel Telecommunications Review*, no. 2, pp. 130-137, 1996.

3. Ackland, B., "The Role of VLSI in Multimedia," *IEEE J. Solid-State Circuits*, vol. 29, no. 4, pp. 381-388, Apr. 1994.

4. Kuroda, I. and Nishitani, T., "Multimedia Processors," *Proc. IEEE*, vol. 86, no. 6, pp. 1203-1221, Jun. 1998.

5. Clemens, J. T, "Silicon Microelectronics Technology," *Bell Labs Technical Journal*, vol. 2, no. 4, pp. 76-102, Fall 1997.

6. Asai, S. and Wada, Y., "Technology Challenges for Integration Near and Below 0.1 Micron [Review]," *Proc. IEEE*, vol. 85, no. 4, pp. 505-520, Apr. 1997.

7. Tsividis, Y., *Mixed Analog-Digital VLSI Devices and Technology: An Introduction*, McGraw-Hill, New York, 1996.

8. Gonzalez, R., Gordon, B.M., and Horowitz, M. A., "Supply and Threshold Voltage Scaling for Low Power CMOS," *IEEE J. Solid-State Circuits*, vol. 32, no. 8, pp. 1210-1216, Aug. 1997.

9. Colinge, J.-P., *Silicon-on-Insulator Technology: Materials to VLSI*, 2nd edition, Kluwer Academic Publishers, Boston, 1997.

10. Assaderaghi, F., Sinitsky, D., Parke, S. A., Bokor, J., Ko, P.K., and Hu, C. M., "Dynamic Threshold-Voltage MOSFET (DTMOS) for Ultra-Low Voltage VLSI," *IEEE Trans. Electron Devices*, vol. 44, no. 3, pp. 414-422, Mar. 1997.

11. Licata, T.J., Colgan, E.G., Harper, J. M. E., and Luce, S. E., "Interconnect Fabrication Processes and the Development of Low-Cost Wiring for CMOS Products," *IBM J. Research & Development*, vol. 39, no. 4, pp. 419-435, Jul. 1995.

12. Itoh, K., Sasaki, K., and Nakagome, Y., "Trends In Low-Power RAM Circuit Technologies," *Proc. IEEE*, vol. 83, no. 4, pp. 524-543, Apr. 1995.

13. Itoh, K., Nakagome, Y., Kimura, S., and Watanabe, T., "Limitations and Challenges of Multigigabit DRAM Chip Design," *IEEE J. Solid-State Circuits*, vol. 32, no. 5, pp. 624-634, May 1997.

14. Kim, K., Hwang, C.-G., and Lee, J. G., "DRAM Technology Perspective for Gigabit Era," *IEEE Trans. Electron Devices*, vol. 45, no. 3, pp. 598-608, Mar. 1998.

15. Cressler, J. D., "SiGe HBT Technology: A New Contender for Si-Based RF and Microwave Circuit Applications," *IEEE Trans. Microwave Theory & Techniques*, vol. 46, no. 5 Part 2, pp. 572-589, May 1998.

16. Hafizi, M., "New Submicron HBT IC Technology Demonstrates Ultra-Fast, Low-Power Integrated Circuits," *IEEE Trans. Electron Devices*, vol. 45, no. 9, pp. 1862-1868, Sept. 1998.

17. Wang, N. L. L., "Transistor Technologies for RFICs in Wireless Applications," *Microwave Journal*, vol. 41, no. 2, pp. 98-110, Feb. 1998.

18. Huang, Q. T., Piazza, F., Orsatti, P., and Ohguro, T.,"The Impact of Scaling Down to Deep Submicron on CMOS RF Circuits," *IEEE J. Solid-State Circuits*, vol. 33, no. 7, pp. 1023-1036, Jul. 1998.

19. Larson, L. E., "Integrated Circuit Technology Options for RFICs — Present Status and Future Directions," *IEEE J. Solid-State Circuits*, vol. 33, no. 3, pp. 387-399, Mar. 1998.

20. Hung, C.-M., Ho, Y. C., Wu, I.-C., and Sin, J. K. O., "High-Q Capacitors Implemented in a CMOS Process for Low-Power Wireless Applications," *IEEE Trans. Microwave Theory & Techniques*, vol. 46, no. 5 Part 1, pp. 505-511, May 1998.

21. Yue, C. P. and Wong, S. S., "On-Chip Spiral Inductors with Patterned Ground Shields for Si-Based RF IC's," *IEEE J. Solid-State Circuits*, vol. 33, no. 5, pp. 743-752, May 1998.

22. Chang, J. Y.-C., Abidi, A. A., and Gaitan, M.,"Large Suspended Inductors on Silicon and Their Use in a 2-mm CMOS RF Amplifier," *IEEE Electron Device Lett.*, vol. 14, pp. 246-248, May 1993.

23. Mohan, N., Undeland, T. M., and Robbins, W. P., *Power Electronics: Converters, Applications, and Design*, 2nd edition, John Wiley & Sons, New York, 1996.

24. Tsui, P. G. Y., Gilbert, P. V., and Sun, S. W., "A Versatile Half-Micron Complementary BiCMOS Technology for Microprocessor-Based Smart Power Applications," *IEEE Trans. Electron Devices*, vol. 42, no. 3, pp. 564-570, Mar. 1995.

25. Chan, W. W. T., Sin, J. K. O., and Wong, S. S., "A Novel Crosstalk Isolation Structure for Bulk CMOS Power ICs," *IEEE Trans. Electron Devices*, vol. 45, no. 7, pp. 1580-1586, Jul. 1998.

26. Baliga, J., "Power Semiconductor Devices for Variable-Frequency Drives," *Proc. IEEE*, vol. 82, no. 8, pp. 1112-1122, Aug. 1994.

27. Mino, M., Yachi, T., Tago, A., Yanagisawa, K., and Sakakibara, K.,"Planar Microtransformer With Monolithically-Integrated Rectifier Diodes For Micro-Switching Converters," *IEEE Trans. Magnetics*, vol. 32, no. 2, pp. 291-296, Mar. 1996.

28. Sunaga, T., Miyatake, H., Kitamura, K., Kogge, P. M., and Retter, E., "A Parallel Processing Chip with Embedded DRAM Macros," *IEEE J. Solid-State Circuits*, vol. 31, no. 10, pp. 1556-1559, Oct. 1996.

29. Watanabe, T., Fujita, R., Yanagisawa, K., Tanaka, H., Ayukawa, K., Soga, M., Tanaka, Y., Sugie, Y., and Nakagome, Y.,"A Modular Architecture for a 6.4-Gbyte/S, 8-Mb DRAM-Integrated Media Chip," *IEEE J. Solid-State Circuits*, vol. 32, no. 5, pp. 635-641, May 1997.

30. Miyano, S., Numata, K., Sato, K., Yabe, T., Wada, M., Haga, R., Enkaku, M., Shiochi, M., Kawashima, Y., Iwase, M., Ohgata, M., Kumagai, J., Yoshida, T., Sakurai, M., Kaki, S., Yanagiya, N., Shinya, H., Furuyama, T., Hansen, P., Hannah, M., Nagy, M., Nagarajan, A., and Rungsea, M., "A 1.6 Gbyte/sec Data Transfer Rate 8 Mb Embedded DRAM," *IEEE J. Solid-State Circuits*, vol. 30, no. 11, pp. 1281-1285, Nov. 1995.

31. Bang, S. H., Choi, J., Sheu, B. J., and Chang, R. C., "A Compact Low-Power VLSI Transceiver for Wireless Communication," *IEEE Trans. Circ. Syst. I—Fundamental Theory & Applications*, vol. 42, no. 11, pp. 933-945, Nov. 1995.

32. Rofougaran, A., Chang, J.G., Rael, J., Chang, J. Y.-C., Rofougaran, M., Chang, P. J., Djafari, M., Ku, M. K., Roth, E. W., Abidi, A. A., and Samueli, H., "A Single-Chip 900-MHz Spread-Spectrum Wireless Transceiver in 1-micron CMOS — Part I: Architecture and Transmitter Design," *IEEE J. Solid-State Circuits*, vol. 33, no. 4, pp. 515-534, Apr. 1998.

33. Rofougaran, A., Chang, G., Rael, J. J., Chang, J. Y.-C., Rofougaran, M., Chang, P. J., Djafari, M., Min, J., Roth, E. W., Abidi, A. A., and Samueli, H., "A Single-Chip 900-MHz Spread-Spectrum Wireless Transceiver in 1-micron CMOS — Part II: Receiver Design," *IEEE J. Solid-State Circuits*, vol. 33, no. 4, pp. 535-547, Apr. 1998.

34. Rofougaran, A., Chang, J. Y. C., Rofougaran, M., and Abidi, A. A., "A 1 GHz CMOS RF Front-End IC for a Direct-Conversion Wireless Receiver," *IEEE J. Solid-State Circuits*, vol. 31, no. 7, pp. 880-889, Jul. 1996.

35. Meyer, R. G. and Mack, W. D., "A 1-GHz BiCMOS RF Front-End IC," *IEEE J. Solid-State Circuits*, vol. 29, no. 3, pp. 350-355, Mar. 1994.

36. Mendis, S., Kemeny, S. E., and Fossum, E. R., "CMOS Active Pixel Image Sensor," *IEEE Trans. Electron Devices*, vol. 41, no. 3, pp. 452-453, Mar. 1994.

37. Fossum, E. R.,"CMOS Image Sensors — Electronic Camera-On-a-Chip," *IEEE Trans. Electron Devices*, vol. 44, no. 10, pp. 1689-1698, Oct. 1997.

38. Pederson, M., Olthuis, W., and Bergveld, P., "High-Performance Condenser Microphone with Fully Integrated CMOS Amplifier and DC-DC Voltage Converter," *IEEE J. Microelectromechanical Systems*, vol. 7, no. 4, pp. 387-394, Dec. 1998.

39. Benard, W. L., Kahn, H., Heuer, A. H., and Huff, M. A., "Thin-Film Shape-Memory Alloy Actuated Micropumps," *IEEE J. Microelectromechanical Systems*, vol. 7, no. 2, pp. 245-251, Jun. 1998.

40. Brassard, G., Chuang, I., Lloyd, S., and Monroe, C., "Quantum Computing," *Proc. of the National Academy of Sciences of the USA*, vol. 95, no. 19, pp. 11032-11033, Sept. 15, 1998.

41. Wallace, R., Price, H., and Breitbeil, F., "Toward a Charge-Transfer Model of Neuromolecular Computing," *Intl J. Quantum Chemistry*, vol. 69, no. 1, pp. 3-10, Jul. 1998.
42. Scarani, V., "Quantum Computing," *American J. of Physics*, vol. 66, no. 11, pp. 956-960, Nov. 1998.
43. Grover, L. K., "Quantum Computing — Beyond Factorization And Search," *Science*, vol. 281, no. 5378, pp. 792-794, Aug. 1998.
44. Gershenfeld, N. and Chuang, I. L., "Quantum Computing with Molecules," *Scientific American*, vol. 278, no. 6, pp. 66-71, Jun. 1998.
45. Conrad, M. and Zauner, K. P., "DNA as a Vehicle for the Self-Assembly Model of Computing," *Biosystems*, vol. 45, no. 1, pp. 59-66, Jan. 1998.
46. Rocha, A. F., Rebello, M. P., and Miura, K., "Toward a Theory of Molecular Computing," *J. Information Sciences*, vol. 106, pp. 123-157, 1998.
47. Kari, L., Paun, G., Rozenberg, G., Salomaa, A., Yu, S., "DNA Computing, Sticker Systems, and Universality," *Acta Informatica*, vol. 35, no. 5, pp. 401-420, May 1998.
48. Forbes, N. A. and Lipton, R. J.,"DNA Computing — A Possible Efficiency Boost for Specialized Problems," *Computers in Physics*, vol. 12, no. 4, pp. 304-306, Jul.-Aug. 1998.
49. Kampfner, R. R.,"Integrating Molecular and Digital Computing — An Information Systems Design Perspective," *Biosystems*, vol. 35, no. 2-3, pp. 229-232, 1995.
50. Rambidi, N. G., "Practical Approach to Implementation of Neural Nets at the Molecular Level," *Biosystems*, vol. 35, no. 2-3, pp. 195-198, 1995.

2

CMOS/BiCMOS Technology

Yasuhiro Katsumata
Tatsuya Ohguro
Kazumi Inoh
Eiji Morifuji
Takashi Yoshitomi
Hideki Kimijima
Hideaki Nii
Toyota Morimoto
Hisayo S. Momose
Kuniyoshi Yoshikawa
Hidemi Ishiuchi
Toshiba Corporation

Hiroshi Iwai
Tokyo Institute of Technology

2.1   Introduction

Silicon LSIs (large-scale integrated circuits) have progressed remarkably in the past 25 years. In particular, complementary metal-oxide-semiconductor (CMOS) technology has played a great role in the progress of LSIs. By downsizing[2] MOS field-effect-transistors (FETs), the number of transistors in a chip increases, and the functionality of LSIs is improved. At the same time, the switching speed of MOSFETs and circuits increases and operation speed of LSIs is improved.

On the other hand, system-on-chip technology has come into widespread use and, as a result, the LSI system requires several functions, such as logic, memory, and analog functions. Moreover, the LSI system sometimes needs an ultra-high-speed logic or an ultra-high-frequency analog function. In some cases, bipolar-CMOS (BiCMOS) technology is very useful.

The first part of this chapter focuses on CMOS technology as the major LSI process technology, including embedded memory technology. The second part, describes BiCMOS technology; and finally, future process technology is introduced.

2.2   CMOS Technology

Device Structure and Basic Fabrication Process Steps

Complementary MOS (CMOS) was first proposed by Wanlass and Sah in 1963.[1] Although the CMOS process is more complex than the NMOS process, it provides both n-channel (NMOS) and p-channel (PMOS) transistors on the same chip, and CMOS circuits can achieve lower power consumption. Consequently, the CMOS process has been widely used as an LSI fabrication process.

Figure 2.1 shows the structure of a CMOS device. Each FET consists of a gate electrode, source, drain, and channel, and gate bias controls carrier flow from source to drain through the channel layer.

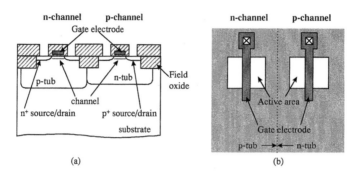

FIGURE 2.1   Structure of CMOS device: (a) cross-sectional view of CMOS, (b) plain view of CMOS.

Figure 2.2 shows the basic fabrication process flow. The first process step is the formation of p tub and n tub (twin tub) in silicon substrate. Because CMOS has two types of FETs, NMOS is formed in p tub and PMOS in n tub.

The isolation process is the formation of field oxide in order to separate each MOSFET active area in the same tub. After that, impurity is doped into channel region in order to adjust the threshold voltage, V_{th}, for each type of FET. The gate insulator layer, usually silicon dioxide (SiO_2), is grown by thermal oxidation, because the interstate density between SiO_2 and silicon substrate is small. Polysilicon is deposited as gate electrode material and gate electrode is patterned by reactive ion etching (RIE).

The gate length, L_g, is the critical dimension because L_g determines the performance of MOSFETs and it should be small in order to improve device performance. Impurity is doped in the source and drain regions of MOSFETs by ion implantation. In this process step, gate electrodes act as a self-aligned mask to cover channel layers. After that, thermal annealing is carried out in order to activate the impurity of diffused layers.

In the case of high-speed LSI, the self-aligned silicide (salicide) process is applied for the gate electrode and source and drain diffused layers in order to reduce parasitic resistance. Finally, the metallization process is carried out in order to form interconnect layers.

Key Process Steps in Device Fabrication

Starting Material

Almost all silicon crystals for LSI applications are prepared by the Czochralski crystal growth method,[2] because it is advantageous for the formation of large wafers. (100) orientation wafers are usually used for MOS devices because their interstate trap density is smaller than those of (111) and (110) orientations.[3] The light doping in the substrate is convenient for the diffusion of tub and reduces the parasitic

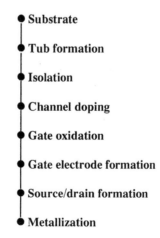

FIGURE 2.2   Basic process flow of CMOS.

capacitance between silicon substrate and tub region. As a starting material, lightly doped ($\sim 10^{15}$ atoms/cm^3) p-type substrate is generally used.

Tub Formation

Figure 2.3 shows the tub structures, which are classified into 6 types: p tub, n tub, twin tub,[4] triple tub, twin tub with buried p$^+$ and n$^+$ layers, and twin tub on p-epi/p$^+$ substrate. In the case of the p tub process, NMOS is formed in p diffusion (p tub) in the n substrate, as shown in Fig. 2.3(a). The p tub is formed by implantation and diffusion into the n substrate at a concentration that is high enough to over-compensate the n substrate.

The other approach is to use an n tub.[5] As shown in Fig. 2.3(b), NMOS is formed in the p substrate.

Figure 2.3(c) shows the twin-tub structure[4] that uses two separate tubs implanted into silicon substrate. In this case, doping profiles in each tub region can be controlled independently, and thus neither type of device suffers from excess doping effect.

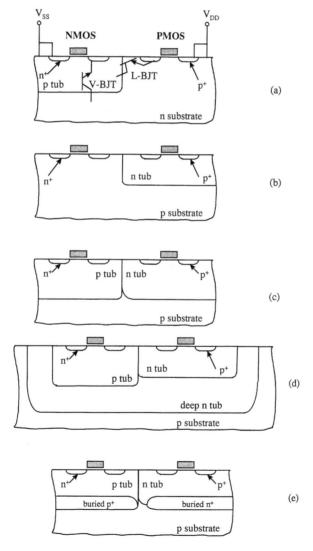

FIGURE 2.3   Tub structures of CMOS: (a) p tub; (b) n tub; (c) twin tub; (d) triple tub; (e) twin tub with buried p$^+$ and n$^+$ layers; and (f) twin tub on p-epi/p$^+$ substrate.

In some cases, such as mixed signal LSIs, a deep n tub layer is sometimes formed optionally, as shown in Fig. 2.3(d), in order to prevent the crosstalk noise between digital and analog circuits. In this structure, both n and p tubs are electrically isolated from the substrate or other tubs on the substrate.

In order to realize high packing density, the tub design rule should be shrunk; however, an undesirable mechanism, the well-known latch-up, might occur.

Latch-up (i.e., the flow of high current between V_{DD} and V_{SS}) is caused by parasitic lateral pnp bipolar (L-BJT) and vertical npn bipolar (V-BJT) transistor actions[6] as shown in Fig. 2.3(a), and it sometimes destroys the functions of LSIs. The collectors of each of these bipolar transistors feed each others' bases and together make up a pnpn thyristor structure. In order to prevent latch-up, it is important to reduce the current gain, h_{FE}, of these parasitic bipolar transistors, and the doping concentration of the tub region should be higher. As a result, device performance might be suppressed because of large junction capacitances.

In order to solve this problem, several techniques have been proposed, such as p^+ or n^+ buried layer under p tub[7] as shown in Fig. 2.3(e), the use of high-dose, high-energy boron p tub implants,[8,9] and the shunt resistance for emitter-base junctions of parasitic bipolar transistors.[7,10,11] It is also effective to provide many well contacts to stabilize the well potential and hence to suppress the latch-up. Recently, substrate with p epitaxial silicon on p^+ substrate can also be used to stabilize the potential for high-speed logic LSIs.[12]

Isolation

Local oxidation of silicon (LOCOS)[13] is a widely used isolation process, because this technique can allow channel-stop layers to be formed self-aligned to the active transistor area. It also has the advantage of recessing about half of the field oxide below the silicon surface, which makes the surface more planar.

Figure 2.4 shows the LOCOS isolation process. First, silicon nitride and pad oxide are etched for the definition of active transistor area. After channel implantation as shown in Fig. 2.4(a), the field oxide is selectively grown, typically to a thickness of several hundred nanometers.

A disadvantage of LOCOS is that involvement of nitrogen in the masking of silicon nitride layer sometimes causes the formation of a very thin nitride layer in the active region, and this often impedes the subsequent growth of gate oxide, thereby causing low gate breakdown voltage of the oxides. In order to prevent this problem, after stripping the masking silicon nitride, a sacrificial oxide is grown and then removed before the gate oxidation process.[14,15]

In addition, the lateral spread of field oxide (bird's beak)[14] poses a problem regarding reduction of the distance between active transistor areas in order to realize high packing density. This lateral spread is suppressed by increasing the thickness of silicon nitride and/or decreasing the thickness of pad oxide. However, there is a tradeoff with the generation of dislocation of silicon.

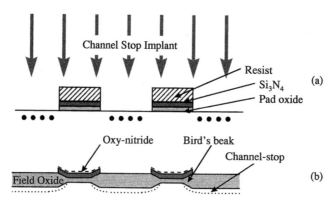

FIGURE 2.4   Process for local oxidation of silicon: (a) after silicon nitride/pad oxide etch and channel-stop implant; (b) after field oxidation, which produces an oxynitride film on nitride.

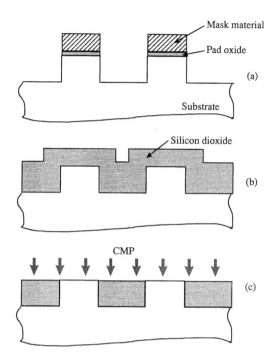

FIGURE 2.5 Process flow of STI: (a) trenches are formed by RIE; (b) filling by deposition of SiO_2; and (c) planarization by CMP.

Recently, shallow trench isolation (STI)[16] has become a major isolation process for advanced CMOS devices. Figure 2.5 shows the process flow of STI. After digging the trench into the substrate by RIE as shown in Fig. 2.5(a), the trench is filled with insulator such as silicon dioxide as shown in Fig. 2.5(b). Finally, by planarization with chemical mechanical polishing (CMP),[17] filling material on the active transistor area is removed, as shown in Fig. 2.5(c).

STI is a useful technique for downsizing not only the distance between active areas, but also the active region itself. However, a mechanical stress problem[18] still remains, and several methods have been proposed[19] to deal with it.

Channel Doping

In order to adjust the threshold voltage of MOSFETs, V_{th}, to that required by a circuit design, the channel doping process is usually required. The doping is carried out by ion implantation, usually through a thin dummy oxide film (10 to 30 nm) thermally grown on the substrate in order to protect the surface from contamination, as shown in Fig. 2.6. This dummy oxide film is removed prior to the gate oxidation. Figure 2.7 shows a typical CMOS structure with channel doping. In this case, n^+ polysilicon gate electrodes are used for both n- and p-MOSFETs and, thus, this type of CMOS is called single-gate CMOS. The role of the channel doping is to enhance or raise the threshold voltage of n-MOSFETs. It is desirable to keep the concentration of p tub lower in order to reduce the junction capacitance of source and drain. Thus, channel doping of p-type impurity — boron — is required. Drain-to-source leakage current in short-channel MOSFETs flows in a deeper path, as shown in Fig. 2.8; this is called the short-channel effect. Thus, heavy doping of the deeper region is effective in suppressing the short-channel effect. This doping is called deep ion implantation.

In the case of p-MOSFET with an n^+ polysilicon gate electrode, the threshold voltage becomes too high in the negative direction if there is no channel doping. In order to adjust the threshold voltage, an ultra-shallow p-doped region is formed by the channel implantation of boron. This p-doped layer is often called a counter-doped layer or buried-channel layer, and p-MOSFETs with this structure are called buried-channel MOSFETs. (On the other hand, MOSFETs without a buried-channel layer are called

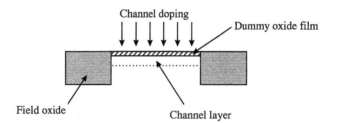

FIGURE 2.6 Channel doping process step.

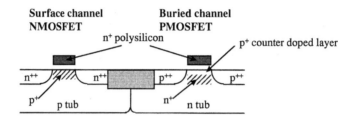

FIGURE 2.7 Schematic cross-section of single-gate CMOS structure.

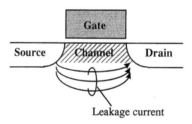

FIGURE 2.8 Leakage current flow in short-channel MOSFET.

surface-channel MOSFETs. n-MOSFETs in this case are the surface-channel MOSFETs.) In the buried-channel case, the short-channel effect is more severe, and, thus, deep implantation of an n-type impurity such as arsenic or phosphorus is necessary to suppress them.

In deep submicron gate length CMOS, it is difficult to suppress the short-channel effect,[20] and thus, a p+-polysilicon electrode is used for p-MOSFETs, as shown in Fig. 2.9. For n-MOSFETs, an n+-polysilicon electrode is used. Thus, this type of CMOS is called dual-gate CMOS. In the case of p+-polysilicon p-MOSFET, the threshold voltage becomes close to 0 V because of the difference in work function between n- and p-polysilicon gate electrode,[21–23] and thus, buried layer is not required. Instead, n-type impurity channel doping such as arsenic is required to raise the threshold voltage slightly in the negative direction. Impurity redistribution during high-temperature LSI manufacturing processes sometimes makes channel profile broader, which causes the short-channel effect. In order to suppress the redistribution, a dopant with a lower diffusion constant, such as indium, is used instead of boron.

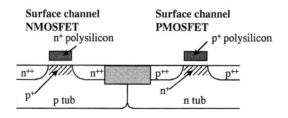

FIGURE 2.9 Schematic cross-section of dual-gate CMOS structure.

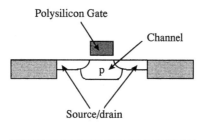

FIGURE 2.10 Localized channel structure.

For the purpose of realizing a high-performance transistor, it is important to reduce junction capacitance. In order to realize lower junction capacitance, a localized diffused channel structure,[24,25] as shown in Fig. 2.10, is proposed. Since the channel layer exists only around the gate electrode, the junction capacitance of source and drain is reduced significantly.

Gate Insulator

The gate dielectric determines several important properties of MOSFETs and thus uniformity in its thickness, low defect density of the film, low fixed charge and interface state density at the dielectric and silicon interface, small roughness at the interface, high reliability of time-dependent dielectric breakdown (TDDB) and hot-carrier induced degradation, and high resistivity to boron penetration (explained in this section) are required. As a consequence of downsizing of MOSFET, the thickness of the gate dielectric has become thinner. Generally, the thickness of the gate oxide is 7 to 8 nm for 0.4-μm gate length MOSFETs, and 5 to 6 nm for 0.25-μm gate length MOSFETs.

Silicon dioxide (SiO_2) is commonly used for gate dielectrics, and can be formed by several methods, such as dry O_2 oxidation,[26] and wet or steam (H_2O) oxidation.[26] The steam is produced by the reaction of H_2 and O_2 ambient in the furnace. Recently, H_2O oxidation has been widely used for gate oxidation because of good controllability of oxide thickness and high reliability.

In the case of the dual-gate CMOS structure shown in Fig. 2.9, boron penetration from the p^+ gate electrode to the channel region through the gate silicon dioxide, which is described in the following section, is a problem. In order to prevent this problem, oxynitride has been used as the gate dielectric material.[27,28] In general, the oxynitride gate dielectric is formed by the annealing process in NH_3, NO (or N_2O) after silicon oxidation, or by direct oxynitridation of silicon in NO (or N_2O) ambient. Figure 2.11 shows the typical nitrogen profile of the oxynitride gate dielectric. Recently, remote plasma nitridation[29,30] has been much studied, and it is reported that the oxynitride gate dielectric grown by the remote plasma method showed better quality and reliability than that grown by the silicon nitridation method.

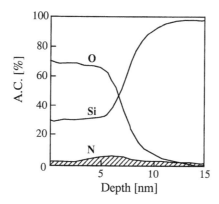

FIGURE 2.11 Oxygen, nitrogen, and silicon concentration profile of oxynitride gate dielectrics measured by AES.

In the regime of a sub-quarter-micron CMOS device, gate oxide thickness is close to the limitation of tunneling current flow, around 3 nm thickness. In order to prevent tunneling current, high κ materials, such as Si_3N_4[31] and Ta_2O_5,[32] are proposed instead of silicon dioxide. In these cases, the thickness of the gate insulator can be kept relatively thick because high κ insulator realizes high gate capacitance, and thus better driving capability.

Gate Electrode

Heavily doped polysilicon has been widely used for gate electrodes because of its resistance to high-temperature LSI fabrication processing. In order to reduce the resistance of the gate electrode, which contributes significantly to RC delay time, silicides of refractory metals have been put on the polysilicon electrode.[33,34] Polycide,[34] the technique of combining a refractory metal silicide on top of doped polysilicon, has the advantage of preserving good electrical and physical properties at the interface between polysilicon and the gate oxide while, at the same time, the sheet resistance of gate electrode is reduced significantly.

For doping the gate polysilicon, ion implantation is usually employed. In the case of heavy doping, dopant penetration from boron-doped polysilicon to the Si substrate channel region through the gate oxide occurs in the high-temperature LSI fabrication process, as shown in Fig. 2.12. (On the other hand, usually, penetration of an n-type dopant [such as phosphorus or arsenic] does not occur.) When the doping of impurities in the polysilicon is not sufficient, the depletion of the gate electrode occurs as shown in Fig. 2.13, resulting in a significant decrease of the drive capability of the transistor, as shown in Fig. 2.14.[35] There is a tradeoff between the boron penetration and the gate electrode depletion, and so thermal process optimization is required.[36]

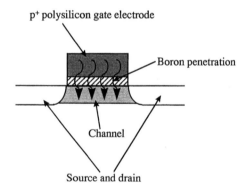

FIGURE 2.12 Dopant penetration from boron-doped polysilicon to silicon substrate channel region.

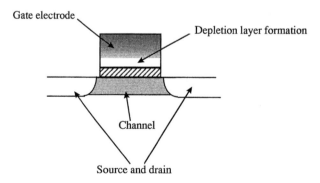

FIGURE 2.13 Depletion of gate electrode in the case that the doping of impurities in the gate electrode is not sufficient.

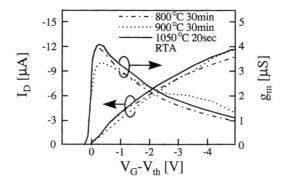

FIGURE 2.14 I_D, $g_m - V_G$ characteristics for various thermal conditions. In the case of 800°C/30 min, a significant decrease in drive capability of transistor occurs because of the depletion of the gate electrode.

Gate length is one of most important dimensions defining MOSFET performance; thus, the lithography process for gate electrode patterning requires high-resolution technology.

In the case of a light-wave source, the g-line (wavelength 436 nm) and the i-line (365 nm) of a mercury lamp were popular methods. Recently, a higher-resolution process, excimer laser lithography, has been used. In the excimer laser process, KrF (248 nm)[37] and ArF (193 nm)[38] have been proposed and developed. For a 0.25-μm gate length electrode, the KrF excimer laser process is widely used in the production of devices. In addition, electron-beam[39–41] and X-ray[42] lithography techniques are being studied for sub-0.1 μm gate electrodes.

For the etching of gate polysilicon, a high-selectivity RIE process is required for selecting polysilicon from SiO_2 because a gate dielectric beneath polysilicon is a very thin film in the case of recent devices.

Source/Drain Formation

Source and drain diffused layers are formed by the ion implantation process. As a consequence of transistor downsizing, at the drain edge (interface of channel region and drain) where reverse biased pn junctions exist, a higher electrical field has been observed. As a result, carriers across these junctions are suddenly accelerated and become hot carriers, which creates a serious reliability problem for MOSFET.[43]

In order to prevent the hot carrier problem, the lightly doped drain (LDD) structure is proposed.[44] The LDD process flow is shown in Fig. 2.15. After gate electrode formation, ion implantation is carried out to make extension layers, and the gate electrode plays the role of a self-aligned mask that covers the channel layer, as shown in Fig. 2.15(b). In general, arsenic is doped for n-type extension of NMOS, and BF_2 for p-type extension of PMOS. To prevent the short-channel effect, the impurity profile of extension layers must be very shallow. Although shallow extension can be realized by ion implantation with low dose, the resistivity of extension layers becomes higher and, thus, MOSFET characteristics degrade. Hence, it is very difficult to meet these two requirements. Also, impurities diffusion in this extension affects the short-channel effect significantly. Thus, it is necessary to minimize the thermal process after forming the extension.

Insulating film, such as Si_3N_4 or SiO_2, is deposited by a chemical vapor deposition method. Then, etching back RIE treatment is performed on the whole wafer; as a result, the insulating film remains only at the gate electrode side, as shown in Fig. 2.15(c). This remaining film is called a sidewall spacer. This spacer works as a self-aligned mask for deep source/drain n^+ and p^+ doping, as shown in Fig. 2.15(d). In general, arsenic is doped for deep source/drain of n-MOSFET, and BF_2 for p-MOSFET. In the dual-gate CMOS process, gate polysilicon is also doped in this process step to prevent gate electrode depletion.

After that, in order to make doped impurities activate electrically and recover from implantation damage, an annealing process, such as rapid thermal annealing (RTA), is carried out.

According to the MOSFET scaling law, when gate length and other dimensions are shrunk by factor k, the diffusion depth also needs to be shrunk by $1/k$. Hence, the diffusion depth of the extension part is required to be especially shallow.

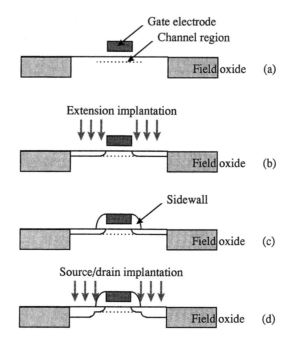

FIGURE 2.15 Process flow of LDD structure: (a) after gate electrode patterning; (b) extension implantation; (c) sidewall spacer formation; and (d) source/drain implantation.

Several methods have been proposed for forming an ultra-shallow junction. For example, very low accelerating voltage implantation, the plasma doping method,[45] and implantation of heavy molecules, such as $B_{10}H_{14}$ for p-type extension,[46] are being studied.

Salicide Technique

As the vertical dimension of transistors is reduced with device downscaling, an increase is seen in sheet resistance — both of the diffused layers, such as source and drain, and the polysilicon films, such as the gate electrode. This is becoming a serious problem in the high-speed operation of integrated circuits.

Figure 2.16 shows the dependence of the propagation delay (t_{pd}) of CMOS inverters on the scaling factor, k, or gate length.[47] These results were obtained by simulations in which two cases were considered. First is the case in which source and drain contacts with the metal line were made at the edge of the diffused layers, as illustrated in the figure inset. In an actual LSI layout, it often happens that the metal contact to the source or drain can be made only to a portion of the diffused layers, since many other signal or power lines cross the diffused layers. The other case is that in which the source and drain contacts cover the entire area of the source and drain layers, thus reducing diffused line resistance. It is clear that without a technique to reduce the diffused line resistance, t_{pd} values cannot keep falling as transistor size is reduced; they will saturate at gate lengths of around 0.25 microns.

In order to solve this problem — the high resistance of shallow diffused layers and thin polysilicon films — self-aligned silicide (salicide) structures for the source, drain, and gate have been proposed, as shown in Fig. 2.17.[48-50]

First, a metal film such as Ti or Co is deposited on the surface of the MOSFET after formation of the polysilicon gate electrode, gate sidewall, and source and drain diffused layers, as shown in Fig. 2.17(b). The film is then annealed by rapid thermal annealing (RTA) in an inert ambient. During the annealing process, the areas of metal film in direct contact with the silicon layer — that is, the source, drain, and gate electrodes — are selectively converted to the silicide, and other areas remain metal, as shown in Fig. 2.17(c). The remaining metal can be etched off with an acid solution such as $H_2O_2 + H_2SO_4$, leaving the silicide self-aligned with the source, drain, and gate electrode, as shown in Fig. 2.17(d).

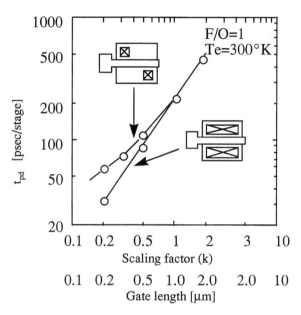

FIGURE 2.16 Dependence of the propagation delay (t_{pd}) of CMOS inverters on the scaling factor, k, or gate length.

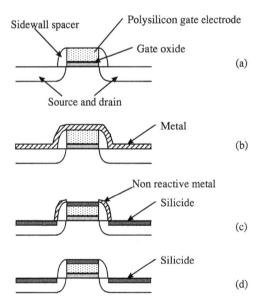

FIGURE 2.17 A typical process flow and schematic cross-section of salicide process: (a) MOSFET formation; (b) metal deposition; (c) silicidation by thermal annealing; and (d) removal of non-reactive metal.

When the salicide process first came into use, furnace annealing was the most popular heat-treatment process[48–50]; however, RTA[51–53] replaced furnace annealing early on, because it is difficult to prevent small amounts of oxidant from entering through the furnace opening, and these degrade the silicide film significantly since silicide metals are easily oxidized. On the other hand, RTA reduces this oxidation problem significantly, resulting in reduced deterioration of the film and consequently of its resistance.

At present, TiSi$_2$[51–53] is widely used as a silicide in LSI applications. However, in the case of ultra-small geometry MOSFETs for VLSIs, use of TiSi$_2$ is subject to several problems. When the TiSi$_2$ is made thick, a large amount of silicon is consumed during silicidation, and this results in problems of junction leakage at the source or drain. On the contrary, if a thin layer of TiSi$_2$ is chosen, agglomeration of the film occurs[54] at higher silicidation temperatures.

On the other hand, CoSi$_2$[55] has a large silicidation temperature window for low sheet resistance; hence, it is expected to be widely used as silicidation material for advanced VLSI applications.[47]

Interconnect and Metallization

Aluminum is widely used as a wiring metal. However, in the case of downsized CMOS, electromigration (EM)[56] and stress migration (SM)[57] become serious problems. In order to prevent these problems, Al-Cu (typically ~0.5 wt % Cu)[58] is a useful wiring material. In addition, ultra-shallow junction for downsized CMOS sometimes needs barrier metal,[58] such as TiN, between the metal and silicon, in order to prevent junction leakage current.

Figure 2.18 shows a cross-sectional view of a multi-layer metallization structure. As a consequence of CMOS downscaling, contact or via aspect ratio becomes larger; and, as a result, filling of contact or via is not sufficient. Hence, new filling techniques, such as W-plug,[59,60] are widely used.

In addition, considering both reliability and low resistivity, Cu is a useful wiring material.[61] In the case when Cu is used, metal thickness can be reduced in order to realize the same interconnect resistance. The reduction of the metal thickness is useful for reducing the capacitance between the dense interconnect wires, resulting in the high-speed operation of the circuit. In order to reduce RC delay of wire in CMOS LSI, not only wiring material but also interlayer material is important. In particular, low-κ material[62] is widely studied.

In the case of Cu wiring, the dual damascene process[63] is being widely studied because it is difficult to realize fine Cu pattern by reactive ion etching. Figure 2.19 shows the process flow of Cu dual damascene metallization. After formation of transistors and contact holes as shown in Fig. 2.19(a), barrier metal, such as TiN, and Cu are deposited as shown in Fig. 2.19(b). By using the CMP planarization process, Cu and barrier metal remains in the contact holes, as shown in Fig. 2.19(c). Insulator, such as silicon dioxide, is deposited and the grooves for first metal wires are formed by reactive ion etching, as shown in Fig. 2.19(d). After the deposition of barrier metal and Cu as shown in Fig. 2.19(e), Cu and barrier metal remain only in the wiring grooves due to use of a planarization process such as CMP, as shown in Fig. 2.19(f).

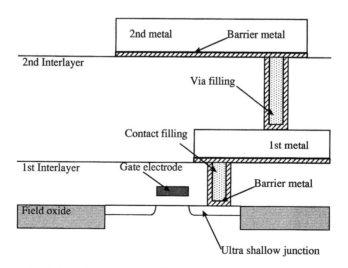

FIGURE 2.18 Cross-sectional view of multi-layer metallization.

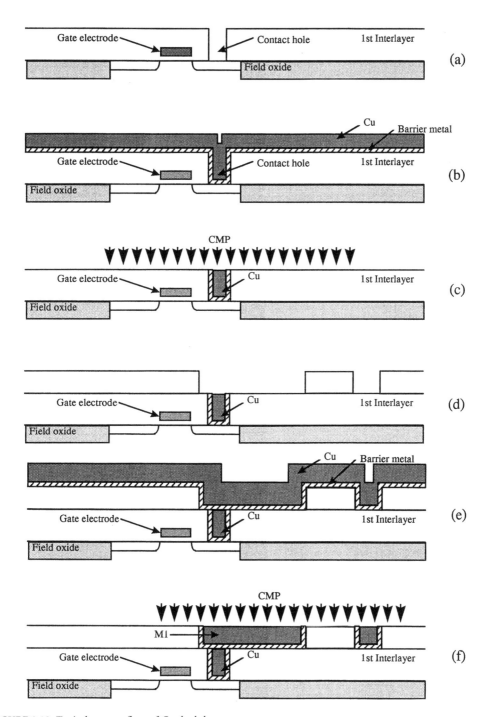

FIGURE 2.19 Typical process flow of Cu dual damascene.

Passive Device for Analog Operation

System-on-chip technology has come into widespread use; and as a result, an LSI system sometimes requires analog functions. In this case, analog passive devices should be integrated,[64] as shown in Fig. 2.20.

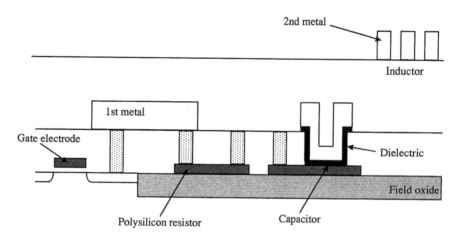

FIGURE 2.20 Various passive devices for analog application.

Resistors and capacitors already have good performance, even for high-frequency applications. On the other hand, it is difficult to realize a high-quality inductor on a silicon chip because of inductance loss in Si substrate, in which the resistivity is lower than that in the compound semiconductor, such as GaAs, substrate. The relatively higher sheet resistance of aluminum wire used for high-density LSI is another problem. Recently, quality of inductor has been improved by using thicker Al or Cu wire[65] and by optimizing the substrate structure.[66]

Embedded Memory Technology

Embedded DRAM

There has been strong motivation to merge DRAM cell arrays and logic circuits into a single silicon chip. This approach makes it possible to realize high bandwidth between memory and logic, low power consumption, and small footprint of the chip.[67] In order to merge logic and DRAM into a single chip, it is necessary to establish process integration for the embedded DRAM. Figure 2.21 shows a typical structure of embedded DRAM. However, the logic process and the DRAM process are not compatible with each other. There are many variations and options in constructing a consistent process integration for the embedded DRAM.

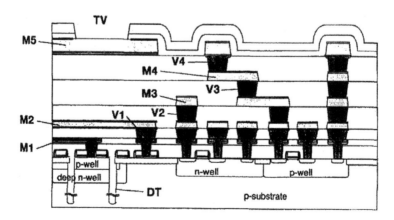

FIGURE 2.21 Schematic cross-section of the embedded DRAM, including DRAM cells and logic MOSFETs.

Trench Capacitor Cell versus Stacked Capacitor Cell

There are two types of DRAM cell structure: stacked capacitor cell[68–73] and trench capacitor cell.[74,75]

In trench cell technology, the cell capacitor process is completed before gate oxidation. Therefore, there is no thermal process due to cell capacitor formation after the MOSFET formation. Another advantage of the trench cell is that there is little height difference between the cell array region and the peripheral circuit region.[76–79]

In the stacked capacitor cell, the height difference high aspect ratio contact holes and difficulty in the planarization process after cell formation. The MOSFET formation steps are followed by the stacked capacitor formation steps, which include high-temperature process steps such as storage node insulator (SiO_2/Si_3N_4) formation, and Si_3N_4 deposition for the self-aligned contact formation. The salicide process for the source and drain of the MOSFETs should be carefully designed to endure the high-temperature process steps. Recently, high-permittivity film for capacitor insulators, such as Ta_2O_5 and BST, has been developed for commodity DRAM and embedded DRAM. The process temperature for Ta_2O_5 and BST is lower than that for SiO_2/Si_3N_4; this means the process compatibility is better with such high-permittivity film.[80–82]

MOSFET Structure

The MOSFET structure in DRAMs is different from that in logic ULSIs. In recent DRAMs, the gate is covered with Si_3N_4 for self-aligned contact process steps in the bit-line contact formation. It is very difficult to apply the salicide process to the gate, source, and drain at the same time. A solution to the problem is to apply the salicide process to the source and drain only. A comparison of the MOSFET structures is shown in Fig. 2.22. Tsukamoto et al.[68] proposed another approach, namely the use of W-bit line layer as the local interconnect in the logic portion.

Gate Oxide Thickness

Generally, DRAM gate oxide thickness is greater than that of logic ULSIs. This is because the maximum voltage of the transfer gate in the DRAM cells is higher than V_{CC}, the power supply voltage. In the logic ULSI, the maximum gate voltage is equal to V_{CC} in most cases. To keep up with the MOSFET performance in logic ULSIs, the oxide thickness of the embedded DRAMs needs to be scaled down further than in the DRAM case. To do so, a highly reliable gate oxide and/or new circuit scheme in the word line biasing, such as applying negative voltage to the cell transfer gate, is required.

Another approach is to use thick gate oxide in the DRAM cell and thin gate oxide in the logic.[83]

Fabrication Cost per Wafer

The conventional logic ULSIs do not need the process steps for DRAM cell formation. On the other hand, most of DRAMs use only two layers of aluminum. This raises wafer cost of the embedded DRAMs. Embedded DRAM chips are used only if the market can absorb the additional wafer cost for some reasons: high bandwidth, lower power consumption, small footprint, flexible memory configuration, lower chip assembly cost, etc.

Next-Generation Embedded DRAM

Process technology for the embedded DRAM with 0.18-μm or 0.15-μm design rules will include state-of-the-art DRAM cell array . and high-performance MOSFETs in the logic circuit. The embedded DRAM

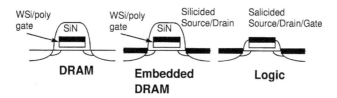

FIGURE 2.22 Typical MOSFET structures for DRAM, embedded DRAM, and logic.

could be a technology driver because the embedded DRAM contains most of the key process steps for DRAM and logic ULSIs.

Embedded Flash Memory Technology[84]

Recently, the importance of embedded flash technology has been increasing and logic chips with non-volatile functions have become indispensable for meeting various market requirements.

Key issues in the selection of an embedded flash cell[85] are (1) tunnel-oxide reliability (damage-less program/erase(P/E) mechanism), (2) process and transistor compatibility with CMOS logic, (3) fast read with low *Vcc*, (4) low power (especially in P/E), (5) simple control circuits, (6) fast program speed, and (7) cell size. This ordering greatly depends on target device specification and memory density, and, in general, is different from that of high-density stand-alone memories. NOR-type flash is essential and EEPROM functionality is also required on the same chip. Figure 2.23 shows the typical device structure of a NOR-type flash memory with logic device.[86]

Process Technology[87]

To realize high-performance embedded flash chips, at least three kinds of gate insulators are required beyond the 0.25-μm regime in order to form flash tunnel oxide, CMOS gate oxide, high voltage transistor gate oxide, and I/O transistor gate oxide. Flash cells are usually made by a stacked gate process. Therefore, it is difficult to achieve less than 150% of the cost of pure logic devices.

The two different approaches to realize embedded flash chips are memory-based and logic-based, as shown in Fig. 2.24.

Memory-based approach is advantageous in that it exploits established flash reliability and yield guaranteed by memory mass production lines, but is disadvantageous for realizing high-performance CMOS transistors due to the additional flash process thermal budget. On the contrary, logic-based approach can use fully CMOS-compatible transistors as they are; but, due to the lack of dedicated mass production lines, great effort is required in order to establish flash cell reliability and performance. Historically, memory-based embedded flash chips have been adopted, but the logic-based chips have

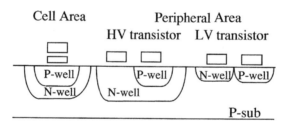

FIGURE 2.23 Device structure schematic view of the NOR flash memories with dual-gate Ti-salicide.

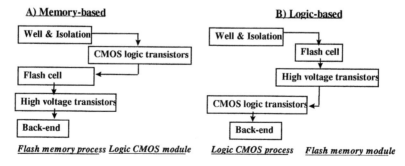

FIGURE 2.24 Process modules.

become more important recently. In general, the number of additional masks required to embed a flash cell into logic chips ranges from 4 to 9.

For high-density embedded flash chips, one transistor stack gate cell using channel hot electron programming and channel FN tunneling erasing will be mainstream. For medium- or low-density, high-speed embedded flash chips, two transistors will be important in the case of using the low power P/E method. From the reliability point of view, a p-channel cell using band-to-band tunneling-induced electron injection[88] and channel FN tunneling ejection is promising since page-programmable EEPROM can also be realized by this mechanism.[85]

2.3   BiCMOS Technology

The development of BiCMOS technology began in the early 1980s. In general, bipolar devices are attractive because of their high speed, better gain, better driving capability, and low wide-band noise properties that allow high-quality analog performance. CMOS is particularly attractive for digital applications because of its low power and high packing density. Thus, the combination would not only lead to the replacement and improvement of existing ICs, but would also provide access to completely new circuits.

Figure 2.25 shows a typical BiCMOS structure.[89] Generally, BiCMOS has a vertical npn bipolar transistor, a lateral pnp transistor, and CMOS on the same chip. Furthermore, if additional mask steps are allowed, passive devices are integrated, as described in the previous section. The main feature of the BiCMOS structure is the existence of a buried layer because bipolar processes require an epitaxial layer grown on a heavily doped n+ subcollector to reduce collector resistance.

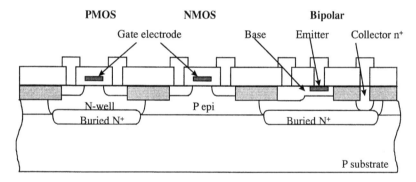

FIGURE 2.25 Cross-sectional view of BiCMOS structure.

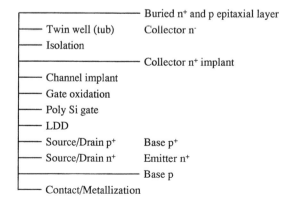

FIGURE 2.26 Typical process flow of BiCMOS device.

Figure 2.26 shows typical process flow for BiCMOS. This is the simplest arrangement for incorporating bipolar devices and a kind of low-cost BiCMOS. Here, the BiCMOS process is completed with minimum additional process steps required to form the npn bipolar device, transforming the CMOS baseline process into a full BiCMOS technology. For this purpose, many processes are merged. The p tub of n-MOSFET shares an isolation of bipolar devices, the n tub of p-MOSFET device is used for the collector, the n^+ source and drain are used for the emitter regions and collector contacts, and also extrinsic base contacts have the p^+ source and drain of PMOS device for common use.

Recently, there have been two significant uses of BiCMOS technology. One is high-performance MPU[90] by using the high driving capability of bipolar transistor; the other is mixed signal products that utilize the excellent analog performance of the bipolar transistor, as shown in Table 2.1.

For the high-performance MPU, merged processes were commonly used, and the mature version of the MPU product has been replaced by CMOS LSI. However, this application has become less popular now with reduction in the supply voltage. Mixed-signal BiCMOS requires high performance, especially with respect to f_T, f_{max}, and low noise figure. Hence, a double polysilicon structure with a silicon[91] or SiGe[92] base with trench isolation technology is used.

The fabrication cost of BiCMOS is a serious problem and, thus, a low-cost mixed-signal BiCMOS process[93] has also been proposed.

2.4 Future Technology

In this section, advanced technologies for realizing future downsized CMOS devices are introduced.

Ultra-Thin Gate Oxide MOSFET

From a performance point of view, ultra-thin gate oxide in a direct-tunneling regime is desirable for future LSIs.[94] In this section, the potential and possibility are discussed.

Figure 2.27 shows a TEM cross-section a 1.5-nm gate oxide. Figure 2.28 shows I_d-V_d characteristics for 1.5-nm gate oxide MOSFETs with various gate lengths. In the long-channel case, unusual electrical characteristics were observed because of the significant tunneling leakage current through the gate oxide.

TABLE 2.1   Recent BiCMOS Structures

Type	Structure	Future
Digital BiCMOS		Simplified bipolar. Merged well or diffused layer. For digital application.
Mixed-signal BiCMOS		Double poly self-aligned bipolar, including Si or SiGe base. Non-merged process. For mixed-signal application.

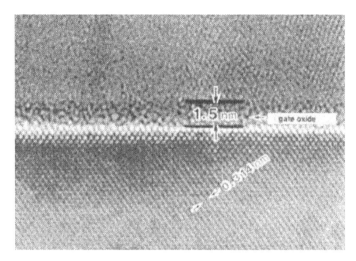

FIGURE 2.27 TEM cross-section of a 1.5-nm gate oxide film. Uniform oxide of 1.5-nm thickness is observed.

However, the characteristics become normal as the gate length is reduced because the gate leakage current decreases in proportion to the gate length and the drain current increases in inverse proportion to the gate length.[95,96] Recently, very high drive currents of 1.8 mA/mm and very high transconductances of more than 1.1 S/mm have been reported using a 1.3-nm gate oxide at a supply voltage of 1.5 V.[97] They also operate well at low power and high speed with a low supply voltage in the 0.5-V range.[98]

Figure 2.29 shows the dependence of cutoff frequency, f_T, of 1.5-nm gate oxide MOSFETs on gate length.[99] Very high cutoff frequencies of more than 150 GHz were obtained at gate lengths in the sub-0.1-μm regime due to the high transconductance. Further, it was confirmed that the high transconductance offers promise of a good noise figure.

Therefore, the MOSFETs with ultra-thin gate oxides beyond the direct-tunneling limit have the potential to enable extremely high-speed digital circuit operation as well as high RF performance in analog applications. Fortunately, the hot-carrier and TDDB reliability of these ultra-thin gate oxides seem to be good.[95,96,100] Thus, ultra-thin gate oxides are likely to be used for such LSIs, for certain application.

In actual applications, even though the leakage current of a single transistor may be very small, the combined leakage of the huge number of transistors in a ULSI circuit poses problems, particularly for battery backup operation.[101,102] There is, however, the possibility of using these direct-tunneling gate oxide MOSFETs only for the smaller number of switches in the critical path determining operation speed. Also, use of a slightly thicker oxide of 2.0 or 2.5 nm would significantly reduce leakage current. The use of these direct-tunneling gate oxide MOSFETs in LSI devices with smaller integration is another possibility.

Epitaxial Channel MOSFET

As the design rule progresses, the supply voltage decreases. In order to obtain high drivability, lower V_{th} has been required under low supply voltage. However, substrate concentration must be higher in order to suppress the short-channel effects. The ideal channel profile for this requirement is that the channel surface concentration is lower to realize lower V_{th}, and the concentration around the extension region is higher in order to suppress the short-channel effects. It is difficult to realize such a channel profile by using ion implantation because the profile is very broad. This requirement can be realized by non-doped epitaxial Si formation on doped Si substrate.[103] Figure 2.30 shows the process flow of MOSFETs with epitaxial Si channel, n channel, and p channel. Although the problem with this structure is the quality of epitaxial Si, degradation of the quality can be suppressed by wet treatment to clean the Si surface, and heating process before epitaxial growth. The zero V_{th} can be realized while suppressing the short-channel effect even when gate length is 0.1 μm. A 20% improvement of drivability can be realized.

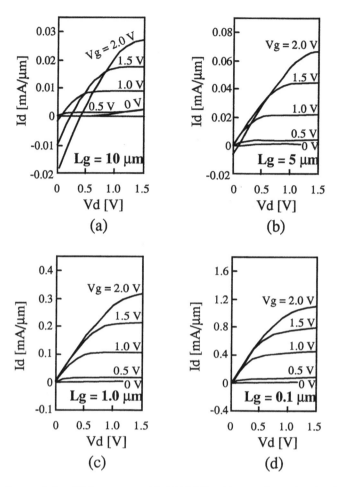

FIGURE 2.28 I_d-V_d characteristics of 1.5-nm gate oxide MOSFETs with several gate lengths: (a) Lg = 10 μm; (b) Lg = 5 μm; (c) Lg = 1.0 μm; and (d) Lg = 0.1 μm.

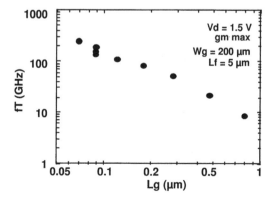

FIGURE 2.29 Dependence of cutoff frequency (f_T) on gate length (L_g) of 1.5-nm gate oxide MOSFETs.

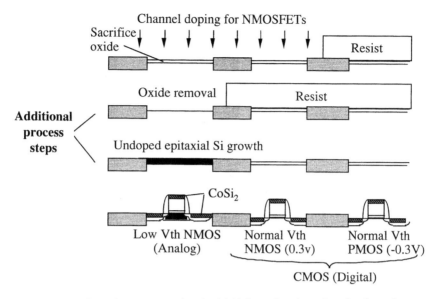

FIGURE 2.30 The process flow of MOSFETs with epitaxial Si channel, n channel, and p channel.

Raised Gate/Source/Drain Structure

In order to realize high drivability of MOSFETs, it is necessary to reduce the resistance under the gate sidewall. However, as the sidewall thickness becomes thinner, the stability of the short-channel effect degrades because the deeper source and drain become closer. If this junction depth becomes shallower, junction leakage degradation occurs because the distance between the bottom of the silicide and the junction becomes shorter. Raised gate/source/drain[104] has been proposed as one way to resolve these problems. The structure is shown in Fig. 2.31.

Even if the junction depth measured from the original Si substrate becomes shallower, the distance between the bottom of silicide and the junction becomes constant using this structure. Additionally, gate resistance becomes lower because the top of the gate electrode is T-shaped. Thus, low gate, source, and drain resistance can be realized while the short-channel effects and junction leakage current degradation are suppressed.

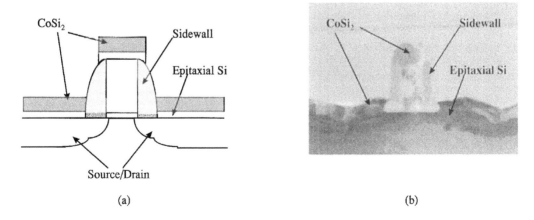

FIGURE 2.31 Structure of raised source/drain/gate FET: (a) schematic cross-section; (b) TEM photograph.

2.5 Summary

This chapter has described CMOS and BiCMOS technology. CMOS is the most important device structure for realizing the future higher-performance devices required for multimedia and other demanding applications. However, certain problems are preventing the downsizing of device dimensions. The chapter described not only conventional technology but also advanced technology that has been proposed with a view to overcoming these problems.

BiCMOS technology is also important, especially for mixed-signal applications. However, CMOS device performance has already been demonstrated for RF applications and, thus, analog CMOS circuit technology will be very important for realizing the production of analog CMOS.

References

1. Wanlass, F. M. and Sah, C. T., "Nanowatt Logic Using Field Effect Metal-Oxide Semiconductor Triode," *IEEE Solid State Circuits Conf.*, p. 32, Philadelphia, 1963.
2. Rea, S. N., "Czochralski Silicon Pull Rate Limits," *Journal of Crystal Growth*, vol. 54, p. 267, 1981.
3. Sze, S. M., *Physics of Semiconductor Devices, 2nd ed.*, Wiley, New York, 1981.
4. Parrillo, L. C., Payne, R. S., Davis, R. E., Reutlinger, G. W., and Field, R. L., "Twin-Tub CMOS — A Technology for VLSI Circuits," *IEEE International Electron Device Meeting 1980*, p. 752, Washington, D.C., 1980.
5. Ohzone, T., Shimura, H., Tsuji, K., and Hirano, "Silicon-Gate n-Well CMOS Process by Full Ion-Implantation Technology," *IEEE Trans. Electron Devices*, vol. ED-27, p. 1789, 1980.
6. Krambeck, R. H., Lee, C. M., and Law, H. F. S., "High-Speed Compact Circuits with CMOS," *IEEE J. Solid-State Circuits*, vol. SC-17, p. 614, 1982.
7. Ochoa, A., Dawes, W., and Estreich, D. B., "Latchup Control in CMOS Integrated Circuits," *IEEE Trans. Nuclear Science*, vol. NS-26(6), p. 5065, 1979.
8. Rung, R. D., Dell'Oca, C. J., and Walker, L. G., "A Retrograde p-Well for Higher Density CMOS," *IEEE Trans. Electron Devices*, vol. ED-28, p. 1115, 1981.
9. Combs, S. R., "Scaleable Retrograde p-Well CMOS Technology," *IEEE International Electron Device Meeting 1981*, p. 346, Washington, D.C., 1981.
10. Schroeder, J. E., Ochoa Jr., A., and Dressendorfer, P. V., "Latch-up Elimination in Bulk CMOS LSI Circuits," *IEEE Trans. Nuclear Science*, vol. NS-27, p. 1735, 1980.
11. Sakai, Y., Hayashida, T., Hashimoto, N., Mimato, O., Musuhara, T., Nagasawa, K., Yasui, T., and Tanimura, N., "Advanced Hi-CMOS Device Technology," *IEEE International Electron Device Meeting 1981*, p. 534, Washington, D.C., 1981.
12. de Werdt, R., van Attekum, P., den Blanken, H., de Bruin, L., op den Buijsch, F., Burgmans, A., Doan, T., Godon, H., Grief, M., Jansen, W., Jonkers, A., Klaassen, F., Pitt, M., van der Plass, P., Stomeijer, A., Verhaar, R., and Weaver, J., "A 1M SRAM with Full CMOS Cells Fabricated in a 0.7 μm Technology," *IEEE International Electron Device Meeting 1987*, p. 532, Washington, D.C., 1987.
13. Apples, J. A., Kooi, E., Paffen, M.M., Schlorje, J. J. H., and Verkuylen, W. H. C. G., "Local Oxidation of Silicon and its Application in Semiconductor Technology," *Philips Research Report*, vol. 25, p. 118, 1970.
14. Shankoff, T. A., Sheng, T. T., Haszko, S. E., Marcus, R. B., and Smith, T. E., "Bird's Beak Configuration and Elimination of Gate Oxide Thinning Produced During Selective Oxidation," *Journal of Electrochemical Society*, vol. 127, p. 216, 1980.
15. Nakajima, S., Kikuchi, K., Minegishi, K., Araki, T., Ikuta, K., and Oda, M., "1 μm 256K RAM Process Technology Using Molybdenum-Polysilicon Gate," *IEEE International Electron Device Meeting 1981*, p. 663, Washington, D.C., 1981.
16. Fuse, G., Ogawa, H., Tateiwa, K., Nakano, I., Odanaka, S., Fukumoto, M., Iwasaki, H., and Ohzone, T., "A Practical Trench Isolation Technology with a Novel Planarization Process," *IEEE International Electron Device Meeting 1987*, p. 732, Washington, D.C., 1987.

17. Perry, K. A., "Chemical Mechanical Polishing: The Impact of a New Technology on an Industry," *1998 Symposium on VLSI Technology, Digest of Technical Papers*, p. 2, Honolulu, 1998.

18. Kuroi, T., Uchida, T., Horita, K., Sakai, M., Itoh, Y., Inoue, Y., and Nishimura, T., "Stress Analysis of Shallow Trench Isolation for 256M DRAM and Beyond," *IEEE International Electron Device Meeting 1998*, p. 141, San Francisco, CA, 1998.

19. Matsuda, S., Sato, T., Yoshimura, H., Sudo, A., Mizushima, I., Tsumashima, Y., and Toyoshima, Y., "Novel Corner Rounding Process for Shallow Trench Isolation utilizing MSTS (Micro-Structure Transformation of Silicon)," *IEEE International Electron Device Meeting 1998*, p. 137, San Francisco, CA, 1998.

20. Hu, G. J. and Bruce, R. H., "Design Trade-off between Surface and Buried-Channel FETs," *IEEE Trans. Electron Devices*, vol. 32, p. 584, 1985.

21. Cham, K. M., D. W. Wenocur, Lin, J., Lau, C. K., and Hu, H.-S., "Submicronmeter Thin Gate Oxide p-Channel Transistors with p^+ Poly-silicon Gates for VLSI Applications," *IEEE Electron Device Letters*, vol. EDL-7, p. 49, 1986.

22. Amm, D. T., Mingam, H., Delpech, P., and d'Ouville, T. T., "Surface Mobility in p^+ and n^+ Doped Polysilicon Gate PMOS Transistors," *IEEE Trans. Electron Devices*, vol. 36, p. 963, 1989.

23. Toriumi, A., Mizuno, T., Iwase, M., Takahashi, M., Niiyama, H., Fukumoto, M., Inaba, S., Mori, I., and Yoshimi, M., "High Speed 0.1 μm CMOS Devices Operating at Room Temperature," *Extended Abstract of 1992 International Conference on Solid State Devices and Materials*, p. 487, Tsukuba, Japan, 1992.

24. Oyamatsu, H., Kinugawa, M., and Kakumu, M., "Design Methodology of Deep Submicron CMOS Devices for 1V Operation," *1993 Symposium on VLSI Technology, Digest of Technical Papers*, p. 89, Kyoto, Japan, 1993.

25. Takeuchi, K., Yamamoto, T., Tanabe, A., Matsuki, T., Kunio, T., Fukuma, M., Nakajima, K., Aizaki, H., Miyamoto, H., and Ikawa, E., "0.15 μm CMOS with High Reliability and Performance," *IEEE International Electron Device Meeting 1993*, p. 883, Washington, D.C., 1993.

26. Ligenza, J. R. and Spitzer, W. G., "The Mechanism for Silicon Oxidation in Steam and Oxygen," *J. Phys. Chem. Solids*, vol. 14, p. 131, 1960

27. Morimoto, T., Momose, H. S., Ozawa, Y., Yamabe, K., and Iwai, H., "Effects of Boron Penetration and Resultant Limitations in Ultra Thin Pure-Oxide and Nitrided-Oxide," *IEEE International Electron Device Meeting 1990*, p. 429, Washington, D.C., 1990.

28. Uchiyama, A., Fukuda, H., Hayashi, T., Iwabuchi, T., and Ohno, S., "High Performance Dual-Gate Sub-Halfmicron CMOSFETs with 6nm-thick Nitrided SiO_2 Films in an N_2O Ambient," *IEEE International Electron Device Meeting 1990*, p. 425, Washington, D.C., 1990.

29. Rodder, M., Chen, I.-C., Hattangaly, S., and Hu, J. C., "Scaling to a 1.0V-1.5V, sub 0.1μm Gate Length CMOS Technology: Perspective and Challenges," *Extended Abstract of 1998 International Conference on Solid State Devices and Materials*, p. 158, Hiroshima, Japan, 1998.

30. Rodder, M., Hattangaly, S., Yu, N., Shiau, W., Nicollian, P., Laaksonen, T., Chao, C. P., Mehrotra, M., Lee, C., Murtaza, S., and Aur, A., "A 1.2V, 0.1μm Gate Length CMOS Technology: Design and Process Issues," *IEEE International Electron Device Meeting 1998*, p. 623, San Francisco, 1998.

31. Khare, M., Guo, X., Wang, X. W., and Ma, T. P., "Ultra-Thin Silicon Nitride Gate Dielectric for Deep-Sub-Micron CMOS Devices," *1997 Symposium on VLSI Technology, Digest of Technical Papers*, p. 51, Kyoto, Japan, 1997.

32. Yagishita, A., Saito, T., Nakajima, K., Inumiya, S., Akasaka, Y., Ozawa, Y., Minamihaba, G., Yano, H., Hieda, K., Suguro, K., Arikado, T., and Okumura, K., "High Performance Metal Gate MOSFETs Fabricated by CMP for 0.1 μm Regime," *IEEE International Electron Device Meeting 1998*, p. 785, San Francisco, CA, 1998.

33. Murarka, S. P., Fraser, D. B., Shinha, A. K., and Levinstein, H. J., "Refractory Silicides of Titanium and Tantalum for Low-Resistivity Gates and Interconnects," *IEEE Trans. Electron Devices*, vol. ED-27, p. 1409, 1980.

34. Geipel, H. J., Jr., Hsieh, N., Ishaq, M. H., Koburger, C. W., and White, F. R., "Composite Silicide Gate Electrode — Interconnections for VLSI Device Technologies," *IEEE Trans. Electron Devices*, vol. ED-27, p. 1417, 1980.

35. Hayashida, H., Toyoshima, Y., Suizu, Y., Mitsuhashi, K., Iwai, H., and Maeguchi, K., "Dopant Redistribution in Dual Gate W-polycide CMOS and its Improvement by RTA," *1989 Symposium on VLSI Technology, Digest of Technical Papers*, p. 29, Kyoto, Japan, 1989.

36. Uwasawa, K., Mogami, T., Kunio, T., and Fukuma, M., "Scaling Limitations of Gate Oxide in p^+ Polysilicon Gate MOS Structure for Sub-Quarter Micron CMOS Devices," *IEEE International Electron Device Meeting 1993*, p. 895, Washington, D.C., 1993.

37. Ozaki, T., Azuma, T., Itoh, M., Kawamura, D., Tanaka, S., Ishibashi, Y., Shiratake, S., Kyoh, S., Kondoh, T., Inoue, S., Tsuchida, K., Kohyama, Y., and Onishi, Y., "A 0.15μm KrF Lithography for 1Gb DRAM Product Using High Printable Patterns and Thin Resist Process," *1998 Symposium on VLSI Technology, Digest of Technical Papers*, p. 84, Honolulu, 1998.

38. Hirukawa, S., Matsumoto, K., and Takemasa, K., "New Projection Optical System for Beyond 150 nm Patterning with KrF and ArF Sources," *Proceedings of 1998 International Symposium on Optical Science, Engineering, and Instrumentation, SPIE's 1998 Annual Meeting*, p. 414, 1998.

39. Triumi, A. and Iwase, M., "Lower Submicrometer MOSFETs Fabricated by Direct EB Lithography," *Extended Abstract of the 19th Conference on Solid State Devices and Materials*, p. 347, Tokyo, Japan, 1987.

40. Liddle, J. A. and Berger, S. D., "Choice of System Parameters for Projection Electron-Beam Lithography: Accelerating Voltage and Demagnification Factor," *Journal of Vacuum and Science Technology*, vol. B10(6), p. 2776, 1992.

41. Nakajima, K., Yamashita, H., Kojima, Y., Tamura, T., Yamada, Y., Tokunaga, K., Ema, T., Kondoh, K., Onoda, N., and Nozue, H., "Improved 0.12μm EB Direct Writing for Gbit DRAM Fabrication," *1998 Symposium on VLSI Technology, Digest of Technical Papers*, p. 34, Honolulu, 1998.

42. Deguchi, K., Miyoshi, K., Ban, H., Kyuragi, H., Konaka, S., and Matsuda, T., "Application of X-ray Lithography with a Single-Layer Resist Process to Subquartermicron LSI Fabrication," *Journal of Vacuum and Science Technology*, vol. B10(6), p. 3145, 1992.

43. Matsuoka, F., Iwai, H., Hayashida, H., Hama, K., Toyoshima, Y., and Maeguchi, K., "Analysis of Hot Carrier Induced Degradation Mode on pMOSFETs," *IEEE Trans. Electron Devices*, vol. ED-37, p. 1487, 1990.

44. Ogura, S., Chang, P. J., Walker, W. W., Critchlow, D. L., and Shepard, J. F., "Design and Characteristics of the Lightly-Doped Drain-Source (LDD) Insulated Gate Field Effect Transistor," *IEEE Trans. Electron Devices*, vol. ED-27, p. 1359, 1980.

45. Ha, J. M., Park, J. W., Kim, W. S., Kim, S. P., Song, W. S., Kim, H. S., Song, H. J., Fujihara, K., Lee, M. Y., Felch, S., Jeong, U., Groeckner, M., Kim, K. H., Kim, H. J., Cho, H. T., Kim, Y. K., Ko, D. H., and Lee, G. C., "High Performance pMOSFET with BF_3 Plasma Doped Gate/Source/Drain and A/D Extension," *IEEE International Electron Device Meeting 1998*, p.639, San Francisco, 1998.

46. Goto, K., Matsuo, J., Sugii, T., Minakata, H., Yamada, I., and Hisatsugu, T., "Novel Shallow Junction Technology Using Decaborone ($B_{10}H_{14}$)," *IEEE International Electron Device Meeting 1996*, p. 435, San Francisco, 1996.

47. Ohguro, T., Nakamura, S., Saito, M., Ono, M., Harakawa, H., Morifuji, E., Yoshitomi, T., Morimoto, T., Momose, H. S., Katsumata, Y., and Iwai, H., "Ultra-shallow Junction and Salicide Technique for Advanced CMOS Devices," *Proceedings of the Sixth International Symposium on Ultralarge Scale Integration Science and Technology, Electrochemical Society*, p. 275, May 1997.

48. Osburn, C. M., Tsai, M. Y., and Zirinsky, S., "Self-Aligned Silicide Conductors in FET Integrated Circuits," *IBM Technical Disclosure Bulletin*, vol. 24, p. 1970, 1981.

49. Shibata, T., Hieda, K., Sato, M., Konaka, M., Dang, R. L. M., and Iizuka, H., "An Optimally Designed Process for Submicron MOSFETs," *IEEE International Electron Device Meeting 1981*, p. 647, Washington, D.C., 1981.

50. Ting, C. Y., Iyer, S. S., Osburn, C. M., Hu, G. J., and Schweighart, A. M., "The Use of TiSi$_2$ in a Self-Aligned Silicide Technology," *Proceedings of 1st International Symposium on VLSI Science and Technology, Electrochemical Society Meeting*, vol. 82(7), p. 224, 1982.

51. Haken, R. A., "Application of the Self-Aligned Titanium Silicide Process to Very Large Scale Integrated N-Metal-Oxide-Semiconductor and Complementary Metal-Oxide-Semiconductor Technologies," *Journal of Vacuum Science and Technology*, vol. B3(6), p. 1657, 1985.

52. Kobayashi, N., Hashimoto, N., Ohyu, K., Kaga, T., and Iwata, S., "Comparison of TiSi$_2$ and WSi$_2$ for Sub-Micron CMOSs," *1986 Symposium on VLSI Technology, Digest of Technical Papers*, p. 49, 1986.

53. Ho, V. Q. and Poulin, D., "Formation of Self-Aligned TiSi$_2$ for VLSI Contacts and Interconnects," *Journal of Vacuum Science and Technology*, vol. A5, p. 1396, 1987.

54. Ting, C. H., d'Heurle, F. M., Iyer, S. S., and Fryer, P. M., "High Temperature Process Limitation on TiSi$_2$," *Journal of Electrochemical Society*, vol. 133(12), p. 2621, 1986.

55. Osburn, C. M., Tsai, M. Y., Roberts, S., Lucchese, C. J., and Ting, C. Y., "High Conductivity Diffusions and Gate Regions Using a Self-Aligned Silicide Technology," *Proceedings of 1st International Symposium on VLSI Science and Technology, Electrochemical Society*, vol. 82-1, p. 213, 1982.

56. Kwork, T., "Effect of Metal Line Geometry on Electromigration Lifetime in Al-Cu Submicron Interconnects," *26th Annual Proceedings of Reliability Physics 1988*, p. 185, 1988.

57. Owada, N., Hinode, K., Horiuchi, M., Nishida, T., Nakata, K., and Mukai, K., "Stress Induced Slit-Like Void Formation in a Fine-Pattern Al-Si Interconnect during Aging Test," *1985 Proceedings of the 2nd International IEEE VLSI Multilevel Interconnection Conference*, p. 173, 1985.

58. Kikkawa, T., Aoki, H., Ikawa, E., and Drynan, J. M., "A Quarter-Micrometer Interconnection Technology Using a TiN/Al-Si-Cu/Al-Si-Cu/TiN/Ti Multilayer Structure," *IEEE Trans. Electron Devices*, vol. ED-40, p. 296, 1993.

59. White, F., Hill, W., Eslinger, S., Payne, E., Cote, W., Chen, B., and Johnson, K., "Damascene Stud Local Interconnect in CMOS Technology," *IEEE International Electron Device Meeting 1992*, p. 301, San Francisco, 1992.

60. Kobayashi, N., Suzuki, M., and Saitou, M., "Tungsten Plug Technology: Substituting Tungsten for Silicon Using Tungsten Hexafluoride," *Extended Abstract of 1988 International Conference on Solid State Devices and Materials*, p. 85, 1988.

61. Cote, W., Costrini, G., Eldlstein, D., Osborn, C., Poindexter, D., Sardesai, V., and Bronner, G., "An Evaluation of Cu Wiring in a Production 64Mb DRAM," *1998 Symposium on VLSI Technology, Digest of Technical Papers*, p. 24, Honolulu, 1998.

62. Loke, A. L. S., Wetzel, J., Ryu, C., Lee, W.-J., and Wong, S. S., "Copper Drift in Low-κ Polymer Dielectrics for ULSI Metallization," *1998 Symposium on VLSI Technology, Digest of Technical Papers*, p. 26, Honolulu, 1998.

63. Wada, J., Oikawa, Y., Katata, T., Nakamura, N., and Anand, M. B., "Low Resistance Dual Damascene Process by AL Reflow Using Nb Liner," *1998 Symposium on VLSI Technology, Digest of Technical Papers*, p. 48, Honolulu, 1998.

64. Momose, H. S., Fujimoto, R., Ohtaka, S., Morifuji, E., Ohguro, T., Yoshitomi, T., Kimijima, H., Nakamura, S., Morimoto, T., Katsumata, Y., Tanimoto, H., and Iwai, H., "RF Noise in 1.5 nm Gate Oxide MOSFETs and the Evaluation of the NMOS LNA Circuit Integrated on a Chip," *1998 Symposium on VLSI Technology, Digest of Technical Papers*, p. 96, Honolulu, 1998.

65. Burghartz, J. N., "Progress in RF Inductors on Silicon — Understanding Substrate Loss," *IEEE International Electron Device Meeting 1998*, p. 523, San Francisco, 1998.

66. Yoshitomi, T., Sugawara, Y., Morifuji, E., Ohguro, T., Kimijima, H., Morimoto, T., Momose, H. S., Katsumata, Y., and Iwai, H., "On-Chip Inductors with Diffused Shield Using Channel-Stop Implant," *IEEE International Electron Device Meeting 1998*, p. 540, San Francisco, 1998.

67. Borel, J., "Technologies for Multimedia Systems on a Chip," *International Solid State Circuit Conference, Digest of Technical Papers*, p. 18, 1997.

68. Tsukamoto, M., Kuroda, H., and Okamoto, Y., "0.25mm W-polycide Dual Gate and Buried Metal on Diffusion Layer (BMD) Technology for DRAM-Embedded Logic Devices," *1997 Symposium on VLSI Technology, Digest of Technical Papers*, p. 23, 1997.

69. Itabashi, K., Tsuboi, S., Nakamura, H., Hashimoto, K., Futoh, W., Fukuda, K., Hanyu, I., Asai. S., Chijimatsu, T., Kawamura, E., Yao, T., Takagi, H., Ohta, Y., Karasawa, T., Iio, H., Onoda, M., Inoue, F., Nomura, H., Satoh, Y., Higashimoto, M., Matsumiya, M., Miyabo, T., Ikeda, T., Yamazaki, T., Miyajima, M., Watanabe, K., Kawamura, S., and Taguchi, M., "Fully Planarized Stacked Capacitor Cell with Deep and High Aspect Ratio Contact Hole for Giga-bit DRAM," *1997 Symposium on VLSI Technology, Digest of Technical Papers*, p. 21, 1997.

70. Kim, K. N., Lee, J. Y., Lee, K. H., Noh, B. H., Nam, S. W., Park, Y. S., Kim, Y. H., Kim, H. S., Kim, J. S., Park, J. K., Lee, K. P., Lee, K. Y., Moon, J. T., Choi, J. S., Park, J. W., and Lee, J. G., "Highly Manufacturable 1Gb SDRAM," *1997 Symposium on VLSI Technology, Digest of Technical Papers*, p. 10, 1997.

71. Kohyama, Y., Ozaki, T., Yoshida, S., Ishibashi, Y., Nitta, H., Inoue, S., Nakamura, K., Aoyama, T., Imai, K., and Hayasaka, N., "A Fully Printable, Self-Aligned and Planarized Stacked Capacitor DRAM Cell Technology for 1Gbit DRAM and Beyond," *1997 Symposium on VLSI Technology, Digest of Technical Papers*, p. 17, 1997.

72. Drynan, J. M., Nakajima, K., Akimoto, T., Saito, K., Suzuki, M., Kamiyama, S., and Takaishi, Y., "Cylindrical Full Metal Capacitor Technology for High-Speed Gigabit DRAMs," *1997 Symposium on VLSI Technology, Digest of Technical Papers*, p. 151, 1997.

73. Takehiro, S., Yamauchi, S., Yoshimura, M., and Onoda, H., "The Simplest Stacked BST Capacitor for the Future DRAMs Using a Novel Low Temperature Growth Enhanced Crystallization," *1997 Symposium on VLSI Technology, Digest of Technical Papers*, p. 153, 1997.

74. Nesbit, L., Alsmeier, J., Chen, B., DeBrosse, J., Fahey, P., Gall, M., Gambino, J., Gerhard, S., Ishiuchi, H., Kleinhenz, R., Mandelman, J., Mii, T., Morikado, M., Nitayama, A., Parke, S., Wong, H., and Bronner, G., "A 0.6μm² 256Mb Trench DRAM Cell with Self-Aligned BuriEd STrap (BEST)," *IEEE International Electron Device Meeting*, p. 627, Washington, D.C., 1993.

75. Bronner, G., Aochi, H., Gall, M., Gambino, J., Gernhardt, S., Hammerl, E., Ho, H., Iba, J., Ishiuchi, H., Jaso, M., Kleinhenz, R., Mii, T., Narita, M., Nesbit, L., Neumueller, W., Nitayama, A., Ohiwa, T., Parke, S., Ryan, J., Sato, T., Takato, H., and Yoshikawa, S., "A Fully Planarized 0.25μm CMOS Technology for 256Mbit DRAM and Beyond," *1995 Symposium on VLSI Technology, Digest of Technical Papers*, p. 15, 1995.

76. Ishiuchi, H., Yoshida, Y., Takato, H., Tomioka, K., Matsuo, K., Momose, H., Sawada, S., Yamazaki, K., and Maeguchi, K., "Embedded DRAM Technologies," *IEEE International Electron Device Meeting*, p. 33, Washington, D.C., 1997.

77. Togo, M., Iwao, S., Nobusawa, H., Hamada, M., Yoshida, K., Yasuzato, N., and Tanigawa, T., "A Salicide-Bridged Trench Capacitor with a Double-Sacrificial-Si₃N₄-Sidewall (DSS) for High-Performance Logic-Embedded DRAMs," *IEEE International Electron Device Meeting*, p. 37, Washington, D.C., 1997.

78. Crowder, S., Stiffler, S., Parries, P., Bronner, G., Nesbit, L., Wille, W., Powell, M., Ray, A., Chen, B., and Davari, B., "Trade-offs in the Integration of High Performance Devices with Trench Capacitor DRAM," *IEEE International Electron Device Meeting*, p. 45, Washington, D.C., 1997.

79. Crowder, S., Hannon, R., Ho, H., Sinitsky, D., Wu, S., Winstel, K., Khan, B., Stiffler, S. R., and Iyer, S. S., "Integration of Trench DRAM into a High-Performance 0.18 μm Logic Technology with Copper BEOL," *IEEE International Electron Device Meeting*, p. 1017, San Francisco, 1998.

80. Yoshida, M., Kumauchi, T., Kawakita, K., Ohashi, N., Enomoto, H., Umezawa, T., Yamamoto, N., Asano, I., and Tadaki, Y., "Low Temperature Metal-based Cell Integration Technology for Gigabit and Embedded DRAMs," *IEEE International Electron Device Meeting*, p. 41, Washington, D.C., 1997.

81. Nakamura, S., Kosugi, M., Shido, H., Kosemura, K., Satoh, A., Minakata, H., Tsunoda, H., Kobayashi, M., Kurahashi, T., Hatada, A., Suzuki, R., Fukuda, M., Kimura, T., Nakabayashi, M., Kojima, M., Nara, Y., Fukano, T., and Sasaki, N., "Embedded DRAM Technology Compatible to the 0.13 μm High-Speed Logics by Using Ru Pillars in Cell Capacitors and Peripheral Vias," *IEEE International Electron Device Meeting*, p. 1029, San Francisco, 1998.

82. Drynan, J. M., Fukui, K., Hamada, M., Inoue, K., Ishigami, T., Kamiyama, S., Matsumoto, A., Nobusawa, H., Sugai, K., Takenaka, M., Yamaguchi, H., and Tanigawa, T., "Shared Tungsten Structures for FEOL/BEOL Compatibility in Logic-Friendly Merged DRAM," *IEEE International Electron Device Meeting*, p. 849, San Francisco, 1998.

83. Togo, M., Noda, K., and Tanigawa, T., "Multiple-Thickness Gate Oxide and Dual-Gate Technologies for High-Performance Logic-Embedded DRAMs," *IEEE International Electron Device Meeting*, p. 347, San Francisco, 1998.

84. Yoshikawa, K., "Embedded Flash Memories — Technology Assessment and Future —," *1999 International Symposium on VLSI Technology, System, and Applications*, p. 183, Taipei, 1999.

85. Yoshikawa, K., "Guide-lines on Flash Memory Cell Selection," *Extended Abstract of 1998 International Conference on Solid State Devices and Materials*, p. 138, 1998.

86. Watanabe, H., Yamada, S., Tanimoto, M., Mitsui, M., Kitamura, S., Amemiya, K., Tanzawa, T., Sakagami, E., Kurata, M., Isobe, K., Takebuchi, M., Kanda, M., Mori, S.,and Watanabe, T., "Novel $0.44\mu m^2$ Ti-Salicide STI Cell Technology for High-Density NOR Flash Memories and High Performance Embedded Application," *IEEE International Electron Device Meeting 1998*, p. 975, San Francisco, 1998.

87. Kuo, C., "Embedded Flash Memory Applications, Technology and Design," *1995 IEDM Short Course: NVRAM Technology and Application, IEEE International Electron Device Meeting*, Washington, D.C., 1995.

88. Ohnakado, T., Mitsunaga, K., Nunoshita, M., Onoda, H., Sakakibara, K., Tsuji, N., Ajika, N., Hatanaka, M., and Miyoshi, H., "Novel Electron Injection Method Using Band-to-Band Tunneling Induced Hot Electron (BBHE) for Flash Memory with a p-channel Cell," *IEEE International Electron Device Meeting*, p. 279, Washington, D.C., 1995.

89. Iwai, H., Sasaki, G., Unno, Y., Niitsu, Y., Norishima, M., Sugimoto, Y., and Kannzaki, K., "$0.8\mu m$ Bi-CMOS Technology with High f_T Ion-Implanted Emitter Bipolar Transistor," *IEEE International Electron Device Meeting 1987*, p. 28, Washington, D. C., 1987.

90. Clark, L. T. and Taylor, G. F., "High Fan-in Circuit Design," *1994 Bipolar/BiCMOS Circuits & Technology Meeting*, p. 27, Minneapolis, MN, 1994.

91. Nii, H., Yoshino, C., Inoh, K., Itoh, N., Nakajima, H., Sugaya, H., Naruse, H., Kataumata, Y., and Iwai, H., "0.3 μm BiCMOS Technology for Mixed Analog/Digital Application System," *1997 Bipolar/BiCMOS Circuits & Technology Meeting*, p. 68, Minneapolis, MN, 1997.

92. Johnson, R. A., Zierak, M. J., Outama, K. B., Bahn, T. C., Joseph, A. J.,Cordero, C. N., Malinowski, J., Bard, K. A., Weeks, T. W., Milliken, R. A., Medve, T. J., May, G. A., Chong, W., Walter, K. M., Tempest, S. L., Chau, B. B., Boenke, M., Nelson, M. W., and Harame, D. L.,"1.8 million Transistor CMOS ASIC Fabricated in a SiGe BiCMOS Technology," *IEEE International Electron Device Meeting 1998*, p. 217, San Francisco, 1998.

93. Chyan, Y.-F., Ivanov, T. G., Carroll, M. S., Nagy, W. J., Chen, A. S., and Lee, K. H., "A 50-GHz 0.25-μm High-Energy Implanted BiCMOS (HEIBiC) Technology for Low-Power High-Integration Wireless-Communication System," *1998 Symposium on VLSI Technology, Digest of Technical Papers*, p. 92, Honolulu, 1998.

94. "The National Technology Roadmap for Semiconductors," Semiconductor Industry Association, 1997.

95. Momose, H. S., Ono, M., Yoshitomi, T., Ohguro, T., Nakamura, S., Saito M., and Iwai, H., "Tunneling Gate Oxide Approach to Ultra-High Current Drive in Small-Geometry MOSFETs," *IEEE International Electron Device Meeting*, p. 593, San Francisco, 1994.

96. Momose, H. S., Ono, M., Yoshitomi, T., Ohguro, T., Nakamura, S., Saito, M., and Iwai, H., "1.5 nm Direct-Tunneling Gate Oxide Si MOSFETs," *IEEE Trans. Electron Devices*, vol. ED-43, p. 1233, 1996.

97. Timp, G., Agarwal, A., Baumann, F. H., Boone, T., Buonanno, M., Cirelli, R., Donnelly, V., Foad, M., Grant, D., Green, M., Gossmann, H., Hillenius, S., Jackson, J., Jacobson, D., Kleiman, R., Kornblit, A., Klemens, F., Lee, J. T.-C., Mansfield, W., Moccio, S., Murrell, A., O'Malley, M., Rosamilia, J., Sapjeta, J., Silverman, P., Sorsch, T.,Tai, W. W., Tennant, D., Vuong, H., and Weir, B., "Low Leakage, Ultra-Thin Gate Oxides for Extremely High Performance sub-100 nm nMOSFETs," *IEEE International Electron Device Meeting*, p. 930, Washington, D.C., 1997.

98. Momose, H. S., Ono, M., Yoshitomi, T., Ohguro, T., Nakamura, S., Saito, M., and Iwai, H., "Prospects for Low-Power, High-Speed MPUs Using 1.5 nm Direct-Tunneling Gate Oxide MOS-FETs," *Journal of Solid-State Electronics*, vol. 41, p. 707, 1997.

99. Momose, H. S., Morifuji, E., Yoshitomi, T., Ohguro, T., Saito, M., Morimoto, T., Katsumata, Y., and Iwai, H., "High-Frequency AC Characteristics of 1.5 nm Gate Oxide MOSFETs," *IEEE International Electron Device Meeting*, p. 105, San Francisco, 1996.

100. Momose, H. S., Nakamura, S., Ohguro, T., Yoshitomi, T., Morifuji, E., Morimoto, T., Katsumata, Y., and Iwai, H., "Study of the Manufacturing Feasibility of 1.5 nm Direct-Tunneling Gate Oxide MOSFETs: Uniformity, Reliability, and Dopant Penetration of the Gate Oxide," *IEEE Trans. Electron Devices*, vol. ED-45, p. 691, 1998.

101. Lo, S.-H., Buchanan, D. A., Taur, Y., and Wang, W., "Quantum-Mechanical Modeling of Electron Tunneling Current from the Inversion Layer of Ultra-Thin-Oxide nMOSFETs," *IEEE Electron Devices Letters*, vol. EDL-18, p. 209, 1997.

102. Sorsch, T., Timp, W., Baumann, F. H., Bogart, K. H. A., Boone, T., Donnelly, V. M., Green, M., Evans-Lutterodt, K., Kim, C. Y., Moccio, S., Rosamilia, J., Sapjeta, J., Silverman, P., Weir B., and Timp, G., "Ultra-Thin, 1.0–3.0 nm, Gate Oxides for High Performance sub-100 nm Technology," *1998 Symposium on VLSI Technology, Digest of Technical Papers*, p. 222, 1998.

103. Ohguro, T., Naruse, N., Sugaya, H., Morifuji, E., Nakamura, S., Yoshitomi, T., Morimoto, T., Momose, H. S., Katsumata, Y., and Iwai, H., "0.18 μm Low Voltage/Low Power RF CMOS with Zero Vth Analog MOSFETs made by Undoped Epitaxial Channel Technique," *IEEE International Electron Device Meeting*, p. 837, Washington, D.C., 1997.

104. Ohguro, T., Naruse, H., Sugaya, H., Kimijima, H., Morifuji, E., Yoshitomi, T., Morimoto, T., Momose, H. S., Katsumata, Y., and Iwai, H., "0.12 μm Raised Gate/Source/Drain Epitaxial Channel NMOS Technology," *IEEE International Electron Device Meeting 1998*, p. 927, San Francisco, 1998.

3

Bipolar Technology

Jan V. Grahn
Mikael Östling
Royal Institute of Technology (KTH)

3.1   Introduction

The development of a bipolar technology for integrated circuits goes hand in hand with the steady improvement in semiconductor materials and discrete components during the 1950s and 1960s. Consequently, silicon bipolar technology formed the basis for the IC market during the 1970s. As circuit dimensions shrink, the MOSFET (or MOS) has gradually taken over as the major technological platform for silicon integrated circuits. The main reasons are the ease of miniaturization and high yield for MOS compared to bipolar technology. However, during the same period of MOS growth, much progress was simultaneously achieved in bipolar technology.[1,2] This is illustrated in Fig. 3.1 where the reported gate delay time for emitter-coupled logic (ECL) is plotted versus year.[2,3] In 1984, the 100 ps/gate limit was broken and, since then, the speed performance has been improved by a factor of ten. The high speed and large versatility of the silicon bipolar transistor still make it an attractive choice for a variety of digital and analog applications.[4]

Apart from high-speed performance, the bipolar transistor is recognized by its excellent analog properties. It features high linearity, superior low- and high-frequency noise behavior, and a very large transconductance.[5] Such properties are highly desirable for many RF applications, both for narrow-band as well as broad-band circuits.[6] The high current drive capability per unit silicon area makes the bipolar transistor suitable for input/output stages in many IC designs (e.g., in fast SRAMs). The disadvantage of bipolar technology is the low transistor density, combined with a large power dissipation. High-performance bipolar circuits are therefore normally fabricated at a modest integration level (MSI/LSI). By using BiCMOS design, the benefits of both MOS and bipolar technology are utilized.[7] One example is mixed analog/digital systems where a high-performance bipolar process is integrated with high-density CMOS. This technology forms a vital part in several system-on-a-chip designs (e.g., for telecommunication circuits).

In this chapter, a brief overview of bipolar technology is given with an emphasis on the integrated silicon bipolar transistor. The information presented here is based on the assumption that the reader is

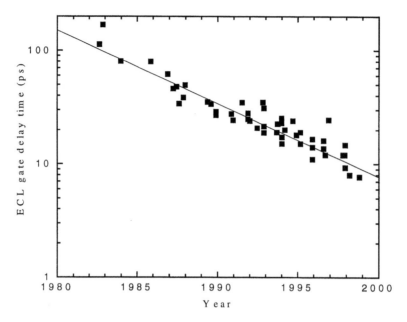

FIGURE 3.1 Reported gate delay time for bipolar ECL circuits vs. year.

familiar with bipolar device fundamentals and basic VLSI process technology. Bipolar transistors are treated in detail in the well-known textbooks by Ashburn[8] and Roulston.[9] The first part of this chapter will outline the general concepts in bipolar process design and optimization (Section 3.2). The second part will present the three generations of integrated devices representing state-of-the-art bipolar technologies for the 1970s, 1980s, and 1990s (Sections 3.3, 3.4, and 3.5, respectively). Finally, some future trends in bipolar technology are outlined.

3.2 Bipolar Process Design

The design of a bipolar process starts with the specification of the application target and its circuit technology (digital or analog). This leads to a number of requirements formulated in device parameters and associated figures-of-merit. These are mutually dependent and must therefore be traded off against each other, making the final bipolar process design a compromise between various conflicting device requirements.

Figures-of-Merit

In the digital bipolar process, the cutoff frequency (f_T) is a well-known figure-of-merit for speed. The f_T is defined for a common-emitter configuration with its output short-circuit when extrapolating the small signal current gain to unity. From a circuit perspective, a more adequate figure-of-merit is the gate delay time (τ_d) measured for a ring-oscillator circuit containing an odd number of inverters.[10] The τ_d can be expressed as a linear combination of the incoming time constants weighted by a factor determined by the circuit topology (e.g., ECL).[10,11] Alternative expressions for τ_d calculations have been proposed.[12] Besides speed, power dissipation can also be a critical issue in densely packed bipolar digital circuits, resulting in the power-delay product as a figure-of-merit.[4]

In the analog bipolar process, the dc properties of the transistor are of utmost importance. This involves minimum values on common-emitter current gain (β), Gummel plot linearity (β_{max}/β) breakdown voltage (BV_{CEO}), and Early voltage (V_A). The product $\beta \times V_A$ is often introduced as a figure-of-merit for the device dc characteristics.[13] Rather than f_T, the maximum oscillation frequency, $f_{max} = \sqrt{f_T/(8\pi R_B C_{BC})}$ is preferred as a figure-of-merit in high-speed analog design, where R_B and C_{BC} denote the total base resistance and the base-collector capacitance, respectively.[14] Alternative figures-of-merit for speed have been proposed in the literature.[15,16] Analog bipolar circuits are often crucially dependent on a certain noise immunity, leading to the introduction of the corner frequency and noise figure as figures-of-merit for low-frequency and high-frequency noise properties, respectively.[17]

Process Optimization

The optimization of the bipolar process is divided between the intrinsic and extrinsic device design. This corresponds to the vertical impurity profile and the horizontal layout of the transistor, respectively.[10] See example in Fig. 3.2, where the device cross-section is also included. It is clear that the vertical profile and horizontal layout are primarily dictated by the given process and lithography constraints, respectively.

Figure 3.3 shows a simple flowchart of the bipolar design procedure. Starting from the specified dc parameters at a given operation point, the doping profiles can be derived. The horizontal layout must be adjusted for minimization of the parasitics. A (speed) figure-of-merit can then be calculated. An implicit relation is thus obtained between the figure-of-merit and the processing parameters.[11,18] In practice, several iterations must be performed in the optimization loop in order to find an acceptable compromise between the device parameters. This procedure is substantially alleviated by two-dimensional process simulations of the device fabrication[19] as well as device simulations of the bipolar transistor.[20,21] For optimization of a large number of device parameters, the strategy is based on screening out the unimportant factors, combined with a statistical approach (e.g., response surface methodology).[22]

Vertical Structure

The engineering of the vertical structure involves the design of the collector, base, and emitter impurity profiles. In this respect, f_T is an adequate parameter to optimize. For a modern bipolar transistor with suppressed parasitics, the maximum f_T is usually determined by the forward transit time of minority carriers through the intrinsic component. The most important f_T tradeoff is against BV_{CEO}, as stated by the Johnson limit for silicon transistors:[23] the product $f_T \times BV_{CEO}$ cannot exceed 200 GHz-V (recently updated to 500 GHz-V).[24]

Collector Region

The vertical n-type collector of the bipolar device in Fig. 3.2 consists of two regions below the p-type base diffusion: a lowly or moderately doped n-type epitaxial (epi) layer, followed by a highly doped n^+-subcollector. The thickness and doping level of the subcollector are non-critical parameters; a high arsenic or antimony doping density between 10^{19} and 10^{20} cm^{-3} is representative, resulting in a sheet resistance of 20 to 40 Ω/sq. In contrast, the design of the epi-layer constitutes a fundamental topic in bipolar process optimization.

To first order, the collector doping in the epi-layer is determined by the operation point (more specifically, the collector current density) of the component (see Fig. 3.3). A normal condition is to have the operation point corresponding to maximum f_T, which typically means a collector current density on the order of 2–4×10^4 A/cm^2. As will be recognized later, bipolar scaling results in increased collector current densities. Above a certain current, there will be a rapid roll-off in current gain as well as cutoff frequency. This is due to high-current effects, primarily the base push-out or Kirk effect, leading to a steep increase in the forward transit time.[25] Since the critical current value is proportional to the collector doping,[26] a minimum impurity concentration for the epi-layer is required, thus avoiding f_T degradation (typically around 10^{17} cm^{-3} for a high-speed device). Usually, the epi-layer is doped only in the intrinsic structure by a selectively implanted collector (SIC) procedure.[27] An example of such a doping profile is seen in Fig. 3.4. Such a collector design permits an improved control over the base-collector junction; that is, shorter base width as well as suppressed Kirk effect. The high collector doping concentration, however, may be a concern for both C_{BC} and BV_{CEO}. The latter value will therefore often set a higher limit on the collector doping value. One way to reduce the electrical field in the junction is to implement a

(a)

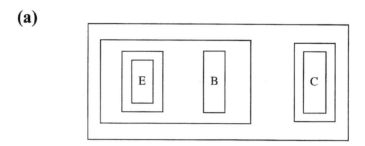

(b)

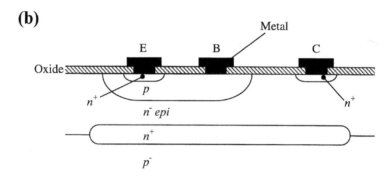

(c)

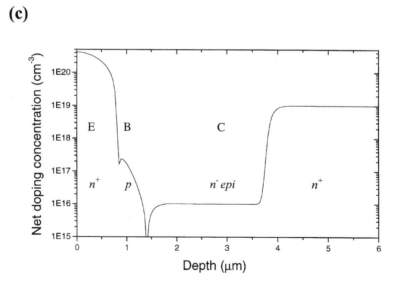

FIGURE 3.2 (a) Layout, (b) cross-section, and (c) simulated impurity profile through emitter window for an integrated bipolar transistor (E = emitter, B = base, C = collector).

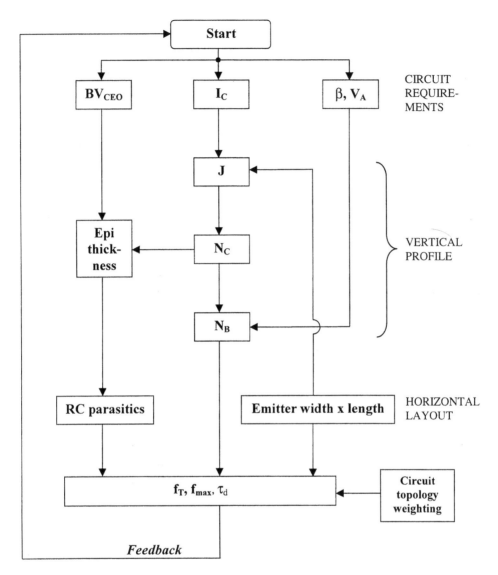

FIGURE 3.3 Generic bipolar device optimization flowchart.

lightly doped collector spacer layer between the heavily doped base and collector regions.[28,29] A retrograde collector profile with a low impurity concentration near the base-collector junction and then increasing toward the subcollector has also been reported to enhance f_T.[30,31]

The thickness of the epi-layer exhibits large variations between different device designs, extending several micrometers in depth for analog bipolar components, whereas a high-speed digital design typically has an epi-layer thickness around 1 μm or below, thus reducing total collector resistance. As a result, the transistor breakdown voltage is sometimes determined by reach-through breakdown (i.e., full depletion penetration of the epi-collector). The thickness of the collector layer can therefore be used as a parameter in determining BV_{CEO}, which in turn is traded off vs. f_T.[32]

In cases where f_{max} is of interest, the collector design must be carefully taken into account. Compared to f_T, the optimum f_{max} is found for thicker and lower doped collector epi-layers.[33,34] The vertical collector design will therefore, to a large extent, determine the tradeoff between f_T and f_{max}.

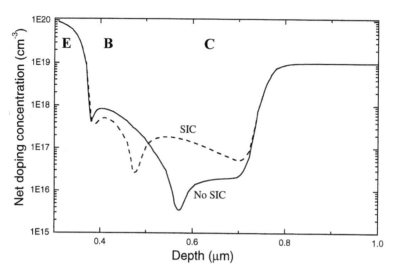

FIGURE 3.4 Simulated vertical impurity profile with and without SIC.

Base Region

The width and peak concentration of the base profile are two of the most fundamental parameters in vertical profile design. The base width W_B is normally in the range 0.1 to 1 μm, whereas a typical base peak concentration lies between 10^{17} and 10^{18} cm^{-3}. The current gain of the transistor is determined by the ratio of the Gummel number in the emitter and base. Usually, a current gain of at least 100 is required for analog bipolar transistors, whereas in digital applications, β around 20 is often acceptable. A normal base sheet resistance (or pinch resistance) for conventional bipolar processes is of the order of 100 Ω/sq., whereas the number for high-speed devices typically is in the interval 1 to 10 kΩ/sq. This is due to the small $W_B < 0.1$ μm necessary for a short base transit time ($\propto W_B^2$). On the other hand, a too narrow base will have a negative impact on f_{max} because of its R_B dependence. As a result, f_{max} exhibits a maximum when plotted against W_B.[35]

The base impurity concentration must be kept high enough to avoid punch-through at low collector voltages; that is, the base-collector depletion layer penetrates across the neutral base. In other words, the base doping level is also dictated by the collector impurity concentration. Punch-through is the ultimate consequence of base width modulation or the Early effect manifested by a finite output resistance in the I_C-V_{CE} transistor characteristic.[36] The associated V_A or the product $\beta \times V_A$ serves as an indicator of the linear properties for the bipolar transistor. The V_A is typically at a relatively high level (>30 V) for analog applications, whereas digital designs often accept relatively low $V_A < 15$ V.

Figure 3.5 demonstrates simulations of current gain versus the base doping for various W_B.[37] It is clearly seen that the base doping interval permitting a high current gain while avoiding punch-through will be pushed to high impurity concentrations for narrow base widths. In addition, Fig. 3.5 points to another limiting factor for high base doping numbers above 5×10^{18} cm^{-3}, namely, the onset of forward-biased tunneling currents in the emitter-base junction[38] leading to non-ideal base current characteristics.[39] It is concluded that the allowable base doping interval will be very narrow for $W_B < 0.1$ μm.

The shape of the base profile has some influence over the device performance. The final base profile is the result of an implantation and diffusion process and, normally, only the peak base concentration is given along with the base width. Nonetheless, there will be an impurity grading along the base profile (see Figs. 3.2 and 3.4), creating a built-in electrical field and thereby adding a drift component for the minority carrier transport.[40] Recent research has shown that for very narrow base transistors, the uniform doping profile is preferable when maximizing f_T.[41,42] This is also valid under high injection conditions in the base.[43] Uniformly doped base profiles are common in advanced bipolar processes using epitaxial techniques for growing the intrinsic base.

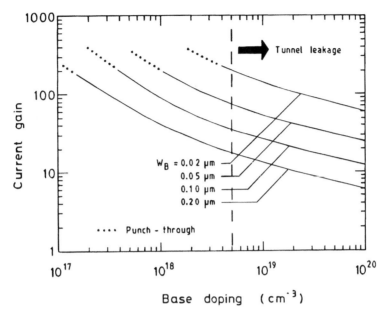

FIGURE 3.5 Simulated current gain vs. base doping density for different base widths ($N_C = 6 \times 10^{16}$ cm^{-3}, $N_E = 10^{20}$ cm^{-3}, and emitter depth 0.20 µm) (after Ref. 37, copyright© 1990, IEEE).

During recent years, base profile design has largely been devoted to implementation of narrow bandgap SiGe in the base. The resulting emitter-base heterojunction allows a considerable enhancement in current gain, which can be traded off against increased base doping, thus substantially alleviating the problem with elevated base sheet resistances typical of high-speed devices.[44] Excellent dc as well as high-frequency properties can be achieved. The position of the Ge profile with respect to the boron profile has been discussed extensively in the literature.[45–47] More details about SiGe heterojunction engineering are found in Chapter 5 by Cressler.

Emitter Region

The conventional metal-contacted emitter is characterized by an abrupt arsenic or phosphorus profile fabricated by direct diffusion or implantation into the base area (see Fig. 3.2).[48] In keeping emitter efficiency close to unity (and thus high current gain), the emitter junction cannot be made too shallow (~1 µm). The emitter doping level lies typically between 10^{20} and 10^{21} cm^{-3} close to the solid solubility limit at the silicon surface, hence providing a low emitter resistance as well as a large emitter Gummel number required for keeping current gain high. Bandgap narrowing, however, will be present in the emitter, causing a reduction in the efficient emitter doping.[49]

When scaling bipolar devices, the emitter junction must be made more shallow to ensure a low emitter-base capacitance. When the emitter depth becomes less than the minority carrier recombination length, the current gain will inevitably degrade. This precludes the use of conventional emitters in a high-performance bipolar technology. Instead, polycrystalline (poly) silicon emitter technology is utilized. By diffusing impurity species from the polysilicon contact into the monocrystalline (mono) silicon, a very shallow junction (< 0.2 µm) is formed; yet gain can be kept at a high level and even traded off against a higher base doping.[50] A gain enhancement factor between 3 and 30 for the polysilicon compared to the monosilicon emitter has been reported (see also Section 3.4).[51,52]

Scaling Rules

The principles for vertical design can be summarized in the bipolar scaling rules formulated by Solomon and Tang;[53,54] see Table 3.1. Since the bipolar transistor is scaled under constant voltage, the current

TABLE 3.1   Bipolar Scaling Rules (Scaling factor $\lambda < 1$)

Parameter	Scaling Factor
Voltage	1
Base width W_B	$\lambda^{0.8}$
Base doping N_B	W_B^{-2}
Current density J	λ^{-2}
Collecting doping	J
Depletion capacitances	λ
Delay	λ
Power	1
Power-delay product	λ

density increases with reduced device dimensions. At medium or high current densities, the vertical structure determines the speed. At low current densities, performance is normally limited by device parasitics. Eventually, tunnel currents or contact resistances constitute a final limit to further speed improvement based on the scaling rules. A solution is to use SiGe bandgap engineering to further enhance device performance without scaling.

Horizontal Layout

The horizontal layout is carried out in order to minimize the device parasitics. Figure 3.6 shows the essential parasitic resistances and capacitances for a schematic bipolar structure containing two base contacts. The various RC constants in Fig. 3.6 introduce time delays. For conventional bipolar transistors, such parasitics often limit device speed. In contrast, the self-alignment technology applied in advanced bipolar transistor fabrication allows for efficient suppression of the parasitics.

In horizontal layout, f_{max} serves as a first-order indicator in the extrinsic optimization procedure because of its dependence on C_{BC} and (total) R_B. These two parasitics are strongly connected to the geometrical layout of the device. The more advanced τ_d calculation takes all major parasitics into account under given load conditions, thus providing good insight into the various time delay contributions of a bipolar logic gate.[55]

From Fig. 3.6, it is seen that the collector resistance is divided into three parts. Apart from the epi-layer and buried layer previously discussed, the collector contact also adds a series resistance. Provided the epi-layer is not too thick, the transistor is equipped with a deep phosphorus plug from the collector contact down to the buried layer, thus reducing the total R_C.

As illustrated in Fig. 3.6, the base resistance is divided into intrinsic (R_{Bi}) and extrinsic (R_{Bx}) components. The former is the pinched base resistance situated directly under the emitter diffusion, whereas the latter constitutes the base regions contacting the intrinsic base. The intrinsic part decreases with the current due to the lateral voltage drop in the base region.[56] At high current densities, this causes current crowding effects at the emitter diffusion edges. This results in a reduced onset for high-current effects in the transistor. The extrinsic base resistance is bias independent and must be kept as small as possible (e.g., by utilizing self-alignment architectures). By designing a device layout with two or more base contacts surrounding the emitter, the final R_B is further reduced at the expense of chip area. Apart from enhancing f_{max}, the R_B reduction is also beneficial for device noise performance.

The layout of the emitter is crucial since the effective emitter area defines the intrinsic device cross-section.[57] The minimum emitter area, within the lithography constraints, is determined by the operational collector current and the critical current density where high-current effects start to occur.[58] Eventually, a tradeoff must be made between the base resistance and device capacitances as a function of emitter geometry; this choice is largely dictated by the device application. Long, narrow emitter stripes, meaning a reduction in the base resistance, are frequently used. The emitter resistance is usually non-critical for conventional devices; however, for polysilicon emitters, the emitter resistance may become a concern in very small-geometry layouts.[3]

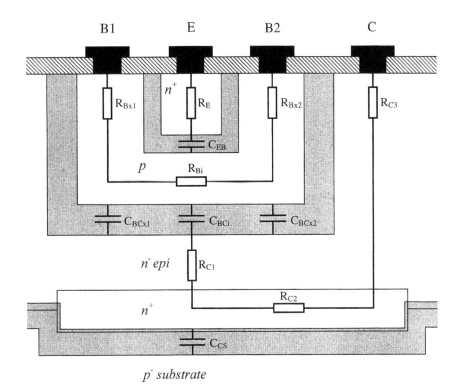

FIGURE 3.6 Schematic view of the parasitic elements in a bipolar transistor equipped with two base contacts. R_E = emitter resistance, R_{Bi} = intrinsic base resistance, R_{Bx} = extrinsic base resistance, R_C = collector resistance, C_{EB} = emitter-base capacitance, C_{BCi} = intrinsic base-collector capacitance, C_{BCx} = extrinsic base-collector capacitance, and C_{CS} = collector-substrate capacitance. Gray areas denote depletion regions. Contact resistances are not shown.

Of the various junction capacitances in Fig. 3.6, the collector-base capacitance is the most significant. This parasitic is also divided into intrinsic (C_{BCi}) and extrinsic (C_{BCx}) contributions. Similar to R_{Bx}, the C_{BCx} is kept low by using self-aligned schemes. For example, the fabrication of a SIC causes an increase only in C_{BCi}, whereas C_{BCx} stays virtually unaffected. The collector-substrate capacitance C_{CS} is one of the minor contributors to f_T; the C_{CS} originates from the depletion regions created in the epi-layer and under the buried layer. C_{CS} will become significant at very high frequencies due to substrate coupling effects.[59]

3.3 Conventional Bipolar Technology

Conventional bipolar technology is based on the device designs developed during the 1960s and 1970s. Despite its age, the basic concept still constitutes a workhorse in many commercial analog processes where ultimate speed and high packing density are not of primary importance. In addition, a conventional bipolar component is often implemented in low-cost BiCMOS processes.

Junction-Isolated Transistors

The early planar transistor technology took advantage of a reverse-biased pn junction in providing the necessary isolation between components. One of the earliest junction-isolated transistors, the so-called triple-diffused process, is simply based on three ion implantations and subsequent diffusion.[60] This device has been integrated into a standard CMOS process using one extra masking step.[61] The triple-diffused bipolar process, however, suffers from a large collector resistance due to the absence of a subcollector, and the npn performance will be low.

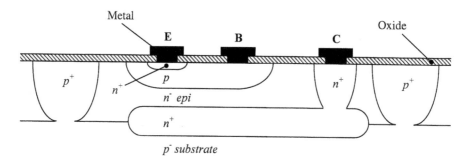

FIGURE 3.7 Cross-section of the buried-collector transistor with junction isolation and collector plug.

By far, the most common junction-isolated transistor is represented by the device cross-section of Fig. 3.7, the so-called buried-collector process.[60] This device is based on the concept previously shown in Fig. 3.2 but with the addition of an n^+-collector plug and isolation. This is provided by the diffused p^+-regions surrounding the transistor. The diffusion of the base and emitter impurities into the epi-layer allows relatively good control of the base width (more details of the fabrication is given in the next section on oxide-isolated transistors).

A somewhat different approach of the buried-collector process is the so-called collector-isolation diffusion.[62] This process requires a p-type epi-layer after formation of the subcollector. An n^+- diffusion serves both as isolation and the collector plug to the buried layer. After an emitter diffusion, the p-type epi-layer will constitute the base of the final device. Compared to the buried-layer collector process, the collector-isolated device concept does not result in very accurate control over the final base width.

The main disadvantage of the junction-isolated transistor is the relatively large chip area occupied by the isolation region, thus precluding the use of such a device in any VLSI application. Furthermore, high-speed operation is ruled out because of the large parasitic capacitances associated with the junction isolation and the relatively deep diffusions involved. Indeed, many of the conventional junction-isolated processes were designed for doping from the gas phase at high temperatures.

Oxide-Isolated Transistors

Oxide isolation permits a considerable reduction in the lateral and vertical dimensions of the buried-layer collector process. The reason is that the base and collector contacts can be extended to the edge of the isolation region. More chip area can be saved by having the emitter walled against the oxide edge. The principal difference between scaling of junction- and oxide-isolated transistors is visualized in Fig. 3.8.[63] The device layouts are Schottky clamped; that is, the base contact extends over the collector region. This hinders the transistor to enter saturation mode under device operation. In Fig. 3.8(b), the effective surface area of the emitter contact has been reduced by a so-called washed emitter approach: since the oxide formed on the emitter window during emitter diffusion has a much higher doping concentration than its surroundings, this particular oxide can be removed by a mask-less wet etching. Hence, the emitter contact becomes self-aligned to the emitter diffusion area.

The process flow including mask layouts for an oxide-isolated bipolar transistor of the buried-layer collector type is shown in Fig. 3.9.[64] After formation of the subcollector by arsenic implantation through an oxide mask in the p^--substrate, the upper collector layer is grown epitaxially on top (Fig. 3.9(a)). The device isolation is fabricated by local oxidation of silicon (LOCOS) or recessed oxide (ROX) process (Figs. 3.9(b) to (d)). The isolation mask in Fig. 3.9(b) is aligned to the buried layer using the step in the silicon (Fig. 3.9(a)) originating from the enhanced oxidation rate for highly doped n^+-silicon compared to the p^--substrate during activation of the buried layer. The ROX is thermally grown (Fig. 3.9(d)) after the boron field implantation (or chan-stop) (Fig. 3.9(c)). This p^+-implant is necessary for suppressing a conducting channel otherwise present under the ROX. The base is then formed by ion implantation of boron or BF$_2$ through a screen oxide (Fig. 3.9(d)); in the simple device of Fig. 3.9, a single base

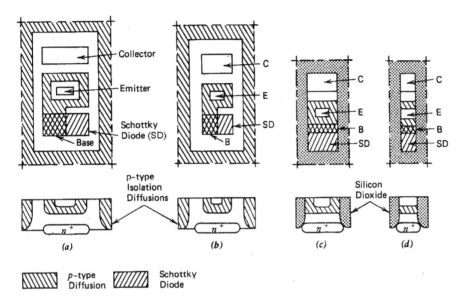

FIGURE 3.8 Device layout and cross-section demonstrating scaling of (a)-(b) junction-isolated and (c)-(d) oxide-isolated bipolar transistors (after Ref. 63, copyright© 1986, Wiley).

implantation is used; in a more advanced bipolar process, the fabrication of the intrinsic and extrinsic base must be divided into one low dose and one high dose implantation, respectively, adding one more mask to the total flow. After base formation, an emitter/base contact mask is patterned in a thermally grown oxide (Fig. 3.9(e)). The emitter is then implanted using a heavy dose arsenic implant (Fig. 3.9(f)). An n+ contact is simultaneously formed in the collector window. After annealing, the device is ready for metallization and passivation.

Apart from the strong reduction in isolation capacitances, the replacement of a junction-isolated process with an oxide-isolated process also adds other high-speed features such as thinner epitaxial layer and shallower emitter/base diffusions. A typical base width is a few 1000 Å and the resulting f_T typically lies in the range of 1 to 10 GHz. The doping of the epitaxial layer is determined by the required breakdown voltage. Further speed enhancement of the oxide-isolated transistor is difficult due to the parasitic capacitances and resistances originating from contact areas and design-rule tolerances related to alignment accuracy.

Lateral pnp Transistors

The conventional npn flow permits the bipolar designer to simultaneously create a lateral pnp transistor. This is made by placing two base diffusions in close proximity to each other in the epi-layer, one of them (pnp-collector) surrounding the other (pnp-emitter) (see Fig. 3.10). In general, the lateral pnp device exhibits poor performance since the base width is determined by lithography constraints rather than vertical base control as in the npn device. In addition, there will be electron injection from the subcollector into the p-type emitter, thus reducing emitter efficiency.

3.4 High-Performance Bipolar Technology

The development of a high-performance bipolar technology for integrated circuits signified a large step forward, both with respect to speed and packing density of bipolar transistors. A representative device cross-section of a so-called double-poly transistor is depicted in Fig. 3.11. The important characteristics for this bipolar technology are the polysilicon emitter contact, the advanced device isolation, and the self-aligned structure. These three features are discussed here with an emphasis on self-alignment where the two basic process flows are outlined — the single-poly and double-poly transistor.

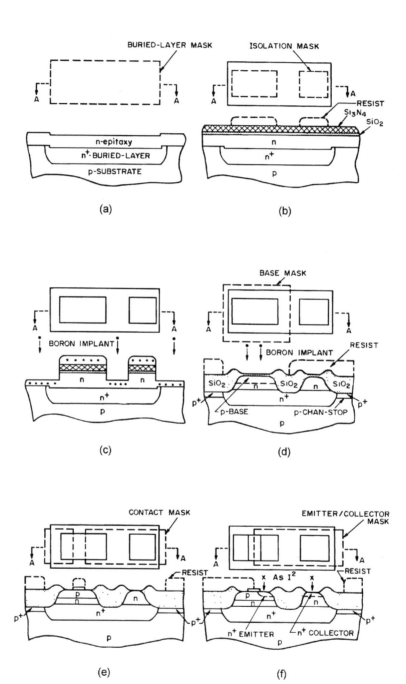

FIGURE 3.9 Layout and cross-section of the fabrication sequence for an oxide-isolated buried-collector transistor (after Ref. 64, copyright© 1983, McGraw-Hill).

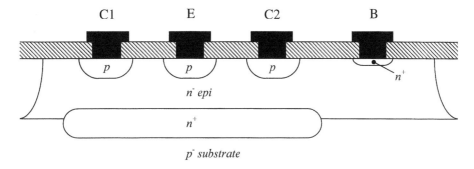

FIGURE 3.10   Schematic cross-section of the lateral pnp transistor.

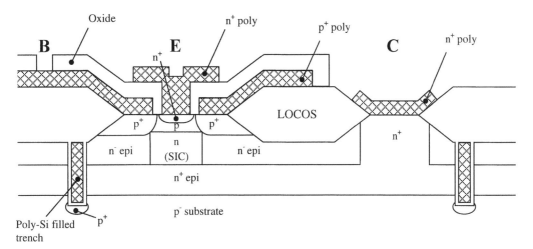

FIGURE 3.11   A double-poly self-aligned bipolar transistor with deep-trench isolation, polysilicon emitter and SIC. Metallization is not shown.

Polysilicon Emitter Contact

The polysilicon emitter contact is fabricated by a shallow diffusion of n-type species (usually arsenic) from an implanted n^+- polysilicon layer into the silicon substrate[65] (see emitter region in Fig. 3.11). The thin oxide sandwiched between the poly- and monosilicon is partially or fully broken up during contact formation. The mechanism behind the improved current gain is strongly related to the details of the interface between the polysilicon layer and the monosilicon substrate.[52] Hence, the cleaning procedure of the emitter window surface before polysilicon deposition must be carefully engineered for process robustness. Otherwise, the average current gain from wafer-to-wafer will exhibit unacceptable variations. The emitter window preparation and subsequent drive-in anneal conditions can also be used in tailoring the process with respect to gain and emitter resistance.

From a fabrication point of view, there are further advantages when introducing polysilicon emitter technology. By implanting into the polysilicon rather than into single-crystalline material, the total defect generation as well as related anomalous diffusion effects are strongly suppressed in the internal transistor after the drive-in anneal. Moreover, the risk for spiking of aluminum during the metallization process, causing short-circuiting of the pn junction, is strongly reduced compared to the conventional contact formation. As a result, some of the yield problems associated with monosilicon emitter fabrication are, to a large extent, avoided when utilizing polysilicon emitter technology.

Advanced Device Isolation

With advanced device isolation, one usually refers to the deep trenches combined with LOCOS or ROX as seen in Fig. 3.11.[66] The starting material before etching is then a double-epitaxial layer (n^+-n) grown on a lowly doped p^--substrate. The deep trench must reach all the way through the double epi-layer, meaning a high-aspect ratio reactive-ion etch. Hence, the trenches will define the extension of the buried layer collector for the transistor.

The main reason for introducing advanced isolation in bipolar technology is the need for a compact chip layout.[67] Quite naturally, the bipolar isolation technology has benefited from the trench capacitor development in the MOS memory area. The deep trench isolation allows bipolar transistors to be designed at the packing density of VLSI.

The fabrication of a deep-trench isolation includes deep-silicon etching, chan-stop p^+-implantation, an oxide/nitride stack serving as isolation, intrinsic polysilicon fill-up, planarization, and cap oxidation.[66] The deep-trench isolation is combined with an ordinary LOCOS or ROX isolation, which is added before or after trench formation. The most advanced isolation schemes take advantage of shallow-trench isolation rather than ordinary LOCOS after the deep-trench process; in this way, a very planar surface with no oxide lateral encroachment ("birds beak") is achieved after the planarization step. The concern regarding stress-induced crystal defects originating from trench etching requires careful attention so as not to seriously affect yield.

Self-Aligned Structures

Advanced bipolar transistors are based on self-aligned structures made possible by polysilicon emitter technology. As a result, the emitter-base alignment is not dependent on the overlay accuracy of the lithography tool. The device contacts can be separated without affecting the active device area. It is also possible to create structures where the base is self-aligned both to the collector and emitter, the so-called sidewall-based contact structure (SICOS).[68] This process, however, has not been able to compete successfully with the self-aligned schemes discussed below.

The self-aligned structures are divided into single-polysilicon (single-poly) and double-polysilicon (double-poly) architectures, as visualized in Fig. 3.12.[69] The double-poly structure refers to the emitter polysilicon and base polysilicon electrode, whereas the single-poly only refers to the emitter polysilicon. From Fig. 3.12, it is seen that the double-poly approach benefits from a smaller active area than the single-poly, manifested in a reduced base-collector capacitance. Moreover, the double-poly transistor in general exhibits a lower base resistance. The double-poly transistor, however, is more complex to fabricate than the single-poly device. On the other hand, by applying inside spacer technology for the double-poly emitter structure, the lithography requirements are not as strict as in the single-poly case where more conventional MOS design rules are used for definition of the emitter electrode.

Single-Poly Structure

The fabrication of a single-poly transistor has been presented in several versions, more or less similar to the traditional MOS flow. An example of a standard single-poly process is shown in Fig. 3.13.[70] After arsenic emitter implantation (Fig. 3.13(a)) and polysilicon patterning, a so-called base-link is implanted using boron ions (Fig. 3.13(b)). Oxide is then deposited and anisotropically etched to form outside spacers, followed by the heavy extrinsic base implantation (Fig. 3.13(c)). Shallow junctions (including emitter diffusion) are formed by rapid thermal annealing (RTA). A salicide or polycide metallization completes the structure (Fig. 3.13(d)).

The intrinsic base does not necessarily need to be formed prior to the extrinsic part. Li et al.[71] have presented a reverse extrinsic-intrinsic base scheme based on a disposable emitter pedestal with spacers. This leads to improved control over the intrinsic base width and a lower surface topography compared to the process represented in Fig. 3.13.

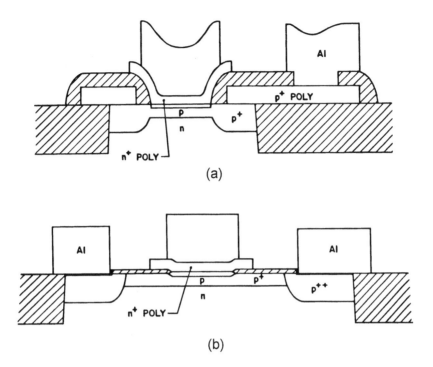

FIGURE 3.12 (a) Double-poly structure and (b) single-poly structure. Buried layer and collector contact are not shown (after Ref. 69, copyright© 1989, IEEE).

Another variation of the single-poly architecture is the so-called quasi-self-aligned process (see Fig. 3.14).[72] A base oxide is formed by thermal oxidation in the active area and an emitter window is opened (Fig. 13.14(a)). Following intrinsic base implantation, the emitter polysilicon is deposited, implanted, and annealed. The polysilicon emitter pedestal is then etched out (Fig. 3.14(b)). The extrinsic base process, junction formation, and metallization are essentially the same as in the single-poly process shown in Fig. 3.13. Note that in Fig. 13.4, the emitter-base formation is self-aligned to the emitter window in the oxide, not to the emitter itself, hence explaining the term quasi-self-aligned. As a result, a higher total base resistance is obtained compared to the standard single-poly process.

The boron implantation illustrated in Fig. 3.13(b) is an example of so-called base-link engineering aimed at securing the electrical contact between the heavily doped p^+-extrinsic base and the much lower doped intrinsic base. Too weak a base link will result in high total base resistance, whereas too strong a base link may create a lateral emitter-base tunnel junction leading to non-ideal base current characteristics.[73] Furthermore, a poorly designed base link jeopardizes matching between individual transistors since the final current gain may vary substantially with the emitter width.

Double-Poly Structure

The double-poly structure originates from the classical IBM structure presented in 1981.[74] Most high-performance commercial processes today are based on double-poly technology. The number of variations are less than for the single-poly, mainly with different aspects on base-link engineering, spacer technology, and SIC formation. One example of a double-poly fabrication is presented in Fig. 3.15. After deposition of the base polysilicon and oxide stack, the emitter window is opened (Fig. 3.15(a)) and thermally oxidized. During this step, p^+-impurities from the base polysilicon diffuse into the monosilicon, thus forming the extrinsic base. In addition, the oxidation repairs the crystal damage caused by the dry etch when opening the emitter window. A thin silicon nitride layer is then deposited, the intrinsic base is implanted using boron, followed by the fabrication of amorphous silicon spacers inside the emitter window (Fig. 3.15(b)). The nitride is exposed to a short dry etch, the spacers are

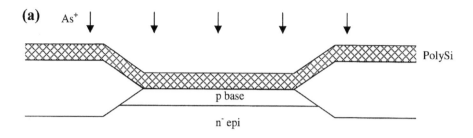

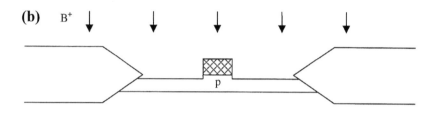

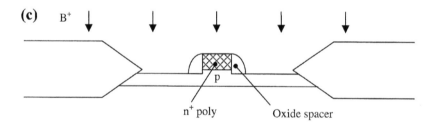

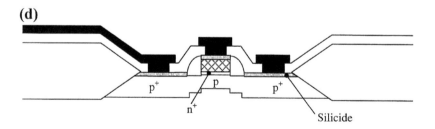

FIGURE 3.13   The single-poly, self-aligned process: (a) polyemitter implantation, (b) emitter etch and base link implantation, (c) oxide spacer formation and extrinsic base implantation, and (d) final device after junction formation and metallization.

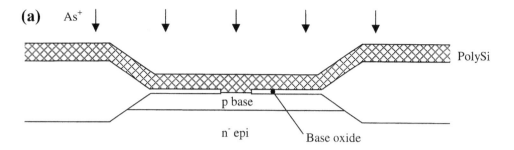

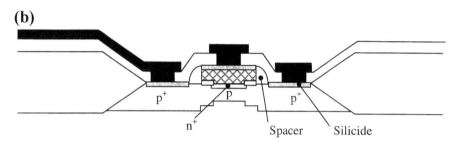

FIGURE 3.14   The single-poly, quasi-self-aligned process: (a) polyemitter implantation, (b) final device.

removed, and the thin oxide is opened up by an HF dip. Deposition and implantation of the polysilicon emitter film is carried out (Fig. 3.15(c)). The structure is patterned and completed by RTA emitter drive-in and metallization (Fig. 3.15(d)). The emitter will thus be fully self-aligned with respect to the base. Note that the inside spacer technology implies that the actual emitter width will be significantly less than the drawn emitter width.

The definition of the polyemitter in the single- and double-poly process inevitably leads to some overetching into the epi-layer, see Figs. 3.13(b) and 3.15(a), respectively. The final recessed region will make control over base-link formation more awkward.[75,76] In fact, the base link will depend both on the degree of overetch as well as the implantation parameters.[77] This situation is of no concern for the quasi-self-aligned process where the etch of the polysilicon emitter stops on the base oxide.

In a modification of the double-poly process, a more advanced base-link technology is proposed.[78] After extrinsic base drive-in and emitter window opening, BF_2-implanted poly-spacers are formed inside the emitter window. The boron is out-diffused through the emitter oxide, thus forming the base link. The intrinsic base is subsequently formed by conventional implantation through the emitter window. New dielectric inside spacers are formed prior to polysilicon emitter deposition, followed by arsenic implantation and emitter drive-in.

Also, vertical pnp bipolar transistors based on the double-poly concept have been demonstrated.[79] Either boron or BF_2 is used for the polyemitter implantation. A pnp device with f_T of 35 GHz has been presented in a classical double-poly structure.[80]

3.5   Advanced Bipolar Technology

This chapter section treats state-of-the-art bipolar technologies reported during the 1990s (but not necessarily put into production). Alongside the traditional down-scaling in design rules, efforts have focused on new innovations in emitter and base electrode fabrication. A key issue has been the integration of epitaxial Si or SiGe intrinsic base into the standard npn process flow. This section concludes with an outlook on the future trends in bipolar technology after the year 2000.

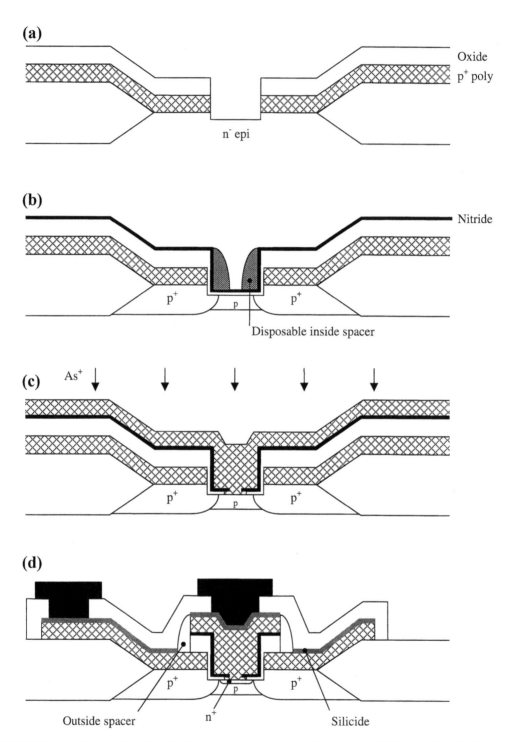

FIGURE 3.15   The double-poly, self-aligned process: (a) emitter window etch, (b) intrinsic base implantation through thin oxide/nitride stack followed by inside spacer formation, (c) polyemitter implantation, (d) final device after emitter drive-in and metallization.

Implanted Base

Today's most advanced commercial processes are specified with an f_T around 30 GHz. The major developments are being carried out using double-poly technology, although new improvements have also been reported for single-poly architectures.[72,81] For double-poly transistors, it was demonstrated relatively early that by optimizing a very low intrinsic base implant energy below 10 keV, devices with an f_T around 50 GHz are possible to fabricate.[82] The emitter out-diffusion is performed by a combined furnace anneal and RTA. In this way, the intrinsic base width is controlled below 1000 Å, whereas the emitter depth is only around 250 Å.

Since ion implantation is associated with a number of drawbacks such as channeling, shadowing effects, and crystal defects, it may be difficult to reach an f_T above 50 to 60 GHz based on such a technology. The intrinsic base implantation has been replaced by rapid vapor deposition using B_2H_6 gas around 900°C.[83] The in-diffused boron profile will form a thin and low-resistive base. Also, the emitter implantation can be removed by utilizing *in situ* doped emitter technology (e.g., AsH_3 gas during polysilicon deposition).[84] Two detrimental effects are then avoided; namely, emitter perimeter depletion and the emitter plug effect.[85] The former effect causes a reduced doping concentration close to the emitter perimeter, whereas the latter implies the plugging of doping atoms in narrow emitter windows causing shallower junctions compared to larger openings on the same chip.

Arsenic came to replace phosphorus as the emitter impurity during the 1970s, mainly because of the emitter push-effect plaguing phosphorus monosilicon emitters. The phosphorus emitter has, however, experienced a renaissance in advanced bipolar transistors by introducing the so-called *in situ* phosphorus doped polysilicon (IDP) emitter.[86] One motivation for using IDP technology is the reduction in final emitter resistance compared to the traditional As polyemitter, in particular for aggressively down-scaled devices with very narrow emitter windows. In addition, the emitter drive-in for an IDP emitter is carried out at a lower thermal budget than the corresponding arsenic emitter due to the difference in diffusivity between the impurity atoms. Using IDP and RVD, very high f_T values (above 60 GHz) have been realized.[83] It has been suggested that the residual stress of the IDP emitter and the interfacial oxide between the poly- and the monosilicon creates a heteroemitter action for the device, thus explaining the high current gains of IDP bipolar transistors.[87]

Base electrode engineering in advanced devices has become an important field in reducing the total base resistance, thus improving f_{max} of the transistor. One straightforward method in lowering the base sheet resistance is by shunting the base polysilicon with an extended silicide across the total base electrode. This has recently been demonstrated in an f_{max} = 60 GHz double-poly process.[88] A still more effective concept is to integrate metal base electrodes.[89] This approach is combined with *in situ* doped boron polysilicon base electrodes as well as an IDP emitter in a double-poly process (see Fig. 3.16). The tungsten electrodes are fully self-aligned using WF_6-selective deposition. The technology, denoted SMI (self-aligned metal IDP), has been applied together with RVD base formation. The bipolar process was shown to produce f_T and f_{max} figures of 100 GHz at a breakdown voltage of 2.5 V.[90]

Epitaxial Base

By introducing epitaxial film growth techniques for intrinsic base formation, the base width is readily controlled on the order of some hundred angstroms. Both selective and non-selective epitaxial growth (SEG and NSEG, respectively) have been reported. One example of a SEG transistor flow is illustrated in Fig. 3.17.[91] Not only the epitaxial base, but also the n⁻-collector is grown using SEG. The p⁺-poly overhangs warrant a strong base link between the SEG intrinsic base and the base electrode. This f_T = 44 GHz process was capable of delivering divider circuits working at 25 GHz.

A natural extension of the Si epitaxy is to apply the previously mentioned SiGe epitaxy, thus creating a heterojunction bipolar transistor (HBT). For example, the transistor process in Fig. 3.17 was later extended to a SiGe process with f_T = 61 GHz and f_{max} = 74 GHz.[92] Apart from high speed, low base resistance is a trademark for many SiGe bipolar processes. For details of the SiGe HBT, the reader is referred to Chapter 5. Here, only some process integration points of view are given of this very important technology for advanced silicon-based bipolar devices during the 1990s.

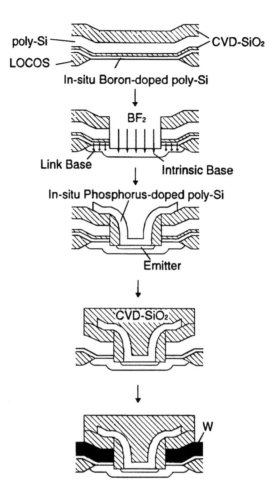

FIGURE 3.16   The self-aligned metal IDP process using selective deposition of tungsten base electrodes (after Ref. 89, copyright© 1997, IEEE).

While the first world records in terms of f_T and f_{max} were broken for non-self-aligned structures or mesa HBTs, planar self-aligned process schemes taking advantage of the benefits using SiGe have subsequently been demonstrated. In recent years, both single-poly and double-poly SiGe transistors exhibiting excellent dc and ac properties have been presented. This also includes the quasi-self-aligned approach where a fully CMOS compatible process featuring an f_{max} of 71 GHz has been shown.[93] A similar concept featuring NSEG, so-called differential epitaxy, has been applied in the design of an HBT with a single-crystalline emitter rather than a polysilicon emitter.[94] This process, however, requires a very low thermal budget because of the high boron and germanium content in the intrinsic base of the transistor. Excellent properties for RF applications are made possible by this approach, such as high f_T and f_{max} around 50 GHz, combined with good dc properties and low noise figures.[95]

In double-poly structures, the extrinsic base is usually deposited prior to SiGe base epitaxy, which is then carried out by SEG (as shown in Fig. 3.17). Several groups report τ_d below 20 ps using this approach. One example of the most advanced bipolar technologies is the SiGe double-poly structure shown in Fig. 3.18.[96] This technology features the SMI base electrode technology, together with SEG of SiGe. The device is isolated by oxide-filled trenches. The reported τ_d was 7.7 ps and the f_{max} was 108 GHz, thus approaching the SiGe HBT mesa record of 160 GHz.[97] A device similar to the one in Fig. 3.18 was also reported to yield an f_T of 130 GHz.[98]

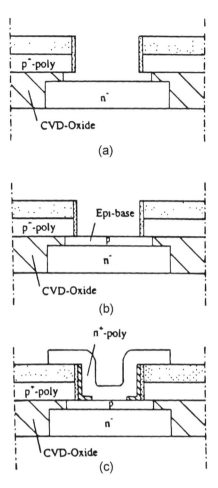

FIGURE 3.17 Process demonstrating selective epitaxial growth: (a) self-aligned formation of p+-poly overhangs, (b) selective epitaxial growth of the intrinsic base, (c) emitter fabrication (after Ref. 91, copyright© 1992, IEEE).

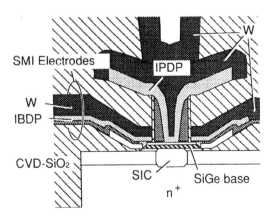

FIGURE 3.18 A state-of-the-art bipolar device featuring SMI electrodes, selectively grown epitaxial SiGe base, *in situ* doped polysilicon emitter/base and oxide-filled trenches (after Ref. 96, copyright© 1998, IEEE).

Future Trends

The shrinking of dimensions in bipolar devices, in particular for digital applications, will proceed one or two generations behind the MOS frontier, leading to a further reduction in τ_d. Several of the concepts reviewed above for advanced bipolar components are expected to be introduced in the next commercial high-performance processes; for example, *in situ* doped emitter and low-resistivity base electrodes. Similar to CMOS, the overall temperature budget must be reduced in processing. Evidently, bipolar technology in the future also continues to benefit from the progress made in CMOS technology; for example, in isolation and back-end processing. CMOS compatibility will be a general requirement for the majority of bipolar process development because of the strong interest in mixed-signal BiCMOS processes.

Advanced isolation technology combining deep and shallow trenches, perhaps on silicon-on-insulator (SOI) or high-resistivity substrates, marks one key trend in future bipolar transistors. Bipolar technology based on SOI substrates may well be accelerated by the current introduction of SOI into CMOS production. However, thermal effects for high current drive bipolar devices must be solved when using SOI. An interesting low-cost alternative insulating substrate for RF-bipolar technology is the silicon-on-anything concept recently presented.[99] In addition, copper metallization will be introduced for advanced bipolars provided that intermetal dielectrics as well as passive components are developed to meet this additional advantage.[100]

The epitaxial base constitutes another important trend where both Si and SiGe are expected to enhance performance in the high-frequency domain, although the introduction may be delayed due to progress in ion-implanted technology. In this respect, SEG technology has yet to prove its manufacturability.

Future Si-based bipolar technology with f_T and f_{max} greater than 100 GHz will continue to play an important role in small-density, high-performance designs. The most important applications are found in communication systems in the range 1 to 10 GHz (wireless telephony and local area networks) and 10 to 70 GHz (microwave and optical-fiber communication systems) where Si and/or SiGe bipolar technologies are expected to seriously challenge existing III-V technologies.[101]

Acknowledgments

We are grateful to G. Malm and M. Linder for carrying out the process simulations. The support from the Swedish High-Frequency Bipolar Technology Consortium is greatly acknowledged.

References

1. Ning, T. H. and Tang, D. D., Bipolar Trends, *Proc. IEEE*, 74, 1669, 1986.
2. Nakamura, T. and Nishizawa, H., Recent progress in bipolar transistor technology, *IEEE Trans. Electron Dev.*, 42, 390, 1995.
3. Warnock, J. D., Silicon bipolar device structures for digital applications: Technology trends and future directions, *IEEE Trans. Electron Dev.*, 42, 377, 1995.
4. Wilson, G. R., Advances in bipolar VLSI, *Proc. IEEE*, 78, 1707, 1990.
5. Barber, H. D., Bipolar device technology challenge and opportunity, *Can. J. Phys.*, 63, 683, 1985.
6. Baltus, P., Influence of process- and device parameters on the performance of portable rf communication circuits, *Proceedings of the 24th European Solid State Device Research Conference*, Hill, C. and Ashburn, P., Eds., 1994, 3.
7. Burghartz, J. N., BiCMOS process integration and device optimization: Basic concepts and new trends, *Electrical Eng.*, 79, 313, 1996.
8. Ashburn, P., *Design and Realization of Bipolar Transistors*, Wiley, Chichester, 1988.
9. Roulston, D. J., *Bipolar Semiconductor Devices*, McGraw-Hill, New York, 1990.
10. Tang, D. D. and Solomon, P. M., Bipolar transistor design for optimized power-delay logic circuits, *IEEE J. Solid-State Circuits*, SC-14, 679, 1979.

11. Chor, E.-F., Brunnschweiler, A., and Ashburn, P., A propagation-delay expression and its application to the optimization of polysilicon emitter ECL processes, *IEEE J. Solid-State Circuits*, 23, 251, 1988.

12. Stork, J. M. C., Bipolar transistor scaling for minimum switching delay and energy dissipation, in *1988 Int. Electron Devices Meeting Tech. Dig.*, 1988, 550.

13. Prinz, E. J. and Sturm, J. C., Current gain — Early voltage products in heterojunction bipolar transistors with nonuniform base bandgaps, *IEEE Electron. Dev. Lett.*, 12, 661, 1991.

14. Kurishima, K., An analytical expression of f_{max} for HBT's, *IEEE Trans. Electron Dev.*, 43, 2074, 1996.

15. Taylor, G. W. and Simmons, J. G., Figure of merit for integrated bipolar transistors, *Solid State Electronics*, 29, 941, 1986.

16. Hurkx, G. A. M., The relevance of f_T and f_{max} for the speed of a bipolar CE amplifier stage, *IEEE Trans. Electron Dev.*, 44, 775, 1997.

17. Larson, L. E., Silicon bipolar transistor design and modeling for microwave integrated circuit applications, in *1996 Bipolar Circuits Technol. Meeting Tech. Dig.*, 1996, 142.

18. Fang, W., Accurate analytical delay expressions for ECL and CML circuits and their applications to optimizing high-speed bipolar circuits, *IEEE J. Solid-State Circuits*, 25, 572, 1990.

19. Silvaco International, 1997: VWF Interactive tools, *Athena User's Manual*, 1997.

20. Silvaco International, 1997: VWF Interactive tools, *Atlas User's Manual*, 1997.

21. Roulston, D. J., *BIPOLE3 User's Manual*, BIPSIM, Inc., 1996.

22. Alvarez, A. R., Abdi, B. L., Young, D. L., Weed, H. D., Teplik, J., and Herald, E. R., Application of statistical design and response surface methods to computer-aided VLSI device design, *IEEE Trans. Comp.-Aided Design*, 7, 272, 1988.

23. Johnson, E. O., Physical limitations on frequency and power parameters of transistors, *RCA Rev.*, 26, 163, 1965.

24. Ng, K. K., Frei, M. R., and King, C. A., Reevaluation of the $f_t BV_{CEO}$ limit in Si bipolar transistors, *IEEE Trans. Electron Dev.*, 45, 1854, 1998.

25. Kirk, C. T., A theory of transistor cut-off frequency falloff at high current densities, *IRE Trans. Electron. Dev.*, ED9, 164, 1962.

26. Roulston, D. J., *Bipolar Semiconductor Devices*, McGraw-Hill, New York, 1990, p. 257.

27. Konaka, S., Amemiya, Y., Sakuma, K., and Sakai, T., A 20 ps/G Si bipolar IC using advanced SST with collector ion implantation, in *1987 Ext. Abstracts 19th Conf. Solid-State Dev. Mater.*, Tokyo, 1987, 331.

28. Lu, P.-F. and Chen, T.-C., Collector-base junction avalanche effects in advanced double-poly self-aligned bipolar transistors, *IEEE Trans. Electron Dev.*, 36, 1182, 1989.

29. Tang, D. D. and Lu, P.-F., A reduced-field design concept for high-performance bipolar transistors, *IEEE Electron. Dev. Lett.*, 10, 67, 1989.

30. Ugajin M., Konaka, S., Yokohama K., and Amemiya, Y., A simulation study of high-speed hetero-emitter bipolar transistors, *IEEE Trans. Electron Dev.*, 36, 1102, 1989.

31. Inou, K. et al., 52 GHz epitaxial base bipolar transistor with high Early voltage of 26.5 V with box-like base and retrograded collector impurity profiles, in *1994 Bipolar Circuits Technol. Meeting Tech. Dig.*, 1994, 217.

32. Ikeda, T., Watanabe, A., Nishio, Y., Masuda, I., Tamba, N., Odaka, M., and Ogiue, K., High-speed BiCMOS technology with a buried twin well structure, *IEEE Trans. Electron Dev.*, 34, 1304, 1987.

33. Kumar, M. J., Sadovnikov, A. D., and Roulston, D. J., Collector design tradeoffs for low voltage applications of advanced bipolar transistors, *IEEE Trans. Electron Dev.*, 40, 1478, 1993.

34. Kumar, M. J. and Datta, K., Optimum collector width of VLSI bipolar transistors for maximum f_{max} at high current densities, *IEEE Trans. Electron Dev.*, 44, 903, 1997.

35. Roulston, D. J. and Hébert, F., Optimization of maximum oscillation frequency of a bipolar transistor, *Solid State Electronics*, 30, 281, 1987.

36. Early, J. M., Effects of space-charge layer widening in junction transistors, *Proc. IRE*, 42, 1761, 1954.

37. Shafi, Z. A., Ashburn, P., and Parker, G., Predicted propagation delay of Si/SiGe heterojunction bipolar ECL circuits, *IEEE J. Solid-State Circuits*, 25, 1268, 1990.

38. Stork, J. M. C. and Isaac, R. D., Tunneling in base-emitter junctions, *IEEE Trans. Electron Dev.*, 30, 1527, 1983.

39. del Alamo, J. and Swanson, R. M., Forward-biased tunneling: A limitation to bipolar device scaling, *IEEE Electron. Dev. Lett.*, 7, 629, 1986.

40. Roulston, D. J., *Bipolar Semiconductor Devices*, McGraw-Hill, New York, 1990, 220 ff.

41. Van Wijnen, P. J. and Gardner, R. D., A new approach to optimizing the base profile for high-speed bipolar transistors, *IEEE Electron. Dev. Lett.*, 4, 149, 1990.

42. Suzuki, K., Optimum base doping profile for minimum base transit time, *IEEE Trans. Electron Dev.*, 38, 2128, 1991.

43. Yuan, J. S., Effect of base profile on the base transit time of the bipolar transistor for all levels of injection, *IEEE Trans. Electron Dev.*, 41, 212, 1994.

44. Ashburn, P. and Morgan, D. V., Heterojunction bipolar transistors, in *Physics and Technology of Heterojunction Devices*, Morgan, D. V. and Williams, R. H., Eds., Peter Peregrinus Ltd., London, 1991, chap. 6.

45. Harame, D. L., Stork, J. M. C., Meyerson, B. S., Hsu, K. Y.-J., Cotte, J., Jenkins, K. A., Cressler, J. D., Restle, P., Crabbé, E. F., Subbana, S., Tice, T. E., Scharf, B. W., and Yasaitis, J. A., Optimization of SiGe HBT technology for high speed analog and mixed-signal applications, in *1993 Int. Electron Devices Meeting Tech. Dig.*, 1993, 71.

46. Prinz, E. J., Garone, P. M., Schwartz, P. V., Xiao, X., and Sturm, J. C., The effects of base dopant outdiffusion and undoped $Si_{1-x}Ge_x$ junction spacer layers in $Si/Si_{1-x}Ge_x/Si$ heterojunction bipolar transistors, *IEEE Electron. Dev. Lett.*, 12, 42, 1991.

47. Hueting, R. J. E., Slotboom, J. W., Pruijmboom, A., de Boer, W. B., Timmering, C. E., and Cowern, N. E. B., On the optimization of SiGe-base bipolar transistors, *IEEE Trans. Electron Dev.*, 43, 1518, 1996.

48. Kerr, J. A. and Berz, F., The effect of emitter doping gradient on f_T in microwave bipolar transistors, *IEEE Trans. Electron Dev.*, ED-22, 15, 1975.

49. Slotboom, J. W. and de Graaf, H. C., Measurement of bandgap narrowing in silicon bipolar transistors, *Solid State Electronics*, 19, 857, 1976.

50. Cuthbertson, A. and Ashburn, P., An investigation of the tradeoff between enhanced gain and base doping in polysilicon emitter bipolar transistors, *IEEE Trans. Electron Dev.*, ED-32, 2399, 1985.

51. Ning, T. H. and Isaac, R. D., Effect on emitter contact on current gain of silicon bipolar devices, *IEEE Trans. Electron Dev.*, ED-27, 2051, 1980.

52. Post, R. C., Ashburn, P., and Wolstenholme, G. R., Polysilicon emitters for bipolar transistors: A review and re-evaluation of theory and experiment, *IEEE Trans. Electron Dev.*, 39, 1717, 1992.

53. Solomon, P. M. and Tang, D. D., Bipolar circuit scaling, in *1979 IEEE International Solid-State Circuits Conference Tech. Dig.*, 1979, 86.

54. Ning, T. H., Tang, D. D., and Solomon, P. M., Scaling properties of bipolar devices, in *1980 Int. Electron Devices Meeting Tech. Dig.*, 1980, 61.

55. Ashburn, P., *Design and Realization of Bipolar Transistors*, Wiley, Chichester, 1988, chap. 7.

56. Lary, J. E. and Anderson, R. L., Effective base resistance of bipolar transistors, *IEEE Trans. Electron Dev.*, ED-32, 2503, 1985.

57. Rein, H.-M., Design considerations for very-high-speed Si-bipolar IC's operating up to 50 Gb/s, *IEEE J. Solid-State Circuits*, 8, 1076, 1996.

58. Schröter, M. and Walkey, D. J., Physical modeling of lateral scaling in bipolar transistors, *IEEE J. Solid-State Circuits*, 31, 1484, 1996.

59. Pfost, M., Rein, H.-M., and Holzwarth, T., Modeling substrate effects in the design of high-speed Si-bipolar IC's, *IEEE J. Solid-State Circuits*, 31, 1493, 1996.

60. Lohstrom, J., Devices and circuits for bipolar (V)LSI, *Proc. IEEE*, 69, 812, 1981.

61. Wolf, S., *Silicon Processing for the VLSI Area*, Vol. 2, Lattice Press, Sunset Beach, 1990, 532-533.

62. *ibid.*, p. 16-17.

63. Muller, R. S. and Kamins, T. I., *Device Electronics for Integrated Circuits*, 2nd ed., Wiley, New York, 1986, 307.

64. Parrillo, L. C., VLSI process integration, in *VLSI Technology*, Sze, S. M., Ed., McGraw-Hill, Singapore, 1983, 449 ff.

65. Ashburn, P., Polysilicon emitter technology, in *1989 Bipolar Circuits Technol. Meeting Tech. Dig.*, 1989, 90.

66. Li, G. P., Ning, T. H., Chuang, C. T., Ketchen, M. B., Tang, D.D., and Mauer, J., An advanced high-performance trench-isolated self-aligned bipolar technology, *IEEE Trans. Electron Dev.*, ED-34, 2246, 1987.

67. Tang, D. D., Solomon, P. M., Isaac, R. D., and Burger, R. E., 1.25 μm deep-groove-isolated self-aligned bipolar circuits, *IEEE J. Solid-State Circuits*, SC-17, 925, 1982.

68. Yano, K., Nakazato, K., Miyamoto, M., Aoki, M., and Shimohigashi, K., A high-current-gain low-temperature pseudo-HBT utilizing a sidewall base-contact structure (SICOS), *IEEE Trans. Electron Dev.*, 10, 452, 1989.

69. Tang, D. D.-L., Chen, T.-C., Chuang, C. T., Cressler, J. D., Warnock, J., Li, G.-P., Polcari, M. R., Ketchen, M. B., and Ning, T. H., The design and electrical characteristics of high-performance single-poly ion-implanted bipolar transistors, *IEEE Trans. Electron Dev.*, 36, 1703, 1989.

70. de Jong, J. L., Lane, R. H., de Groot, J. G., and Conner, G. W., Electron recombination at the silicided base contact of an advanced self-aligned polysilicon emitter, in *1988 Bipolar Circuits Technol. Meeting Tech. Dig.*, 1988, 202.

71. Li, G. P., Chen, T.-C., Chuang, C.-T., Stork, J. M. C., Tang, D. D., Ketchen, M. B., and Wang, L.-K., Bipolar transistor with self-aligned lateral profile, *IEEE Electron. Dev. Lett.*, EDL-8, 338, 1987.

72. Niel, S., Rozeau, O., Ailloud, L., Hernandez, C., Llinares, P., Guillermet, M., Kirtsch, J., Monroy, A., de Pontcharra, J., Auvert, G., Blanchard, B., Mouis, M., Vincent, G., and Chantre, A., A 54 GHz f_{max} implanted base 0.35 μm single-polysilicon bipolar transistor, in *1997 Int. Electron Devices Meeting Tech. Dig.*, 1997, 807.

73. Tang, D. D., Chen, T.-C., Chuang, C.-T., Li, G. P., Stork, J. M. C., Ketchen, M. B., Hackbarth, E., and Ning, T. H., Design considerations of high-performance narrow-emitter bipolar transistors, *IEEE Electron. Dev. Lett.*, EDL-8, 174, 1987.

74. Ning, T. H., Isaac, R. D., Solomon, P. M., Tang, D. D.-L., Yu, H.-N., Feth, G. C., and Wiedmann, S. K., Self-aligned bipolar transistors for high-performance and low-power-delay VLSI, *IEEE Trans. Electron Dev.*, ED-28, 1010, 1981.

75. Chantre, A., Festes, G., Giroult-Matlakowski, G., and Nouailhat, An investigation of nonideal base currents in advanced self-aligned "etched-polysilicon" emitter bipolar transistors, *IEEE Trans. Electron Dev.*, 38, 1354, 1991.

76. Sun, S. W., Denning, D., Hayden, J. D., Woo, M., Fitch, J. T., and Kaushik, V., A nonrecessed-base, self-aligned bipolar structure with selectively deposited polysilicon emitter, *IEEE Trans. Electron Dev.*, 39, 1711, 1992.

77. Chuang, C.-T., Li, G. P., and Ning, T. H., Effect of off-axis implant on the characteristics of advanced self-aligned bipolar transistors, *IEEE Electron. Dev. Lett.*, EDL-8, 321, 1987.

78. Hayden, J. D., Burnett, J. D., Pfiester, J. R., and Woo, M. P., A new technique for forming a shallow link base in a double polysilicon bipolar transistor, *IEEE Trans. Electron Dev.*, 41, 63, 1994.

79. Maritan, C. M. and Tarr, N. G., Polysilicon emitter p-n-p transistors, *IEEE Trans. Electron Dev.*, 36, 1139, 1989.

80. Warnock, J., Lu, P.-F., Cressler, J. D., Jenkins, K. A., and Sun, J. Y. C., 35 GHz/35 psec ECL pnp technology, in *1990 Int. Electron Devices Meeting Tech. Dig.*, 1990, 301.

81. Chantre, A., Gravier, T., Niel, S., Kirtsch, J., Granier, A., Grouillet, A., Guillermet, M., Maury, D., Pantel, R., Regolini, J. L., and Vincent, G., The design and fabrication of 0.35 μm single-polysilicon self-aligned bipolar transistors, *Jpn. J. Appl. Phys.*, 37, 1781, 1998.

82. Warnock, J., Cressler, J. D., Jenkins, K. A., Chen, T.-C., Sun, J. Y.-C., and Tang, D. D., 50-GHz self-aligned silicon bipolar transistors with ion-implanted base profiles, *IEEE Electron. Dev. Lett.*, 11, 475, 1990.

83. Uchino, T., Shiba, T., Kikuchi, T., Tamaki, Y., Watanabe, A., and Kiyota, Y., Very-high-speed silicon bipolar transistors with *in situ* doped polysilicon emitter and rapid vapor-phase doping base, *IEEE Trans. Electron Dev.*, 42, 406, 1995.

84. Burghartz, J. N., Megdnis, A. C., Cressler, J. D., Sun, J. Y.-C., Stanis, C. L., Comfort, J. H., Jenkins, K. A., and Cardone, F., Novel *in-situ* doped polysilicon emitter process with buried diffusion source (BDS), *IEEE Electron. Dev. Lett.*, 12, 679, 1991.

85. Burghartz, J. N., Sun, J. Y.-C., Stanis, C. L., Mader, S. R., and Warnock, J. D., Identification of perimeter depletion and emitter plug effects in deep-submicrometer, shallow-junction polysilicon emitter bipolar transistors, *IEEE Trans. Electron Dev.*, 39, 1477, 1992.

86. Shiba, T., Uchino, T., Ohnishi, K., and Tamaki, Y., *In situ* phosphorus-doped polysilicon emitter technology for very high-speed small emitter bipolar transistors, *IEEE Trans. Electron Dev.*, 43, 889, 1996.

87. Kondo, M., Shiba, T., and Tamaki, Y., Analysis of emitter efficiency enhancement induced by residual stress for *in situ* phosphorus-doped emitter transistors, *IEEE Trans. Electron Dev.*, 44, 978, 1997.

88. Böck, J., Meister, T. F., Knapp, H., Aufinger, K., Wurzer, M., Gabl, R., Pohl, M., Boguth, S., Franosch, M., and Treitinger, L., 0.5 μm / 60 GHz f_{max} implanted base Si bipolar technology, in *1998 Bipolar Circuits Technol. Meeting Tech. Dig.*, 1998, 160.

89. Onai, T., Ohue, E., Tanabe, M., and Washio, K., 12-ps ECL using low-base-resistance Si bipolar transistor by self-aligned metal/IDP technology, *IEEE Trans. Electron Dev.*, 44, 2207, 1997.

90. Kiyota, Y., Ohue, E., Washio, K., Tanabe, M., and Inade, T., Lamp-heated rapid vapor-phase doping technology for 100-GHz Si bipolar transistors, in *1996 Bipolar Circuits Technol. Meeting Tech. Dig.*, 1996, 173.

91. Meister, T. F., Stengl, R., Meul, H. W., Packan, P., Felder, A., Klose, H., Schreiter, R., Popp, J., Rein, H. M., and Treitinger, L., Sub-20 ps silicon bipolar technology using selective epitaxial growth, in *1992 Int. Electron Devices Meeting Tech. Dig.*, 1992, 401.

92. Meister, T. F., Schäfer, H., Franosch, M., Molzer, W., Aufinger, K., Scheler, U., Walz, C., Stolz, M., Boguth, S., and Böck, J., SiGe base bipolar technology with 74 GHz f_{max} and 11 ps gate delay, in *1995 Int. Electron Devices Meeting Tech. Dig.*, 1995, 739.

93. Chantre, A., Marty, M., Regolini, J. L., Mouis, M., de Pontcharra, J., Dutartre D., Morin, C., Gloria, D., Jouan, S., Pantel, R., Laurens, M., and Monroy, A., A high performance low complexity SiGe HBT for BiCMOS integration, in *1998 Bipolar Circuits Technol. Meeting Tech. Dig.*, 1998, 93.

94. Schüppen, A., König, U., Gruhle, A., Kibbel, H., and Erben, U., The differential SiGe-HBT, *Proceedings of the 24th European Solid State Device Research Conference,* Hill, C. and Ashburn, P., Eds., 1994, 469.

95. Schüppen, A., Dietrich, H., Seiler, U., von der Ropp, H., and Erben, U., A SiGe RF technology for mobile communication systems, *Microwave Engineering Europe*, June 1998, 39.

96. Ohue, E., Oda, K., Hayami, R., and Washio, K., A 7.7 ps CML using selective-epitaxial SiGe HBTs, *1998 Bipolar Circuits Technol. Meeting Tech. Dig.*, 1998, 97.

97. Schüppen, A., Erben, U., Gruhle, A., Kibbel, H., Schumacher, H., and König, U., Enhanced SiGe heterojunction bipolar transistors with 160 GHz-f_{max}, *1995 Int. Electron Devices Meeting Tech. Dig.*, 1995, 743.

98. Oda, K., Ohue, E., Tanabe, M., Shimamoto, H., Onai, T., and Washio, K., 130-GHz f_T SiGe HBT technology, in *1997 Int. Electron Devices Meeting Tech. Dig.*, 1997, 791.

99. Dekker, R., Baltus, P., van Deurzen, M., v.d. Einden, W., Maas, H., and Wagemans, A., An ultra low-power RF bipolar technology on glass, *1997 Int. Electron Devices Meeting Tech. Dig.*, 1997, 921.

100. Hashimoto, T., Kikuchi, T., Watanabe, K., Ohashi, N., Saito, N., Yamaguchi, H., Wada, S., Natsuaki, N., Kondo, M., Kondo, S., Homma, Y., Owada, N., and Ikeda, T., A 0.2 μm bipolar-CMOS technology on bonded SOI with copper metallization for ultra high-speed processors, *1998 Int. Electron Devices Meeting Tech. Dig.*, 1998, 209.

101. König, U., SiGe & GaAs as competitive technologies for RF-applications, *1998 Bipolar Circuits Technol. Meeting Tech. Dig.*, 1998, 87.

4

Silicon on Insulator Technology

Sorin Cristoloveanu
Institut National Polytechnique
de Grenoble

4.1 Introduction

Silicon on insulator (SOI) technology (more specifically, silicon on sapphire) was originally invented for the niche of radiation-hard circuits. In the last 20 years, a variety of SOI structures have been conceived with the aim of dielectrically separating, using a buried oxide (Fig. 4.1(b)), the active device volume from the silicon substrate.[1] Indeed, in an MOS transistor, only the very top region (0.1–0.2 μm thick, i.e., less than 0.1% of the total thickness) of the silicon wafer is useful for electron transport and device operation, whereas the substrate is responsible for detrimental, parasitic effects (Fig. 4.1(a)).

More recently, the advent of new SOI materials (Unibond, ITOX) and the explosive growth of portable microelectronic devices have attracted considerable attention on SOI for the fabrication of low-power (LP), low-voltage (LV), and high-frequency (HF) CMOS circuits.

The aim of this chapter is to overview the state-of-the-art of SOI technologies, including the material synthesis (Section 4.2), the key advantages of SOI circuits (Section 4.3), the structure and performance of typical devices (Section 4.4), and the operation modes of fully depleted (Section 4.5) and partially depleted SOI MOSFETs (Section 4.6). Section 4.7 is dedicated to short-channel effects. The main challenges that SOI is facing, in order to successfully compete with bulk-Si in the commercial arena, are critically discussed in Section 4.8.

4.2 Fabrication of SOI Wafers

Many techniques, more or less mature and effective, are available for the synthesis of SOI wafers.[1]

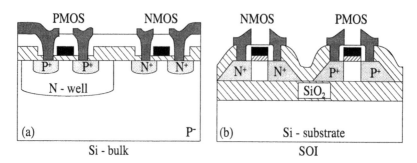

FIGURE 4.1   Basic architecture of MOS transistors in (a) bulk silicon and (b) SOI.

Silicon on Sapphire

Silicon on sapphire (SOS, Fig. 4.2(a_1)) is the initial member of SOI family. The epitaxial growth of Si films on Al_2O_3 gives rise to small silicon islands that eventually coalesce. The interface transition region contains crystallographic defects due to the lattice mismatch and Al contamination from the substrate. The electrical properties suffer from lateral stress, in-depth inhomogeneity of SOS films, and defective transition layer.[2]

SOS has recently undergone a significant lifting: larger wafers and thinner films with higher crystal quality. This improvement is achieved by *solid-phase epitaxial regrowth*. Silicon ions are implanted to amorphise the film and erase the memory of damaged lattice and interface. Annealing allows the epitaxial regrowth of the film, starting from the "seeding" surface towards the Si–Al_2O_3 interface. The result is

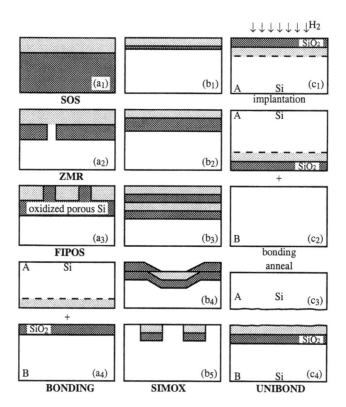

FIGURE 4.2   SOI family: (a) SOS, ZMR, FIPOS, and wafer bonding, (b) SIMOX variants, (c) UNIBOND processing sequence.

visible in terms of higher carrier mobility and lifetime; 100-nm thick SOS films with good quality have recently been grown on 6-in. wafers.[3]

Thanks to the "infinite" thickness of the insulator, SOS looks promising for the integration of RF and radiation-hard circuits.

ELO and ZMR

The *epitaxial lateral overgrowth* (ELO) method consists of growing a single-crystal Si film on a seeded and, often, patterned oxide (Fig. 4.2(a_2)). Since the epitaxial growth proceeds in both lateral and vertical directions, the ELO process requires a post-epitaxy thinning of the Si film. Alternatively, poly-silicon can be deposited directly on SiO_2; subsequently, *zone melting recrystallization* (ZMR) is achieved by scanning high-energy sources (lamps, lasers, beams, or strip heaters) across the wafer. The ZMR process can be seeded or unseeded; it is basically limited by the lateral extension of single-crystal regions, free from grain subboundaries and associated defects. ELO and ZMR are basic techniques for the integration of 3-D stacked circuits.

FIPOS

The FIPOS method (*full isolation by porous oxidized silicon*) makes use of the very large surface-to-volume ratio (10^3 cm^2 per cm^3) of porous silicon which is, thereafter, subject to selective oxidation (Fig. 4.2(a_3)). The critical step is the conversion of selected p-type regions of the Si wafer into porous silicon, via anodic reaction. FIPOS may enlighten Si technology because there are prospects, at least from a conceptual viewpoint, for combining electroluminescent porous Si devices with fast SOI–CMOS circuits.

SIMOX

In the last decade, the dominant SOI technology was SIMOX (*separation by implantation of oxygen*). The buried oxide (BOX) is synthesized by internal oxidation during the deep implantation of oxygen ions into a Si wafer. Annealing at high temperature (1320°C, for 6 h) is necessary to recover a suitable crystalline quality of the film. High current implanters (100 mA) have been conceived to produce 8-in. wafers with good thickness uniformity, low defect density (except threading dislocations: 10^4–10^6 cm^{-2}), sharp Si–SiO$_2$ interface, robust BOX, and high carrier mobility.[4]

The family of SOI structures is presented in Figure 4.2(b):

- Thin and thick Si films fabricated by adjusting the implant energy.
- Low-dose SIMOX: a dose of 4×10^{17} O$^+$/cm^2 and an additional oxygen-rich anneal for enhanced BOX integrity (ITOX process) yield a 0.1-μm thick BOX (Fig. 4.2(b_1)).
- Standard SIMOX obtained with 1.8×10^{18} O$^+$/cm^2 implant dose, at 190 keV and 650°C; the thicknesses of the Si film and BOX are roughly 0.2 μm and 0.4 μm, respectively (Fig. 4.2(b_2)).
- Double SIMOX (Fig. 4.2(b_3)), where the Si layer sandwiched between the two oxides can serve for interconnects, wave guiding, additional gates, or electric shielding.
- Laterally-isolated single-transistor islands (Fig. 4.2(b_4)), formed by implantation through a patterned oxide.
- Interrupted oxides (Fig. 4.2(b_5)) which can be viewed as SOI regions integrated into a bulk Si wafer.

Wafer Bonding

Wafer bonding (WB) and etch-back stands as a rather mature SOI technology. An oxidized wafer is mated to another SOI wafer (Fig. 4.2(a_4)). The challenge is to drastically thin down one side of the bonded structure in order to reach the targeted thickness of the silicon film. Etch-stop layers can be achieved by doping steps (P$^+$/P$^-$, P/N) or porous silicon (Eltran process).[5] The advantage of wafer bonding is to provide unlimited combinations of BOX and film thicknesses, whereas its weakness comes from the difficulty to produce ultra-thin films with good uniformity.

UNIBOND

A recent, revolutionary bonding-related process (UNIBOND) uses the deep implantation of hydrogen into an oxidized Si wafer (Fig. 4.2(c_1)) to generate microcavities and thus circumvent the thinning problem.[6] After bonding wafer *A* to a second wafer *B* and subsequent annealing to enhance the bonding strength (Fig. 4.2(c_2)), the hydrogen-induced microcavities coalesce. The two wafers separate, not at the bonded interface but at a depth defined by the location of hydrogen microcavities. This mechanism, named *Smart-cut*, results in a rough SOI structure (Fig. 4.2(c_4)). The process is completed by touch-polishing to erase the surface roughness.

The extraordinary potential of the Smart-cut approach comes from several distinct advantages: (1) the etch-back step is avoided, (2) the second wafer (Fig. 4.2(c_3)) being recyclable, UNIBOND is a single-wafer process, (3) only conventional equipment is needed for mass production, (4) relatively inexpensive 12-in. wafers are manufacturable, and (5) the thickness of the silicon film and/or buried oxide can be adjusted to match most device configurations (ultra-thin CMOS or thick-film power transistors and sensors). The defect density in the film is very low, the electrical properties are excellent, and the BOX quality is comparable to that of the original thermal oxide. The Smart-cut process is adaptable to a variety of materials: SiC or III–V compounds on insulator, silicon on diamond, etc. Smart-cut can be used to transfer already fabricated bulk-Si CMOS circuits on glass or on other substrates.

4.3   Generic Advantages of SOI

SOI circuits consist of single-device islands dielectrically isolated from each other and from the underlying substrate (Fig. 4.1(b)). The lateral isolation offers more compact design and simplified technology than in bulk silicon; there is no need of wells or interdevice trenches. In addition, the vertical isolation renders the *latch-up* mechanisms impossible.

The source and drain regions extend down to the buried oxide; thus, the junction surface is minimized. This implies reduced leakage currents and junction capacitances, which further translates into improved speed, lower power, and wider temperature range of operation.

The limited extension of drain and source regions allows SOI devices to be less affected by short-channel effects, originated from 'charge sharing' between gate and junctions. Besides the outstanding tolerance of transient radiation effects, SOI MOSFETs experience a lower electric-field peak than in bulk Si and are potentially more immune to hot carrier damage.

It is in the highly competitive domain of LV/LP circuits, operated with one-battery supply (0.9–1.5 V), that SOI can express its entire potential. A small gate voltage gap is suited to switch a transistor from off- to on-state. SOI offers the possibility to achieve a quasi-ideal subthreshold slope (60 mV/decade at room temperature); hence, a threshold voltage shrunk below 0.3 V. Low leakage currents limit the *static* power dissipation, as compared to bulk Si, whereas the *dynamic* power dissipation is minimized by the combined effects of low parasitic capacitances and reduced voltage supply.

Two arguments can be given to outline unequivocally the advantage of SOI over bulk Si:

- Operation at similar *voltage* consistently shows about 30% increase in performance, whereas operation at similar *low-power* dissipation yields as much as 300% performance gain in SOI. It is believed, at least in the SOI community, that SOI circuits of generation (*n*) and bulk-Si circuits from the *next* generation (*n* + 1) perform comparably.

- Bulk Si technology does attempt to mimic a number of features that are natural in SOI: the double-gate configuration is reproduced by processing surrounded-gate vertical MOSFETs on bulk Si, full depletion is approached by tailoring a low-high step doping, and the dynamic-threshold operation is borrowed from SOI.

The problem for SOI is that such an enthusiastic list of merits did not perturb the fantastic progress and authority of bulk Si technology. There has been no room or need so far for an alternative technology such as SOI. However, the SOI community remains confident that the SOI advantages together with the predictable approach of bulk-Si limits will be enough for SOI to succeed soon.

4.4 SOI Devices

CMOS Circuits

High-performance SOI CMOS circuits, compatible with LV/LP and high-speed ULSI applications, have been repeatedly demonstrated on submicron devices. Quarter-micron ring oscillators showed delay times of 14 ps/stage at 1.5 V[7] and of 45 ps/stage at 1V.[8] PLL operated at 2.5 V and 4 GHz dissipate 19 mW only.[8] Microwave SOS MOSFETs, with T-gate configuration, had 66-MHz maximum frequency and low noise figure.[3]

More complex SOI circuits, with direct impact on mainstream microelectronics, have also been fabricated: 0.5 V–200 MHz microprocessor,[9] 4 Mb SRAM,[10] 16 Mb and 1 Gb DRAM,[11] etc.[1,12] Several companies (IBM, Motorola, Sharp) have announced the imminent commercial deployment of 'SOI-enhanced' PC processors and mobile communication devices.

CMOS SOI circuits show capability of successful operation at temperatures higher than 300°C: the leakage currents are much smaller and the threshold voltage is less temperature sensitive (≈ 0.5mV/°C for fully depleted MOSFETs) than in bulk Si.[13] In addition, many SOI circuits are radiation-hard, able to sustain doses above 10 Mrad.

Bipolar Transistors

As a consequence of the small film thickness, most bipolar transistors have a lateral configuration. The implementation of BiCMOS technology on SOI has resulted in devices with a cutoff frequency above 27 GHz.[14] Hybrid MOS–bipolar transistors with increased current drive and transconductance are formed by connecting the gate to the floating body (or base); the MOSFET action governs in strong inversion whereas, in weak inversion, the bipolar current prevails.[12]

Vertical bipolar transistors have been processed in thick-film SOI (wafer bonding or epitaxial growth over SIMOX). An elegant solution for thin-film SOI is to replace the buried collector by an inversion layer activated by the back gate.[12]

High-Voltage Devices

Lateral double-diffused MOSFETs (DMOS), with long drift region, were fabricated on SIMOX and showed 90 V/1.3A capability.[15] Vertical DMOS can be accommodated in thicker wafer-bonding SOI.

The SIMOX process offers the possibility to synthesize locally a buried oxide ('interrupted' SIMOX, Fig. 4.2(b_5)). Therefore, a vertical power device (DMOS, IGBT, UMOS, etc.), located in the bulk region of the wafer, can be controlled by a low-power CMOS/SOI circuit (Fig. 4.3(a)). A variant of this concept is the 'mezzanine' structure, which served for the fabrication of a 600 V/25A smart-power device.[16] Double SIMOX (Fig. 4.2(b_3)) has also been used to combine a power MOSFET with a double-shielded high-voltage lateral CMOS and an intelligent low-voltage CMOS circuit.[17]

Innovative Devices

Most innovative devices make use of special SOI features, including the possibility to (1) combine bulk Si and SOI on a single chip (Fig. 4.3(a)), (2) adjust the thickness of the Si overlay and buried oxide, and (3) implement additional gates in the buried oxide (Fig. 4.3(b)), by ELO process or by local oxidation of the sandwiched Si layer in double SIMOX (Fig. 4.2(b_3)).

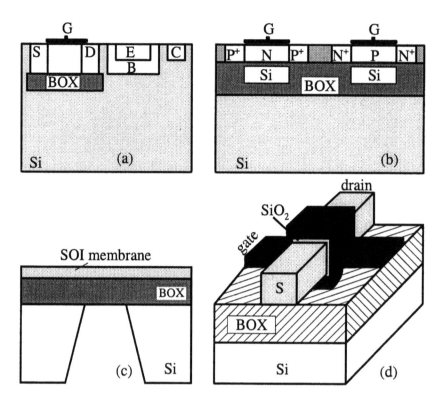

FIGURE 4.3 Examples of innovative SOI devices: (a) combined bipolar (or high power) bulk-Si transistor with low-voltage SOI CMOS circuits, (b) dual-gate transistors, (c) pressure sensor, and (d) gate all-around (GAA) MOSFET.

SOI is an ideal material for microsensors because the Si/BOX interface gives a perfect etch-stop mark, making it possible to fabricate very thin membranes (Fig. 4.3(c)). Transducers for detection of pressure, acceleration, gas flow, temperature, radiation, magnetic field, etc. have successfully been integrated on SOI.[1,16]

The feasibility of three-dimensional circuits has been demonstrated on ZMR structures. For example, an image-signal processor is organized in three levels: photodiode arrays in the upper SOI layer, fast A/D converters in the intermediate SOI layer, and arithmetic units and shift registers in the bottom bulk Si level.[18]

The *gate all-around* (GAA) transistor of Fig. 4.3(d), based on the concept of volume inversion, is fabricated by etching a cavity into the BOX and wrapping the oxidized transistor body into a poly-Si gate.[12] Similar devices include the Delta transistor[19] and various double-gate MOSFETs.

The family of SOI devices also includes optical waveguides and modulators, microwave transistors integrated on high-resistivity SIMOX, twin-gate MOSFETs, and other exotic devices.[1,12] They do not belong to science fiction: the devices have already been demonstrated in terms of technology and functionality... even if most people still do not believe that they can operate.

4.5 Fully Depleted SOI Transistors

In SOI MOSFETs (Fig. 4.1(b)), inversion channels can be activated at both the front Si–SiO$_2$ interface (via gate modulation V_{G_1}) and back Si–BOX interface (via substrate, back-gate bias V_{G_2}).

Full depletion means that the depletion region covers the entire transistor body. The depletion charge is constant and cannot extend according to the gate bias. A better coupling develops between the gate bias and the inversion charge, leading to enhanced drain current. In addition, the front- and back-surface potentials become coupled too. The coupling factor is roughly equal to the thickness ratio between gate oxide and buried oxide. The electrical characteristics of one channel vary remarkably with the bias applied

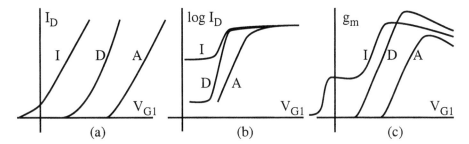

FIGURE 4.4 Generic front-channel characteristics of a fully depleted n-channel SOI MOSFET for accumulation (A), depletion (D), and inversion (I) at the back interface: (a) $I_D(V_{G_1})$ curves in strong inversion, (b) log $I_D(V_{G_1})$ curves in weak inversion, and (c) transconductance $g_m(V_{G_1})$ curves.

to the opposite gate. Due to *interface coupling*, the front-gate measurements are all reminiscent of the back-gate bias and quality of the buried oxide and interface.

Totally new $I_D(V_G)$ relations apply to fully depleted SOI–MOSFETs whose complex behavior is controlled by both gate biases. The typical characteristics of the front-channel transistor are schematically illustrated in Fig. 4.4, for three distinct bias conditions of the back interface (inversion, depletion, and accumulation), and will be explained next.

Threshold Voltage

The lateral shift of $I_D(V_{G_1})$ curves (Fig. 4.4(a)) is explained by the linear variation of the front-channel threshold voltage, $V_{T_1}^{dep}$, with back-gate bias. This *potential coupling* causes $V_{T_1}^{dep}$ to decrease linearly, with increasing V_{G_2}, between two plateaus corresponding, respectively, to accumulation and inversion at the back interface:[20]

$$V_{T_1}^{dep} = V_{T_1}^{acc} - \frac{C_{si}C_{ox_2}(V_{G_2} - V_{G_2}^{acc})}{C_{ox_1}(C_{ox_2} + C_{si} + C_{it_2})} \tag{4.1}$$

where $V_{T_1}^{acc}$ is the threshold voltage when the back interface is accumulated

$$V_{T_1}^{acc} = \Phi_{fb_1} + \frac{C_{ox_1} + C_{si} + C_{it_1}}{C_{ox_1}}2\Phi_F - \frac{Q_{si}}{2C_{ox_1}} \tag{4.2}$$

and $V_{G_2}^{acc}$ is given by

$$V_{G_2}^{acc} = \Phi_{fb_2} - \frac{C_{si}}{C_{ox_2}}2\Phi_F - \frac{Q_{si}}{2C_{ox_2}} \tag{4.3}$$

In the above equations, C_{si}, C_{ox}, and C_{it} are the capacitances of the fully depleted film, oxide, and interface traps, respectively; Q_{si} is the depletion charge, Φ_F is the Fermi potential, and Φ_{fb} is the flat-band potential. The subscripts 1 and 2 hold for the front- or the back-channel parameters and can be interchanged to account for the variation of the back-channel threshold voltage V_{T_2} with V_{G_1}.

The difference between the two plateaus, $\Delta V_{T_1} = (C_{si}/C_{ox_1})2\Phi_F$, slightly depends on doping, whereas the slope does not. We must insist on the polyvalence of Eqs.(4.1) to (4.3) as compared to the simple case of bulk Si MOSFETs (or partially depleted MOSFETs), where

$$V_{T_1} = \Phi_{fb_1} + \left(1 + \frac{C_{it_1}}{C_{ox_1}}\right)2\Phi_F + \frac{\sqrt{4q\varepsilon_{si}N_A\Phi_F}}{C_{ox_1}} \tag{4.4}$$

The extension to p-channels or accumulation-mode SOI–MOSFETs is also straightforward.[1]

In fully depleted MOSFETs, the threshold voltage decreases in thinner films (i.e., reduced depletion charge) until quantum effects arise and lead to the formation of a 2-D subband system. In ultra-thin films ($t_{si} \leq 10$ nm), the separation between the ground state and the bottom of the conduction band increases with reducing thickness: a V_T rebound is then observed.[21]

Subthreshold Slope

For depletion at the back interface, the subthreshold slope (Fig. 4.4(b)) is very steep and the subthreshold *swing S* is given by:[22]

$$S_1^{dep} = 2.3 \frac{kT}{q} \left(1 + \frac{C_{it_1}}{C_{ox_1}} + \alpha_1 \frac{C_{si}}{C_{ox_1}} \right)$$ (4.5)

The interface coupling coefficient α_1

$$\alpha_1 = \frac{C_{ox_2} + C_{it_2}}{C_{si} + C_{ox_2} + C_{it_2}} < 1$$ (4.6)

accounts for the influence of back interface traps C_{it_2} and buried oxide thickness C_{ox_2} on the front channel current.[22]

In the ideal case, where $C_{it_{1,2}} \cong 0$ and the buried oxide is much thicker than both the film and the gate oxide (i.e., $\alpha_1 \cong 0$), the swing approaches the theoretical limit $S_1^{dep} \cong 60$ mV/decade at 300K. Accumulation at the back interface does decouple the front inversion channel from back interface defects but, in turn, makes α_1 tend to unity (as in bulk–Si or partially depleted MOSFETs), causing an overall degradation of the swing.

It is worth noting that the above simplified analysis and equations are valid only when the buried oxide is thick enough, such that substrate effects occurring underneath the BOX can be overlooked. The capacitances of the BOX and Si substrate are actually connected in series. Therefore, the swing may depend, essentially for thin buried oxides, on the density of the traps and surface charge (accumulation, depletion, or inversion) at the *third* interface: BOX–Si substrate. The general trend is that the subthreshold slope improves for thinner silicon films and thicker buried oxides.

Transconductance

For strong inversion and ohmic region of operation, the front-channel drain current and transconductance are given by

$$I_D = \frac{C_{ox_1} W V_D}{L} \cdot \frac{\mu_1}{1 + \theta_1 (V_{G_1} - V_{T_1}(V_{G_2}))} \cdot (V_{G_1} - V_{T_1}(V_{G_2}))$$ (4.7)

$$g_{m_1} = \frac{C_{ox_1} W V_D}{L} \cdot \frac{\mu_1}{[1 + \theta_1 (V_{G_1} - V_{T_1}(V_{G_2}))]^2}$$ (4.8)

where μ_1 is the mobility of front-channel carriers, and θ_1 is the mobility attenuation coefficient.

The complexity of the transconductance curves in fully depleted MOSFETs (Fig. 4.4(c)) is explained by the influence of the back gate bias via $V_{T_1}(V_{G_2})$. The effective mobility and transconductance peak are maximum for depletion at the back interface, due to combined effects of reduced vertical field and series resistances.

An unusual feature is the distortion of the transconductance (curve I, Fig. 4.4(c)), which reflects the possible activation of the back channel, far before the inversion charge build-up is completed at the front channel.[23] While the front interface is still depleted, increasing V_{G_1} reduces the back threshold voltage and eventually opens the *back* channel. The plateau of the front-channel transconductance (Fig. 4.4(c)) can be used to derive directly the back-channel mobility.

Volume Inversion

In thin and low-doped films, the simultaneous activation of front and back channels induces by continuity (i.e., *charge coupling*) the onset of *volume inversion*.[24] Unknown in bulk Si, this effect enables the inversion charge to cover the whole film. Self-consistent solutions of Poisson and Schrödinger equations indicate that the maximum density of the inversion charge is reached in the middle of the film. This results in increased current drive and transconductance, attenuated influence of interface defects (traps, fixed charges, roughness), and reduced $1/f$ noise.

Double-gate MOSFETs (*DELTA* and *GAA* transistors), designed to take full advantage from volume inversion, also benefit from reduced short-channel effects (V_T drop, punch-through, DIBL, hot-carrier injection, etc.), and are therefore very attractive, if not unique, devices for down-scaling below 30-nm gate length.

Defect Coupling

In fully depleted MOSFETs, carriers flowing at one interface may sense the presence of defects located at the opposite interface. *Defect coupling* is observed as an apparent degradation of the front-channel properties, which is actually induced by the buried oxide damage. This unusual mechanism is notorious after back interface degradation via radiation or hot-carrier injection (see also Fig. 4.7 in Section 4.7).

4.6 Partially Depleted SOI Transistors

In partially depleted SOI MOSFETs, the depletion charge controlled by one or both gates does not extend from an interface to the other. A neutral region subsists and, therefore, the interface coupling effects are disabled. When the body is grounded (via independent body contacts or body-source ties), partially depleted SOI transistors behave very much like bulk–Si MOSFETs and most of the standard $I_D(V_G, V_D)$ equations and design concepts apply. If body contacts are not supplied, so-called *floating-body* effects arise, leading to detrimental consequences, which will be explained next.

The *kink* effect is due to majority carriers, generated by impact ionization, that collect in the transistor body. The body potential is raised, which reduces the threshold voltage. This feedback gives rise to extra drain current (kink) in $I_D(V_D)$ characteristics (Fig. 4.5(a)), which is annoying in analog circuits.

In weak inversion and for high drain bias, a similar positive feedback (increased inversion charge → more impact ionization → body charging → threshold voltage lowering) is responsible for negative resistance regions, hysteresis in log $I_D(V_G)$ curves, and eventually latch (loss of gate control, Fig. 4.5(b)).

The floating body may also induce transient effects. A drain current *overshoot* is observed when the gate is turned on (Fig. 4.5(c)). Majority carriers are expelled from the depletion region and collect in the neutral body, increasing the potential. Equilibrium is reached through electron-hole recombination, which eliminates the excess majority carriers, making the drain current decrease gradually with time. A reciprocal *undershoot* occurs when the gate is switched from strong to weak inversion: the current now increases with time (Fig. 4.5(d)) as the majority carrier generation allows the depletion depth to shrink gradually. In short-channel MOSFETs, the transient times are dramatically reduced because of the additional contribution of source and drain junctions to establish equilibrium.

An obvious solution to alleviate floating-body effects is to sacrifice chip space for designing body contacts. The problem is that, in ultra-thin films with large sheet resistance, the body contacts are far

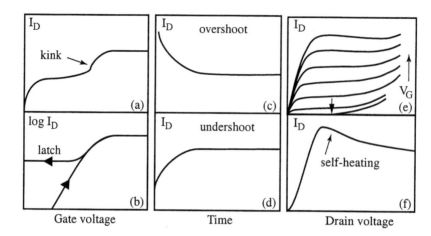

FIGURE 4.5 Parasitic effects in partially depleted SOI MOSFETs: (a) kink in $I_D(V_D)$ curves, (b) latch in $I_D(V_G)$ curves, (c) drain current overshoot, (d) current undershoot, (e) premature breakdown, and (f) self-heating.

from ideal. Their intrinsic resistance does not allow the body to be perfectly grounded and may generate additional noise. A floating body is then preferable to a poor body contact.

An exciting partially depleted device is the *dynamic-threshold* DT-MOS transistor. It is simply configured by interconnecting the gate and the body. As the gate voltage increases in weak inversion, the simultaneous raise in body potential makes the threshold voltage decrease. DT-MOSFETs achieve perfect gate-charge coupling, maximum subthreshold slope, and enhanced current, which are attractive features for LV/LP circuits.

4.7   Short-Channel Effects

In both fully and partially depleted MOSFETs with submicron length, the source–body junction can easily be turned on. The inherent activation of the lateral bipolar transistor has favorable (extra current flow in the body) or detrimental (premature breakdown, Fig. 4.5(e)) consequences. The breakdown voltage is evaluated near threshold, where the bipolar action prevails. The breakdown voltage is especially lowered for n-channels, shorter devices, thinner films, and higher temperatures. As expected, the impact ionization rate and related floating-body effects are attenuated at high temperature. However, the bipolar gain increases dramatically with temperature and accentuates the bipolar action: lower breakdown and latch voltages.[13]

Another concern is *self-heating*, induced by the power dissipation of short-channel MOSFETs and exacerbated by the poor thermal conductivity of the surrounding SiO_2 layers. Self-heating is responsible for mobility degradation, threshold voltage shift, and negative differential conductance shown in Fig. 4.5(f). The temperature rise can exceed 100 to 150°C in SOI, which is far more than in bulk Si.[25] Electromigration may even be initiated by the resulting increase in interconnect temperature. Thin buried oxides (≤100 nm) and thicker Si films (≥100 nm) are suitable when self-heating becomes a major issue.

A familiar short-channel effect is the threshold voltage roll-off due to charge sharing between the gate and source and drain terminals. The key parameters in SOI are the doping level, film thickness, and BOX thickness.[26] Ultra-thin, fully depleted MOSFETs show improved performance in terms of both V_T roll-off and drain-induced barrier lowering (DIBL) as compared to partially depleted SOI or bulk Si transistors (Fig. 4.6(a)).[27] The worst case happens when the film thickness corresponds to the transition between full and partial depletion. An additional origin of V_T roll-off in fully depleted MOSFETs is the field penetration into the buried oxide. An obvious solution is again the use of relatively thin buried oxides.

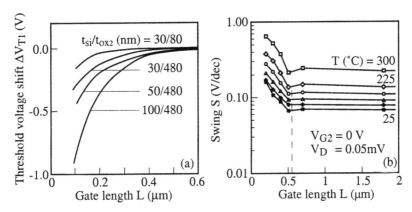

FIGURE 4.6 Typical short-channel effects in fully depleted SOI MOSFETs: (a) threshold voltage roll-off for different thicknesses of film and buried oxide[26] and (b) subthreshold swing degradation below 0.2-μm channel length for various temperatures.[13]

A degradation of the subthreshold swing is observed in very short ($L \leq 0.1$–$0.5\,\mu$m), fully depleted MOSFETs (Fig. 4.6(b)). Two effects are involved: (1) conventional charge sharing, and (2) a surprising non-uniform coupling effect. We have seen that the subthreshold swing is minimum for depletion and increases for inversion at the back interface. In very short transistors, the lateral profile of the back interface potential can be highly inhomogeneous: from depletion in the middle of the channel to weak inversion near the channel ends, due to the proximity of source and drain regions. This localized weak inversion region explains the degradation of the swing.[13]

The transconductance is obviously improved in deep submicron transistors. Velocity saturation occurs as in bulk silicon. The main short-channel limitation of the transconductance comes from series resistance effects.

The lifetime of submicron MOSFETs is affected by hot-carrier injection into the gate oxide(s). The degradation mechanisms are more complex in SOI than in bulk-Si, due to the presence of two oxides, two channels, and related coupling mechanisms. For example, in Fig. 4.7, the front-channel threshold voltage is monitored during back channel stress. The shift ΔV_{T_1}, measured for $V_{G_2} = 0$ (depleted back interface), would imply that many defects are being generated at the front interface. Such a conclusion is totally negated by measurements performed with $V_{G_2} = -40V$: the influence of buried-oxide defects is now masked by the accumulation layer and indeed the apparent front-interface damage disappears ($\Delta V_{T_1} \cong 0$).[28]

In n-channels, the defects are created at the interface where the electrons flow; exceptionally, injection into the opposite interface may arise when the transistor is biased in the breakdown region. Although the device lifetime is relatively similar in bulk Si and SOI, the influence of stressing bias is different: SOI MOSFETs degrade less than bulk Si MOSFETs for $V_G \cong V_D/2$ (i.e., for maximum substrate current) and more for $V_G \cong V_T$ (i.e., enhanced hole injection). Device aging is accelerated by accumulating the back interface.[28]

In p-channels, the key mechanism involves the electrons generated by front-channel impact ionization, which become trapped into the buried oxide. An apparent degradation of the front interface again occurs via coupling.[28]

4.8 SOI Challenges

SOI stands already as a pretty mature technology. However, there are still serious challenges in various domains — fundamental and device physics, technology, device modeling, and circuit design — before SOI will become fully competitive in the commercial market.

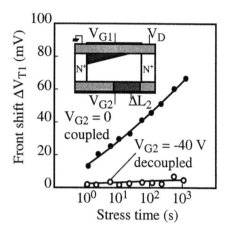

FIGURE 4.7   Front-channel threshold voltage shift during back-channel stress in a SIMOX MOSFET.[28] The apparent degradation of the gate oxide disappears when the stress-induced defects in the buried oxide are masked by interface accumulation ($V_{G_2} = -40$ V).

The minimum dimensions thus far achieved for SOI MOSFETs are: 70 nm length, 10 nm width (quantum wires), and 1 to 2 nm thickness.[21] When these features will be cumulative in a single transistor, the body volume ($\leq 10^{-18}$ cm^3!) will contain 10^4–10^5 silicon atoms and 0 to 1 defects. The body doping (10^{17}–10^{18} cm^{-3}) will be provided by a unique impurity, whose location may become important. Moreover, quantum transport phenomena are already being observed in ultra-thin SOI transistors. It is clear that new physical concepts, ideas, and modeling tools will be needed to account for minimum-size mechanisms in order to take advantage of them.

As far as the technology is concerned, a primary challenge is the mass-production of SOI wafers with large diameter (≥ 12 in.), low defect content, and reasonable cost (2 to 3 times higher than for bulk-Si wafers). The thickness uniformity of the silicon layer is especially important for fully depleted MOSFETs because it governs the fluctuations of the threshold voltage. It is predictable that several SOI technologies will not survive, except for special niches.

There is a demand for appropriate characterization techniques, either imported from other semiconductors or entirely conceived for SOI.[1] Such a pure SOI technique is the pseudo-MOS transistor (Ψ–MOS-FET).[29] Ironically, it behaves very much like the MOS device that Shockley attempted to demonstrate 50 years ago but, at that time, he didn't have the chance to know about SOI. The inset of Figure 4.8 shows that the Si substrate is biased as a gate and induces a conduction channel (inversion or accumulation) at the film–oxide interface. Source and drain probes are used to measure $I_D(V_G)$ characteristics. The Ψ–MOSFET does not require any processing; hence, valuable information is directly available: quality of the film, interface and oxide, electron/hole mobilities, and lifetime.

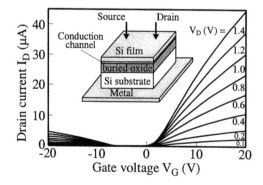

FIGURE 4.8   Pseudo-MOSFET transistor and $I_D(V_G)$ characteristics in SOI.

Full CMOS processing must address typical SOI requirements such as the series resistance reduction in ultra-thin MOSFETs (via local film oxidation, elevated source and drain structures, etc.), the lowering of the source–body barrier by source engineering (silicidation, Si–Ge, etc.), the control of the parasitic bipolar transistor, and the limitation of self-heating effects. It is now clear that the best of SOI is certainly not achievable by simply using a very good bulk-Si technology. For example, double-gate SOI MOSFETs deserve special processing and design.

According to process engineers and circuit designers, partially depleted SOI MOSFETs are more user friendly as they maintain the flavor of bulk-Si technology. On the other hand, fully depleted transistors have superior capability; they need to be domesticated in order to become more tolerant to short-channel effects. A possible solution, which requires further investigation, is the incorporation of a ground plane underneath the buried oxide.

Advanced modeling is required for correct transcription of the transistor behavior, including the transient effects due to body charging and discharging, floating-body mechanisms, bipolar transistor, dual-gate operation, quantum effects, self-heating, and short-channel limitations. Based on such physical models, compact models should then be conceived for customized simulation and design.

It is obvious that SOI does need SOI-dedicated CAD libraries. This implies a substantial amount of work which, in turn, will guarantee that the advantages and peculiar constraints of SOI devices are properly accounted for. The optimum configuration of memories, microprocessors, DSP, etc., will most likely be different in SOI as compared to bulk. Not only can SOI afford to combine fully/partially depleted, low/high power, and DT-MOSFETs in a single chip, but also the basic mechanisms of operation may differ.

4.9 Conclusion

For the next millennium, SOI offers the opportunity to integrate high-performance and/or innovative devices that can push away the present frontiers of the CMOS down-scaling. SOI will play a significant role in the future of microelectronics if subsisting problems can be rapidly solved. The short-term prospects of SOI-based microelectronics will also closely depend on the penetration rate of LV/LP SOI circuits into the market. Not only does SOI offer enhanced performance, but also most of SOI disadvantages (self-heating, hot carriers, early breakdown, etc.) tend to disappear for operation at low voltage.

A key challenge is associated with industrial strategy, which must be oriented to overcome the bulk-Si monocultural barrier. Designers, process engineers, and managers are extremely busy loading the bulk-Si machine. When, eventually, they can afford to take a careful look at the assets of SOI technology, they will realize the immediate and long-term benefits offered in terms of performance and scaling extensions.

This is so because SOI is not a totally different technology — it is just a metamorphosis of silicon.

References

1. Cristoloveanu, S. and Li, S. S., *Electrical Characterization of SOI Materials and Devices*, Kluwer, Norwell, 1995.
2. Cristoloveanu, S., "Silicon films on sapphire," *Rep. Prog. Phys.*, **3**, 327, 1987.
3. Johnson, R. A., de la Houssey, P. R., Chang, C. E., Chen, P.–F., Wood, M. E., Garcia, G. A., Lagnado, I., and Asbeck, P. M., "Advanced thin-film silicon-on-sapphire technology: microwave circuit applications," *IEEE Trans. Electron Devices*, **45**, 1047, 1998.
4. Cristoloveanu, S., "A review of the electrical properties of SIMOX substrates and their impact on device performance," *J. Electrochem. Soc.*, **138**, 3131, 1991.

5. Sato, N., Ishii, S., Matsumura, S., Ito, M., Nakayama, J., and Yonehara, T., "Reduction of crystalline defects to 50/cm^2 in epitaxial layers over porous silicon for Eltran," *IEEE Int. SOI Conf.*, Stuart, FL, 1998.

6. Burel, M., "Silicon on insulator material technology," *Electronics Lett.*, **31**, 1201, 1995.

7. Chen, J., Parke, S., King, J., Assaderaghi, F., Ko, P., and Hu, C., "A high-speed SOI technology with 12 ps/18 ps gate delay operating at 5V/1.5V," *IEDM Techn. Dig.*, 35, 1992.

8. Tsuchiya, T., Ohno, T., and Kado, Y., "Present status and potential of subquarter-micron ultra-thin-film CMOS/SIMOX technology," *SOI Technology and Devices*, Electrochem. Soc., Pennington, 401, 1994.

9. Fuse, T. et al., "A 0.5V 200MHz 1–stage 32b ALU using a body bias controlled SOI pass-gate logic," *ISSCC Techn. Digest*, 286, 1997.

10. Schepis, D. J., et al., "A 0.25 μm CMOS SOI technology and its application to 4 Mb SRAM," *IEDM Techn. Dig.*, 587, 1997.

11. Koh, Y.–H. et al., "1 Gigabit SOI DRAM with fully bulk compatible process and body-contacted SOI MOSFET structure," *IEDM Techn. Dig.*, 579, 1997.

12. Colinge, J.–P., *SOI Technology: Materials to VLSI* (2nd ed.), Kluwer, Boston, 1997.

13. Cristoloveanu, S., and Reichert, G., "Recent advances in SOI materials and device technologies for high temperature," in 1998 High Temperature Electronic Materials, Devices and Sensors Conference, *IEEE Electron Devices Soc.*, 86, 1998.

14. Hiramoto, T., Tamba, N., Yoshida M., et al., "A 27 GHz double polysilicon bipolar technology on bonded SOI with embedded 58 μm^2 CMOS memory cell for ECL–CMOS SRAM applications," *IEDM Techn. Dig.*, 39, 1992.

15. O'Connor, J. M., Luciani, V. K., and Caviglia, A. L., "High voltage DMOS power FETs on thin SOI substrates," *IEEE Int. SOI Conf. Proc.*, 167, 1990.

16. Vogt, H., "Advantages and potential of SOI structures for smart sensors," *SOI Technology and Devices*, Electrochem. Soc., Pennington, 430, 1994.

17. Ohno, T., Matsumoto, S., and Izumi, K., "An intelligent power IC with double buried-oxide layers formed by SIMOX technology," *IEEE Trans. Electron Devices*, **40**, 2074, 1993.

18. Nishimura, T., Inoue, Y., Sugahara, K., Kusonoki, S., Kumamoto, T., Nakagawa, S., Nakaya, M., Horiba, Y., and Akasaka, Y., "Three dimensional IC for high performance image signal processor," *IEDM Dig.*, 111, 1987.

19. Hisamoto, D., Kaga, T., and Takeda, E., "Impact of the vertical SOI 'DELTA' structure on planar device technology," *IEEE Trans. Electron Dev.*, **38**, 1419, 1991.

20. Lim, H.–K. and Fossum, J. G., "Threshold voltage of thin-film silicon on insulator (SOI) MOSFETs," *IEEE Trans. Electron Dev.*, **30**, 1244, 1983.

21. Ohmura, Y., Ishiyama, T., Shoji, M., and Izumi, K., "Quantum mechanical transport characteristics in ultimately miniaturized MOSFETs/SIMOX," *SOI Technology and Devices*, Electrochem. Soc., Pennington, 199, 1996.

22. Mazhari, B., Cristoloveanu, S., Ioannou D. E., and Caviglia, A. L., "Properties of ultra-thin wafer-bonded silicon on insulator MOSFETs," *IEEE Trans. Electron Dev.*, **ED–38**, 1289, 1991.

23. Ouisse, T., Cristoloveanu, S., and Borel, G., "Influence of series resistances and interface coupling on the transconductance of fully depleted silicon-on-insulator MOSFETs," *Solid-State Electron.*, **35**, 141, 1992.

24. Balestra, F., Cristoloveanu, S., Bénachir, M., Brini, J., and Elewa, T., "Double-gate silicon on insulator transistor with volume inversion: a new device with greatly enhanced performance," *IEEE Electron Device Lett.*, **8**, 410, 1987.

25. Su, L. T., Goodson, K. E., Antoniadis, D. A., Flik, M. I., and Chung, J. E., "Measurement and modeling of self-heating effects in SOI n-MOSFETs," *IEDM Dig.*, 111, 1992.

26. Ohmura, Y., Nakashima, S., Izumi, K., and Ishii, T., "0.1-μm-gate, ultra-thin film CMOS device using SIMOX substrate with 80-nm-thick buried oxide layer," *IEDM Techn. Dig.*, 675, 1991.

27. Balestra, F. and Cristoloveanu, S., "Special mechanisms in thin-film SOI MOSFETs," *Microelectron. Reliab.*, **37**, 1341, 1997.
28. Cristoloveanu, S., "Hot-carrier degradation mechanisms in silicon-on-insulator MOSFETs," *Microelectron. Reliab.*, **37**, 1003, 1997.
29. Cristoloveanu, S. and Williams, S., "Point contact pseudo-MOSFET for *in-situ* characterization of as-grown silicon on insulator wavers," *IEEE Electron Device Lett.*, **13**, 102, 1992.

5

SiGe Technology

John D. Cressler
Auburn University

5.1 Introduction

The concept of bandgap engineering has been used for many years in compound semiconductors such as gallium arsenide (GaAs) and indium phosphide (InP) to realize a host of novel electronic devices. A bandgap-engineered transistor is compositionally altered in a manner that improves a specific device metric (e.g., speed). A transistor designer might choose, for instance, to make a bipolar transistor that has a GaAs base and collector region, but which also has a AlGaAs emitter. Such a device has electrical properties that are inherently superior to what could be achieved using a single semiconductor. In addition to simply combining two different materials (e.g., AlGaAs and GaAs), bandgap engineering often involves compositional grading of materials within a device. For instance, one might choose to vary the Al content in an AlGaAs/GaAs transistor from a mole fraction of 0.4 to 0.6 across a given distance within the emitter.

Device designers have long sought to combine the bandgap engineering techniques enjoyed in compound semiconductors technologies with the fabrication maturity, high yield, and hence low cost associated with conventional silicon (Si) integrated circuit manufacturing. Epitaxial silicon-germanium (SiGe) alloys offer considerable potential for realizing viable bandgap-engineered transistors in the Si material system. This is exciting because it potentially allows Si electronic devices to achieve performance levels that were once thought impossible, and thus dramatically extends the number of high-performance circuit applications that can be addressed using Si technology. This chapter reviews the recent progress in both SiGe heterojunction bipolar transistor (HBT) technology and SiGe field effect transistor (FET) technology.

5.2 SiGe Strained Layer Epitaxy

Si and Ge, being chemically compatible elements, can be intermixed to form a stable alloy. Unfortunately, however, the lattice constant of Si is about 4.2% smaller than that of Ge. The difficulties associated with realizing viable SiGe bandgap-engineered transistors can be traced to the problems encountered in growing high-quality, defect-free epitaxial SiGe alloys in the presence of this lattice mismatch. For electronic applications, it is essential to obtain a SiGe film that adopts the same lattice constant as the underlying Si substrate with perfect alignment across the growth interface. In this case, the resultant SiGe alloy is under compressive strain. This strained SiGe film is thermodynamically stable only under a narrow range of conditions that depend on the film thickness and the effective strain (determined by

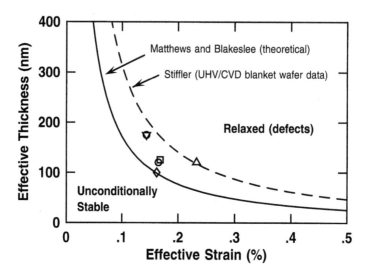

FIGURE 5.1   Effective thickness versus effective strain for SiGe strained layer epitaxy. Shown are a theoretical stability curve (Matthews-Blakeslee), and an empirical curve for UHV/CVD SiGe epitaxy. The symbols represent actual films used in UHV/CVD SiGe HBTs.

the Ge concentration).[1] The critical thickness below which the grown film is unconditionally stable depends reciprocally on the effective strain (Fig. 5.1). Thus, for practical electronic device applications, SiGe alloys must be thin (typically <100–200 nm) and contain only modest amounts of Ge (typically <20–30%). It is essential for electronic devices that the SiGe films remain thermodynamically stable so that conventional Si fabrication techniques such as high-temperature annealing, oxidation, and ion implantation can be employed without generating defects.

From an electronic device viewpoint, the property of the strained SiGe alloy that is most often exploited is the reduction in bandgap with strain and Ge content (roughly 75 meV per 10% Ge).[2] This band offset appears mostly in the valence band, which is particularly useful for realizing n-p-n bipolar transistors and p-channel FETs.[3] While these band offsets are modest compared to those that can be achieved in III–V semiconductors, the Ge content can be compositionally graded to produce local electric fields that aid carrier transport. For instance, in a SiGe HBT, the Ge content might be graded from 0 to 15% across distances as short as 50 to 60 nm, producing built-in drift fields as large as 15 to 20 kV · cm^{-1}. Such fields can rapidly accelerate the carriers to scattering-limited velocity (1 × 10^7 cm · s^{-1}), thereby improving the transistor frequency response. Another benefit of using SiGe strained layers is the enhancement in carrier mobility. This advantage will be exploited in SiGe channel FETs, as discussed below.

Epitaxial SiGe strained layers on Si substrates can be successfully grown today by a number of different techniques, including molecular beam epitaxy (MBE), ultra-high vacuum/chemical vapor deposition (UHV/CVD), rapid-thermal CVD (RTCVD), atmospheric pressure CVD (APCVD), and reduced pressure CVD (RPCVD). Each growth technique has advantages and disadvantages, but it is generally agreed that UHV/CVD[4] has a number of appealing features for the commercialization of SiGe integrated circuits. The features of UHV/CVD that make it particularly suitable for manufacturing include: (1) batch processing on up to 16 wafers simultaneously, (2) excellent doping and thickness control on large (e.g., 200 mm) wafers, (3) very low background oxygen and carbon concentrations, (4) compatibility with patterned wafers and hence conventional Si bipolar and CMOS fabrication techniques, and (5) the ability to compositionally grade the Ge content in a controllable manner across short distances. The experimental results presented in this chapter are based on the UHV/CVD growth technique as practiced at IBM Corporation, and are representative of the state-of-the-art in SiGe technology.

5.3 The SiGe Heterojunction Bipolar Transistor (HBT)

The SiGe HBT is by far the most mature Si-based bandgap-engineered electronic device.[5] The first SiGe HBT was reported in 1987,[6,7] and began commercial production in 1998. Significant steps along the path to manufacturing included the first demonstration of high-frequency (75 GHz) operation of a SiGe HBT in a non-self-aligned structure in early 1990.[8] This result garnered much attention worldwide because the performance of the SiGe HBT was roughly twice what a state-of-the-art Si BJT could achieve (Fig. 5.2). The first fully integrated, self-aligned SiGe HBT technology was demonstrated later in 1990,[9] the first fully integrated 0.5-μm SiGe BiCMOS technology (SiGe HBT + Si CMOS) in 1992,[10] and SiGe HBTs with frequency response above 100 GHz in 1993 and 1994.[11,12] A number of research laboratories around the world have demonstrated robust SiGe HBT technologies, including IBM,[13–17] Daimler-Benz/TEMIC,[18–21] NEC,[22–24] and Siemens.[25] More recent work has begun to focus on practical SiGe HBT circuits for radio-frequency (RF) and microwave applications.[26–28]

Because the intent in SiGe technology is to combine bandgap engineering with conventional Si fabrication techniques, most SiGe HBT technologies appear very similar to conventional Si bipolar technologies. A typical device cross-section is shown in Fig. 5.3. This SiGe HBT has a planar, self-aligned structure with a conventional polysilicon emitter contact, silicided extrinsic base, and deep- and shallow-trench isolation. A 3-5 level, chemical-mechanical-polishing (CMP) planarized, W-stud, AlCu CMOS metallization scheme is used. The extrinsic resistive and capacitive parasitics are intentionally minimized to improve the maximum oscillation frequency (f_{max}) of the transistor. Observe that the Ge is introduced only into the thin base region of the transistor, and is deposited with a thickness and Ge content that ensures the film is thermodynamically stable (examples of real SiGe HBT profiles are shown as symbols in stability space in Fig. 5.1). The *in situ* boron-doped, graded SiGe base is deposited across the entire wafer using the ultra-high vacuum/chemical vapor deposition (UHV/CVD) technique. In areas that are not covered by oxide, the UHV/CVD film consisting of an intrinsic-Si/strained boron-doped SiGe/intrinsic-Si stack is deposited as a perfect single-crystal layer on the Si substrate. Over the oxide, the deposited layer is polycrystalline (poly) and will serve either as the extrinsic base contact of the SiGe HBT, the poly-on-oxide resistor, or the gate electrode of the Si CMOS devices. The metallurgical base and single-crystal emitter widths range from 75 to 90 nm and 25 to 35 nm, respectively. A masked phosphorous implant is used to tailor the intrinsic collector profile for optimum frequency response at high current densities. A conventional deep-trench/shallow-trench bipolar isolation scheme is used, as well as a conventional

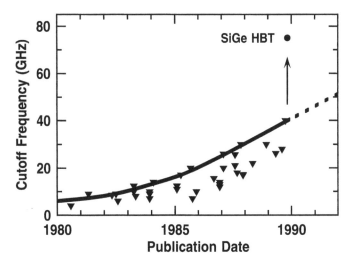

FIGURE 5.2 Transistor cutoff frequency as a function of publication date comparing Si BJT performance and the first SiGe HBT result.

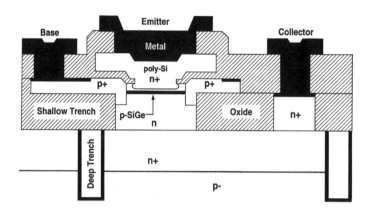

FIGURE 5.3   Schematic cross-section of a self-aligned UHV/CVD SiGe HBT.

arsenic-doped polysilicon emitter layer. This approach ensures that the SiGe HBT is compatible with commonly used (low-cost) bipolar/CMOS fabrication processes. A typical doping profile measured by secondary ion mass spectroscopy (SIMS) of the resultant SiGe HBT is shown in Fig. 5.4.

The smaller base bandgap of the SiGe HBT can be exploited in three major ways, and is best illustrated by examining an energy band diagram comparing a SiGe HBT with a Si BJT (Fig. 5.5). First, note the reduction in base bandgap at the emitter–base junction. The reduction in the potential barrier at the emitter–base junction in a SiGe HBT will exponentially increase the collector current density and, hence, current gain ($\beta = J_C/J_B$) for a given bias voltage compared to a comparably designed Si BJT. Compared to a Si BJT of identical doping profile, this enhancement in current gain is given by:

$$\frac{J_{C,\,SiGe}}{J_{C,\,Si}} = \frac{\beta_{SiGe}}{\beta_{Si}} = \gamma\eta\frac{\Delta E_{g,\,Ge}(grade)/kT \; e^{\Delta E_{g,\,Ge}(0)/kT}}{1 - e^{-\Delta E_{g,\,Ge}(grade)/kT}} \tag{5.1}$$

where $\gamma = N_C N_V$ (SiGe)/$N_C N_V$(Si) is the ratio of the density-of-states product between SiGe and Si, and $\eta = D_{nb}$(SiGe)/D_{nb}(Si) accounts for the differences between the electron and hole mobilities in the base. The position dependence of the band offset with respect to Si is conveniently expressed as a bandgap grading term ($\Delta E_{g,Ge}(grade) = \Delta E_{g,Ge}(Wb) - \Delta E_{g,Ge}(0)$). As can be seen in Fig. 5.6, which compares the measured Gummel characteristics for two identically constructed SiGe HBTs and Si BJTs, these theoretical expectations are clearly borne out in practice.

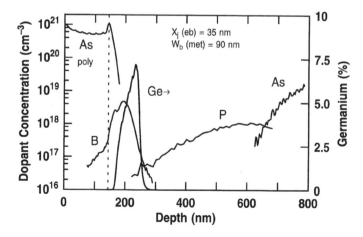

FIGURE 5.4   Secondary ion mass spectroscopy (SIMS) doping profile of a graded-base SiGe HBT.

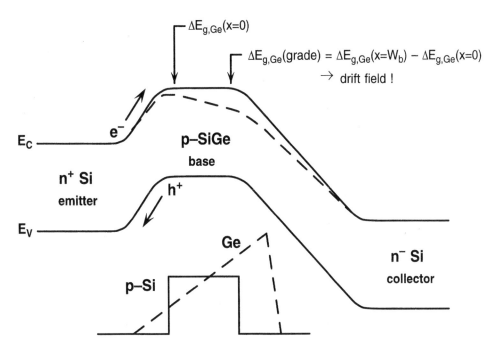

FIGURE 5.5 Energy band diagram for both a Si BJT and a graded-base SiGe HBT.

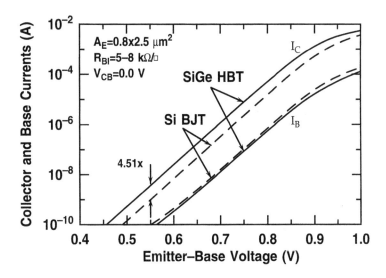

FIGURE 5.6 Measured current/voltage characteristics of both SiGe HBT and Si BJT with a comparable doping profile.

Secondly, if the Ge content is graded across the base region of the transistor, the conduction band edge becomes position dependent (refer to Fig. 5.5), inducing an electric field in the base that accelerates the injected electrons. The base transit time is thereby shortened and the frequency response of the transistor is improved according to:

$$\frac{\tau_{b,SiGe}}{\tau_{b,Si}} = \frac{f_{T,Si}}{f_{T,SiGe}} = \frac{2}{\eta}\left(\frac{kT}{\Delta E_{g,Ge}(grade)}\right)\left[1 - \frac{1 - e^{-\Delta E_{g,Ge}(grade)/kT}}{\Delta E_{g,Ge}(grade)/kT}\right] \tag{5.2}$$

Figure 5.7 compares the measured unity gain cutoff frequency (f_T) of a SiGe HBT and a comparably constructed Si BJT, and shows that an improvement in peak f_T of 1.7X can be obtained with relatively modest Ge profile grading (0–7.5% in this case). More recent studies demonstrate that peak cutoff frequencies in excess of 100 GHz can be obtained using more aggressively designed (although still stable) Ge profiles[12] (see Fig. 5.8).

The final advantage of using a graded Ge profile in a SiGe HBT is the improvement in the output conductance of the transistor, an important analog design metric. For a graded-base SiGe HBT, the Early voltage (a measure of output conductance) increases exponentially compared to a Si BJT of comparable doping, according to:

$$\frac{V_{A,\,SiGe}}{V_{A,\,Si}} = e^{\Delta E_{g,\,Ge}(grade)/kT}\left[\frac{1 - e^{-\Delta E_{g,\,Ge}(grade)/kT}}{(\Delta E_{g,\,Ge}(grade)/kT)}\right] \tag{5.3}$$

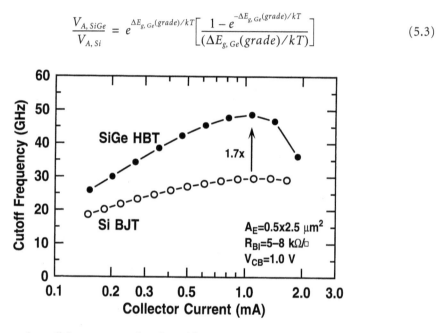

FIGURE 5.7 Measured cutoff frequency as a function of bias current for both SiGe HBT and Si BJT with a comparable doping profile.

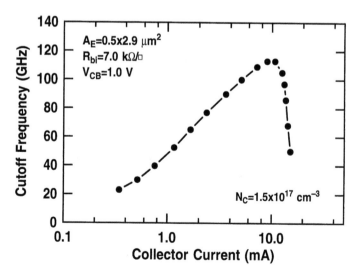

FIGURE 5.8 Cutoff frequency as a function of bias current for an aggressively designed UHV/CVD SiGe HBT.

In essence, the position dependence of the bandgap in the graded-base SiGe HBT weights the base profile toward the collector region, making it harder to deplete the base with collector-base bias, hence yielding a larger Early voltage. A transistor with a high Early voltage has a very flat common-emitter output characteristic, and hence low output conductance. For the device shown in Fig. 5.6, the Early voltage increases from 18 V in the Si BJT to 53 V in the SiGe HBT, a 3X improvement.

SiGe HBTs have been successfully integrated with conventional high-performance Si CMOS to realize a SiGe BiCMOS technology.[15] The SiGe HBT and ac performance in this SiGe BiCMOS technology is shown in Fig. 5.9, and is not compromised by the addition of the CMOS devices. Table 5.1 shows the suite of resultant elements in this SiGe BiCMOS technology. Two SiGe HBTs are available, one with a reduced collector implant and hence higher BV_{CEO} (5.3 V vs. 3.3 V) that is suitable for RF power applications. Table 5.2 gives the typical SiGe HBT device parameters at 300K.

Bandgap-engineered SiGe HBTs have other attractive features that make them ideal candidates for certain circuit applications. For instance, Si BJT technology is well known to have superior low-frequency noise properties compared to compound semiconductor technologies. Low-frequency noise is often a major limitation for RF and microwave systems because it directly limits the spectral purity of the

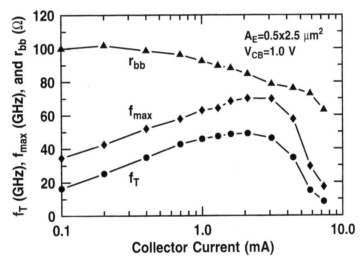

FIGURE 5.9 Cutoff frequency, maximum oscillation frequency, and small-signal base resistance for a small-geometry, state-of-the-art SiGe HBT.

TABLE 5.1   Elements in IBM's SiGe BiCMOS Technology

Element	Characteristics
Standard SiGe HBT	47 GHz f_T at BV_{CEO} = 3.3V
High-breakdown SiGe HBT	28 GHz f_T at BV_{CEO} = 5.3V
Si CMOS	0.36 mm L_{eff} for 2.5 V V_{DD}
Gated lateral pnp	1.0 GHz f_T
Polysilicon resistor	342 $\Omega/\square$
Ion-implanted resistor	1,600 $\Omega/\square$
Thin-oxide decoupling capacitor	1.52 fF/μm^2
MIM precision capacitor	0.70 fF/μm^2
Inductor (6-turn)	10 nH with Q = 10 at 1.0 GHz
Schottky barrier diode	213 mV at 100 μA
p-i-n diode	790 mV at 100 μA
Varactor diode	810 mV at 100 μA
ESD diode	1.2 kV

TABLE 5.2 Typical Parameters for a SiGe HBT with $A_E = 0.5 \times 2.5 \ \mu m^2$. All ac Parameters Were Measured at $V_{CB} = 1.0V$ and f_{max} Was Extracted Using MAG

Parameter	Standard SiGe HBT	High-BVCEO SiGe HBT
Peak β	113	97
V_A (V)	61	132
βV_A (V)	6,893	12,804
Peak f_T (GHz)	48	28
rbb at peak f_T (Ω)	80	N/A
Peak f_{max} (GHz)	69	57
BV_{CEO} (V)	3.3	5.3
BV_{EBO} (V)	4.2	4.1
Peak $f_T \times BV_{CEO}$ (GHz V)	158	143

transmitted signal. Recent work suggests that SiGe HBTs have low-frequency properties as good as or better than Si BJTs, superior to that obtained in AlGaAs/GaAs HBTs and Si CMOS (Fig. 5.10).[29,30] The broadband (RF) noise in SiGe HBTs is competitive with GaAs MESFET technology and superior to Si BJTs. In addition, SiGe HBTs have recently been shown to be very robust with respect to ionizing radiation, an important feature for space-based electronic systems.[31,32] Finally, cooling enhances all of the advantages of a SiGe HBT. In striking contrast to a Si BJT, which strongly degrades with cooling, the current gain, Early voltage, cutoff frequency, and maximum oscillation frequency (f_{max}) all improve significantly as the temperature drops.[33–35] This means that the SiGe HBT is well suited for operation in the cryogenic environment (e.g., 77K), historically the exclusive domain of Si CMOS and III–V compound semiconductor technologies. Cryogenic electronics is in growing use in both military and commercial applications such as space-based satellites, high-sensitivity instrumentation, high-T_C superconductors, and future cryogenic computers.

5.4 The SiGe Heterojunction Field Effect Transistor (HFET)

The effective carrier mobility (μ_{eff}) is the fundamental parameter that limits the speed of field effect transistors (FETs), and SiGe bandgap engineering can be used in two principal ways to significantly improve the mobility and, hence, the speed of the device. First, the valence band offset associated with SiGe can be used to spatially confine carriers such that they are effectively removed from the Si/SiO_2 interface. The surface roughness scattering associated with the Si/SiO_2 interface degrades the mobility

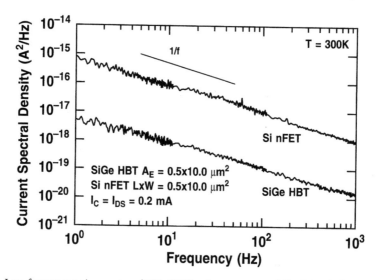

FIGURE 5.10 Low-frequency noise spectra of SiGe HBT and a conventional Si nFET of identical geometry.

in a conventional MOSFET, particularly at high gate bias. If, in addition, the holes are confined to a region of the semiconductor that is intentionally left undoped, the result is a reduction in ionized impurity scattering and, hence, further increase in mobility.[2] This is exactly the approach taken in bandgap-engineered compound semiconductor FETs known as HEMTs (high electron mobility transistor). Second, the strain associated with SiGe epitaxy is known to lift the degeneracy of the light and heavy hole valence bands, resulting in a reduced hole effective mass and hence higher hole mobility.[3] Because it is the hole mobility that is improved in strained SiGe, and the valence band offset can be used to confine these holes, the p-channel FET (pFET) is the logical device to pursue with SiGe bandgap engineering. A SiGe pFET with an improved hole mobility is particularly desirable in CMOS technology because the conventional Si pFET has a mobility that is about a factor of two lower than the Si nFET mobility. This mobility asymmetry between pFET and nFET in conventional Si CMOS technology requires a doubling of the pFET gate width, and thus a serious real estate penalty, to obtain proper switching characteristics in the CMOS logic gate. If SiGe can be used to equalize the pFET and nFET mobilities, then substantial area advantages can be realized in CMOS logic gates, and tighter packing densities achieved.

It is interesting to note that despite the fact that the SiGe HBT is the most commercially mature SiGe device, the first SiGe FET (actually a HEMT structure) predates the first SiGe HBT by one year, having been first reported in 1986.[36,37] A number of different SiGe pFET designs have been successfully demonstrated[38–43] with improvements in mobility as high as 50% at 0.25-μm gate lengths.[40] Figure 5.11 shows perhaps the simplest configuration of a SiGe pFET, which consists of a SiGe hole channel buried underneath a thin Si cap layer and the conventional thermal oxide.[44,45] In this case, the entire device is fabricated on a silicon-on-sapphire substrate to improve microwave performance. Significant mobility advantage can be realized over a comparable Si pFET (Fig. 5.12).

Because a complementary circuit configuration offers many advantages from a power dissipation standpoint, the realization of n-channel SiGe devices is highly desirable. Strictly speaking, this is not possible in strained SiGe on Si substrates because only a valence band offset is available and the electrons cannot be confined as in the SiGe pFET. Fortunately, however, recent work[46–51] using strained Si on relaxed SiGe layers has proven particularly promising because it provides a conduction band offset and enhanced electron mobility compared to Si.

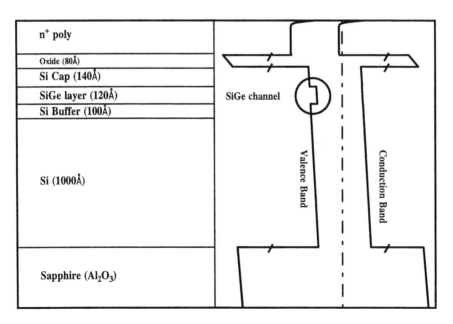

FIGURE 5.11   Schematic cross-section of SiGe pFET on silicon-on-sapphire (SOS).

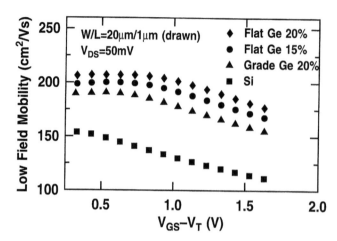

FIGURE 5.12 Effective mobility as a function of gate drive comparing SiGe pFETs on SOS and conventional Si pFETs on SOS.

The fabrication of strained Si nFETs on relaxed SiGe layers is, in general, more complicated than fabricating strained SiGe pFETs in Si substrates. For the strained Si nFET, a graded SiGe buffer layer is used to reduce and confine the dislocations associated with the growth of relaxed SiGe layers.[46] Using this technique, both strained Si nFETs and strained SiGe pFETs can be jointly fabricated to form a Si-based heterojunction CMOS (HCMOS) technology. Figure 5.13 shows a schematic cross-section of such a Si/SiGe HCMOS technology,[51] and represents the state-of-the-art in the field of Si/SiGe HCMOS. Observe that the conducting channels of both transistors are grown in a single step (using UHV/CVD in this case), and electron and hole confinement occurs in the strained SiGe and strained Si for the pFET and nFET, respectively. In this technology, both the pFET and nFET are realized in a planar structure, and are expected to show substantial improvements in performance over conventional Si CMOS. With layer and doping profile optimization, the parasitic surface channel (which degrades mobility) can be minimized. Simulation results of anticipated circuit performance indicate that substantial improvements (4 to 6X) in power-delay product over conventional Si CMOS can be obtained using Si/SiGe HCMOS technology at 0.2-μm effective gate length.

5.5   Future Directions

The future developments in SiGe electronic devices will likely follow two paths. The first path will be toward increased integration and scaling of existing SiGe technologies. There is already a clear trend toward integrating SiGe HBTs with conventional Si CMOS to form a SiGe BiCMOS technology.[14,15] Eventually, one would like to merge SiGe HBTs, Si CMOS, Si/SiGe HCMOS, and SiGe photonic devices on the same chip to provide a complete solution for future RF and microwave electronic and optoelectronic transceivers containing precision analog functions (mixers, LNAs, VCOs, DACs, ADCs, power amps, etc.), digital signal processing and computing functions (traditionally done in CMOS), as well as integrated photonic detectors and possibly transmitters.

The second path in SiGe electronic devices will be to pursue new Si-based material systems that offer the promise of lattice-matching to Si substrates. The intent here is to remove the stability constraints that SiGe device designers currently face, and that limit the useful range of Ge content in practical SiGe devices. The most promising of these new materials is silicon-germanium-carbon (SiGeC). The lattice constant of C is larger than Si and thus can be used to reduce the strain in a SiGe film. Theoretically, it would only take 1% C to lattice match a 9% Ge film to a Si substrate, 2% C for 18% Ge, etc. While research is just beginning on the growth of device-quality SiGeC films, and the properties of the resultant films have not been firmly established, the SiGeC material system clearly offers exciting possibilities for

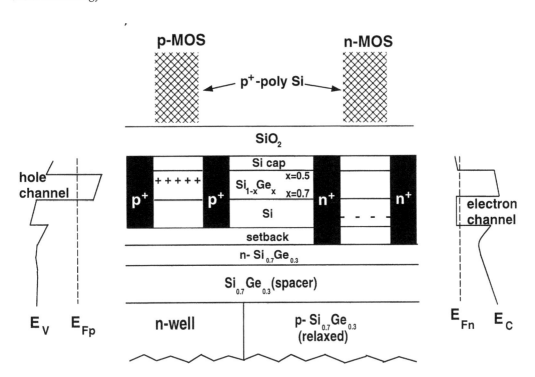

FIGURE 5.13 Schematic device cross-section of SiGe CMOS technology, consisting of a strained SiGe pFET and a strained Si nFET (after Ref. 51).

the future evolution of SiGe technology. In addition, it has recently been shown that low concentrations of C can serve to dramatically reduce boron diffusion in conventional SiGe HBTs.[52] This has the potential to allow much more aggressive SiGe HBT profiles to be realized with stable SiGe strained layers. More research is required to quantify the impact of C on the device electrical characteristics, although initial studies appear promising.

Acknowledgments

The author would like to thank D.L. Harame, B.S. Meyerson, S. Subbanna, D. Ahlgren, M. Gilbert, K. Ismail, and the members of the SiGe team at IBM Corporation, as well as the past and present members of the Auburn University SiGe research group (A. Joseph, D. Richey, L. Vempati, S. Mathew, J. Roldán, G. Bradford, G. Niu, B. Ansley, K. Shivaram, G. Banerjee, S. Zhang, S. Salmon, and U. Gogineni) for their contributions to this work. The support of the Alabama Microelectronics Science and Technology Center, ONR, DARPA, NRL, U.S. Army SSDC, Navy NCCOSC, Navy Crane, and MRC are gratefully acknowledged.

References

1. Matthews, J. W. and Blakeslee, A. E., *Journal of Crystal Growth*, **27**, 118,1974.
2. People, R., *IEEE Journal of Quantum Electronics*, **22**, 1696, 1986.
3. Meyerson, B. S., *Proceedings of the IEEE*, **80**, 1592, 1992.
4. Meyerson, B. S., *Applied Physics Letters*, **48**, 797, 1986.
5. Cressler, J. D., *IEEE Spectrum*, 49, March 1995.
6. Iyer, S. S., Patton, G. L., Delage, S. L., Tiwari, S., and Stork, J. M. C., in *Technical Digest of the IEEE International Electron Device Meeting*, 874, 1987.
7. Patton, G. L., Iyer, S. S., Delage, S. L., Tiwari, S., and Stork, J. M. C., *IEEE Electron Device Letters*, **9**, 165, 1988.

8. Patton, G. L., Comfort, J. H., Meyerson, B. S., Crabbé, E. F., Scilla, G. J., de Fresart, E., Stork, J. M. C., Sun, J. Y.-C., Harame, D. L., and Burghartz, J., *IEEE Electron Device Letters*, **11**, 171, 1990.

9. Comfort, J. H., Patton, G. L., Cressler, J. D., Lee, W., Crabbé, E. F., Meyerson, B. S., Sun, J. Y.-C., Stork, J. M. C., Lu, P.-F., Burghartz, J. N., Warnock, J., Scilla, G., Toh, K.-Y., D'Agostino, M., Stanis, C., and Jenkins, K., in *Technical Digest of the International Electron Device Meeting*, 21, 1990.

10. Harame, D. L., Crabbé, E. F., Cressler, J. D., Comfort, J. H., Sun, J. Y.-C., Stiffler, S. R., Kobeda, E., Gilbert, M., Malinowski, J., Dally, A. J., Ratanaphanyarat, S., Saccamango, M. J., Rausch, W., Cotte, J., Chu, C., and Stork, J. M. C., in *Technical Digest of the International Electron Device Meeting*, 19, 1992.

11. Schuppen, A., Gruhle, A., Erben, U., Kibbel, H., and Konig, U., in *Technical Digest of the International Electron Device Meeting*, 377, 1994.

12. Crabbé, E. F., Meyerson, B. S., Stork, J. M. C., and Harame, D. L., in *Technical Digest of the International Electron Device Meeting*, 83, 1993.

13. Harame, D. L., Stork, J. M. C., Meyerson, B. S., Hsu, K. Y.-J., Cotte, J., Jenkins, K. A., Cressler, J. D., Restle, P., Crabbé, E. F., Subbanna, S., Tice, T. E., Scharf, B. W., and Yasaitis, J. A., in *Technical Digest of the International Electron Device Meeting*, 71, 1993.

14. Harame, D. L., Schonenberg, K., Gilbert, M., Nguyen-Ngoc, D., Malinowski, J., Jeng, S.-J., Meyerson, B. S., Cressler, J. D., Groves, R., Berg, G., Tallman, K., Stein, K., Hueckel, G., Kermarrec, C., Tice, T., Fitzgibbons, G., Walter, K., Colavito, D., Houghton, T., Greco, N., Kebede, T., Cunningham, B., Subbanna, S., Comfort, J. H., and Crabbé, E. F., in *Technical Digest of the International Electron Device Meeting*, 437, 1994.

15. Nguyen-Ngoc, D., Harame, D. L., Malinowski, J. C., Jeng, S.-J., Schonenberg, K. T., Gilbert, M. M., Berg, G., Wu, S., Soyuer, M., Tallman, K. A., Stein, K. J., Groves, R. A., Subbanna, S., Colavito, D., Sunderland, D. A., and Meyerson, B. S., in *Proceedings of the Bipolar/BiCMOS Circuits and Technology Meeting*, 89, 1995.

16. Harame, D. L., Comfort, J. H., Cressler, J. D., Crabbé, E. F., Sun, J. Y.-C., Meyerson, B. S., and Tice, T., *IEEE Transactions on Electron Devices*, 42, 455, 1995.

17. Harame, D. L., Comfort, J. H., Cressler, J. D., Crabbé, E. F., Sun, J. Y.-C., Meyerson, B. S., and Tice, T., *IEEE Transactions on Electron Devices*, **42**, 469, 1995.

18. Gruhle, A., *Journal of Vacuum Science and Technology B*, **11**, 1186, 1993.

19. Gruhle, A., Kibbel, H., Konig, U., Erben, U., and Kasper, E., *IEEE Electron Device Letters*, **13**, 206, 1992.

20. Schuppen, A., in *Technical Digest of the International Electron Device Meeting*, 743, 1995.

21. Schuppen, A., Dietrich, H., Gerlach, S., Arndt, J., Seiler, U., Gotzfried, A., Erben, U., and Schumacher, H., in *Proceedings of the Bipolar/BiCMOS Circuits and Technology Meeting*, 130, 1996.

22. Sato, F., Takemura, H., Tashiro, T., Hirayama, T., Hiroi, M., Koyama, K., and Nakamae, M., in *Technical Digest of the International Electron Device Meeting*, 607, 1992.

23. Sato, F., Hashimoto, T., Tatsumi, T., and Tasshiro, T., *IEEE Transactions on Electron Devices*, **42**, 483, 1995.

24. Sato, F., Tezuka, H., Soda, M., Hashimoto, T., Suzaki, T., Tatsumi, T., Morikawa, T., and Tashiro, T., in *Proceedings of the Bipolar/BiCMOS Circuits and Technology Meeting*, 158, 1995.

25. Meister, T. F., Schafer, H., Franosch, M., Molzer, W., Aufinger, K., Scheler, U., Walz, C., Stolz, M., Boguth, S., and Bock, J., in *Technical Digest of the International Electron Device Meeting*, 739, 1995.

26. Soyuer, M., Burghartz, J., Ainspan, H., Jenkins, K. A., Xiao, P., Shahani, A., Dolan, M., and Harame, D. L., in *Proceedings of the Bipolar/BiCMOS Circuits and Technology Meeting*, 169, 1996.

27. Schumacher, H., Erben, U., Gruhle, A., Kibbelk, H., and Konig, U., in *Proceedings of the Bipolar/BiCMOS Circuits and Technology Meeting*, 186, 1995.

28. Sato, F., Hashimoto, T., Tatsumi, T., Soda, M., Tezuka, H., Suzaki, T., and Tashiro, T., in *Proceedings of the Bipolar/BiCMOS Circuits and Technology Meeting*, 82, 1995.

29. Cressler, J. D., Vempati, L., Babcock, J. A., Jaeger, R. C., and Harame, D. L., *IEEE Electron Device Letters*, **17**, 13, 1996.

30. Vempati, L., Cressler, J. D., Babcock, J. A., Jaeger, R. C., and Harame, D. L., *IEEE Journal of Solid-State Circuits*, **31**, 1458, 1996.

31. Babcock, J. A., Cressler, J. D., Vempati, L., Clark, S. D., Jaeger, R. C., and Harame, D. L., *IEEE Electron Device Letters*, **16**, 351, 1995.

32. Babcock, J. A., Cressler, J. D., Vempati, L., Jaeger, R. C., and Harame, D. L., *IEEE Transactions on Nuclear Science*, **42**, 1558, 1995.

33. Cressler, J. D., Crabbé, E. F., Comfort, J. H., Sun, J.Y.-C., and Stork, J. M. C., *IEEE Electron Device Letters*, **15**, 472, 1994.

34. Cressler, J. D., Comfort, J. H., Crabbé, E. F., Patton, G. L., Stork, J. M. C., Sun, J. Y.-C., and Meyerson, B. S., *IEEE Transactions on Electron Devices*, **40**, 525, 1993.

35. Cressler, J. D., Crabbé, E. F., Comfort, J. H., Stork, J. M. C., and Sun, J. Y.-C., *IEEE Transactions on Electron Devices*, **40**, 542, 1993.

36. Pearsall, T. P. and Bean, J. C., *IEEE Electron Device Letters*, **7**, 308, 1986.

37. Daembkes, H., Herzog, H.-J., Jorke, H., Kibbel, H., and Kasper, E., in *Technical Digest of the International Electron Device Meeting*, 768, 1986.

38. Wang, P. J., Meyerson, B. S., Fang, F. F., Nocera, J., and Parker, B., *Applied Physics Letters*, **55**, 2333, 1989.

39. Nayak, D. K., Woo, J. C. S., Park, J. S., Wang, K. L., and McWilliams, K. P., *IEEE Electron Device Letters*, **12**, 154, 1991.

40. Kesan, V. P., Subbanna, S., Restle, P. J., Tejwani, M. J., Altken, J. M., Iyer, S. S., and Ott, J. A., in *Technical Digest of the International Electron Device Meeting*, 25, 1991.

41. Verdonckt-Vanderbroek, S., Crabbé, E. F., Meyerson, B. S., Harame, D. L., Restle, P. J., Stork, J. M. C., Megdanis, A. C., Stanis, C. L., Bright, A. A., Kroesen, G. M. W., and Warren, A. C., *IEEE Electron Device Letters*, **12**, 447, 1991.

42. Verdonckt-Vanderbroek, S., Crabbé, E. F., Meyerson, B. S., Harame, D. L., Restle, P. J., Stork, J. M. C., and Johnson, J. B., *IEEE Transactions on Electron Devices*, **41**, 90, 1994.

43. Garone, P. M., Venkataraman, V., and Sturm, J. C., *IEEE Electron Device Letters*, **12**, 230, 1991.

44. Mathew, S., Niu, G., Dubbelday, W., Cressler, J. D., Ott, J., Chu, J., Mooney, P., Kavanaugh, K., Meyerson, B., and Lagnado, I., in *Technical Digest of the IEEE International Electron Devices Meeting*, 815, 1997.

45. Mathew, S., Ansley, W., Dubbelday, W., Cressler, J., Ott, J., Chu, J., Meyerson, B., Kavanaugh, K., Mooney, P., and Lagnado, I., in *Technical Digest of the IEEE Device Research Conference*, 130, 1997.

46. Meyerson, B. S., Uram, K., and LeGoues, F., *Applied Physics Letters*, **53**, 2555, 1988.

47. Ismail, K., Meyerson, B. S., Rishton, S., Chu, J., Nelson, S., and Nocera, J., *IEEE Electron Device Letters*, **13**, 229, 1992.

48. Ismail, K., Nelson, S. F., Chu, J. O., and Meyerson, B. S., *Applied Physics Letters*, **63**, 660, 1993.

49. Welser, J., Hoyt, J. L., and Gibbons, J. F., *IEEE Electron Device Letters*, **15**, 100, 1994.

50. Ismail, K., Chu, J. O., and Meyerson, B. S., *Applied Physics Letters*, **64**, 3124, 1994.

51. Sadek, A., Ismail, K., Armstrong, M. A., Antoniadis, D. A., and Stern, F., *IEEE Transactions on Electron Devices*, **43**, 1224, 1996.

52. Lanzerotti, L., Sturm, J., Stach, E., Hull, R., Buyuklimanli, T., and Magee, C., in *Technical Digest of the IEEE International Electron Device Meeting*, 249, 1996.

6

SiC Technology

Philip G. Neudeck
*NASA Glenn Research Center at
Lewis Field*

6.1   Introduction

Silicon carbide (SiC)-based semiconductor electronic devices and circuits are presently being developed for use in high-temperature, high-power, and/or high-radiation conditions under which conventional semiconductors cannot adequately perform. Silicon carbide's ability to function under such extreme conditions is expected to enable significant improvements to a far-ranging variety of applications and systems. These range from greatly improved high-voltage switching[1-4] for energy savings in public electric power distribution and electric motor drives, to more powerful microwave electronics for radar and communications,[5-7] to sensors and controls for cleaner-burning more fuel-efficient jet aircraft and automobile engines. In the particular area of power devices, theoretical appraisals have indicated that SiC power MOSFETs and diode rectifiers would operate over higher voltage and temperature ranges, have superior switching characteristics, and yet have die sizes nearly 20 times smaller than correspondingly rated silicon-based devices.[8] However, these tremendous theoretical advantages have yet to be realized in experimental SiC devices, primarily due to the fact that SiC's relatively immature crystal growth and device fabrication technologies are not yet sufficiently developed to the degree required for reliable incorporation into most electronic systems.[9]

This chapter briefly surveys the SiC semiconductor electronics technology. In particular, the differences (both good and bad) between SiC electronics technology and well-known silicon VLSI technology are highlighted. Projected performance benefits of SiC electronics are highlighted for several large-scale applications. Key crystal growth and device-fabrication issues that presently limit the performance and capability of high-temperature and/or high-power SiC electronics are identified.

6.2   Fundamental SiC Material Properties

SiC Crystallography: Important Polytypes and Definitions

Silicon carbide occurs in many different crystal structures, called polytypes. A comprehensive introduction to SiC crystallography and polytypism can be found in Ref. 10. Despite the fact that all SiC polytypes chemically consist of 50% carbon atoms covalently bonded with 50% silicon atoms, each SiC polytype has its own distinct set of electrical semiconductor properties. While there are over 100 known polytypes of SiC, only a few are commonly grown in a reproducible form acceptable for use as an electronic semiconductor. The most common polytypes of SiC presently being developed for electronics are 3C-SiC, 4H-SiC, and 6H-SiC. 3C-SiC, also referred to as β-SiC, is the only form of SiC with a cubic crystal lattice structure. The non-cubic polytypes of SiC are sometimes ambiguously referred to as α-SiC. 4H-SiC and 6H-SiC are only two of many possible SiC polytypes with hexagonal crystal structure. Similarly, 15R-SiC is the most common of many possible SiC polytypes with a rhombohedral crystal structure.

Because some important electrical device properties are non-isotropic with respect to crystal orientation, lattice site, and surface polarity, some further understanding of SiC crystal structure and terminology is necessary. As discussed much more thoroughly in Ref. 10, different polytypes of SiC are actually composed of different stacking sequences of Si-C bilayers (also called Si-C double layers), where each single Si-C bilayer can simplistically be viewed as a planar sheet of silicon atoms coupled with a planar sheet of carbon atoms. The plane formed by a bilayer sheet of Si and C atoms is known as the basal plane, while the crystallographic c-axis direction, also known as the stacking direction or the [0001] direction, is defined normal to Si-C bilayer plane. Figure 6.1 schematically depicts the stacking sequence of the 6H-SiC polytype, which requires six Si-C bilayers to define the unit cell repeat distance along the c-axis [0001] direction. The [1100] direction depicted in Fig. 6.1 is often referred to as the a-axis direction. The silicon atoms labeled "h" or "k" in Figure 6.1 denote Si-C double layers that reside in "quasi-hexagonal" or "quasi-cubic" environments with respect to their immediately neighboring above and below bilayers. SiC is a polar semiconductor across the c-axis, in that one surface normal to the c-axis is terminated with silicon atoms, while the opposite normal c-axis surface is terminated with carbon atoms. As shown in Fig. 6.1, these surfaces are typically referred to as "silicon face" and "carbon face" surfaces, respectively.

SiC Semiconductor Electrical Properties

Owing to the differing arrangements of Si and C atoms within the SiC crystal lattice, each SiC polytype exhibits unique fundamental electrical and optical properties. Some of the more important semiconductor electrical properties of the 3C, 4H, and 6H silicon carbide polytypes are given in Table 6.1. Much more detailed electrical properties can be found in Refs. 11 to 13 and references therein. Even within a given polytype, some important electrical properties are non-isotropic, in that they are a strong function of crystallographic direction of current flow and applied electric field (e.g., electron mobility for 6H-SiC). Dopants in SiC can incorporate into energetically inequivalent quasi-hexagonal (h) C-sites or Si-sites, or quasi-cubic (k) C-sites or Si-sites (only Si-sites are h or k labeled in Fig. 6.1). While all dopant ionization energies associated with various dopant incorporation sites should normally be considered for utmost accuracy, Table 6.1 lists only the shallowest ionization energies of each impurity.

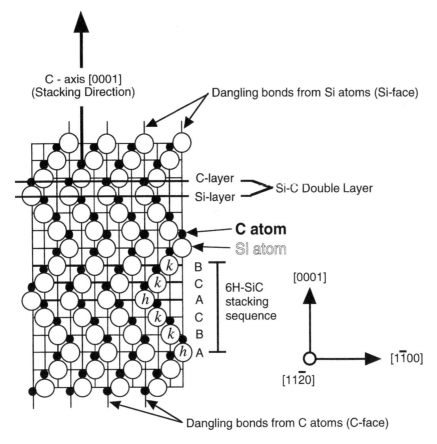

FIGURE 6.1 Schematic cross-section {(1120) plane} of the 6H-SiC polytype. (Modified from Ref. 10. With permission.)

TABLE 6.1 Comparison of Selected Important Semiconductors of Major SiC Polytypes with Silicon and GaAs

Property	Silicon	GaAs	4H-SiC	6H-SiC	3C-SiC
Bandgap (eV)	1.1	1.42	3.2	3.0	2.3
Relative dielectric constant	11.9	13.1	9.7	9.7	9.7
Breakdown field $N_D = 10^{17}$ cm^{-3} (MV/cm)	0.6	0.6	//c-axis: 3.0	// c-axis: 3.2 $\perp$ c-axis: >1	>1.5
Thermal conductivity (W/cm-K)	1.5	0.5	3–5	3–5	3–5
Intrinsic carrier concentration (cm^{-3})	10^{10}	1.8×10^{6}	~10^{-7}	~10^{-5}	~10
Electron mobility @ $N_D = 10^{16}$ cm^{-3} (cm^2/V-s)	1200	6500	//c-axis: 800 $\perp$ c-axis: 800	//c-axis: 60 $\perp$ c-axis: 400	750
Hole mobility @ $N_A = 10^{16}$ cm^{-3} (cm^2/V-s)	420	320	115	90	40
Saturated electron velocity (10^7 cm/s)	1.0	1.2	2	2	2.5
Donor dopants & shallowest ionization energy (meV)	P: 45 As: 54	Si: 5.8	N: 45 P: 80	N: 85 P: 80	N: 50
Acceptor dopants & shallowest ionization energy (meV)	B: 45	Be, Mg, C: 28	Al: 200 B: 300	Al: 200 B: 300	Al: 270
1998 Commercial wafer diameter (cm)	30	15	5	5	None

Note: Data compiled from Refs. 11–13, 15, and references therein. (With permission.)

For comparison, Table 6.1 also includes comparable properties of silicon and GaAs. Because silicon is the semiconductor employed in most commercial solid-state electronics, it is the yardstick against which other semiconductor materials must be evaluated. To varying degrees, the major SiC polytypes exhibit advantages and disadvantages in basic material properties compared to silicon. The most beneficial inherent material superiorities of SiC over silicon listed in Table 6.1 are its exceptionally high breakdown electric field, wide bandgap energy, high thermal conductivity, and high carrier saturation velocity. The electrical device performance benefits that each of these properties enable are discussed in the next section, as are system-level benefits enabled by improved SiC devices.

6.3 Applications and Benefits of SiC Electronics

Two of the most beneficial advantages that SiC-based electronics offer are in the areas of high-temperature device operation and high-power device operation. The specific SiC device physics that enables high-temperature and high-power capabilities will be examined first, followed by several examples of revolutionary system-level performance improvements these enhanced capabilities enable.

High-Temperature Device Operation

The wide bandgap energy and low intrinsic carrier concentration of SiC allow SiC to maintain semiconductor behavior at much higher temperatures than silicon, which in turn permits SiC semiconductor device functionality at much higher temperatures than silicon. As discussed in basic semiconductor physics textbooks,[14,15] semiconductor electronic devices function in the temperature range where intrinsic carriers are negligible so that conductivity is controlled by intentionally introduced dopant impurities. Furthermore, the intrinsic carrier concentration n_i is a fundamental prefactor to well-known equations governing undesired junction reverse-bias leakage currents.[15–18] As temperature increases, intrinsic carriers increase exponentially so that undesired leakage currents grow unacceptably large, and eventually at still higher temperatures, the semiconductor device operation is overcome by uncontrolled conductivity as intrinsic carriers exceed intentional device dopings. Depending on specific device design, the intrinsic carrier concentration of silicon generally confines silicon device operation to junction temperatures less than 300°C. The much smaller intrinsic carrier concentration of SiC theoretically permits device operation at junction temperatures exceeding 800°C; 600°C SiC device operation has been experimentally demonstrated on a variety of SiC devices (Section 6.6).

High-Power Device Operation

The high breakdown field and high thermal conductivity of SiC, coupled with high operational junction temperatures, theoretically permit extremely high-power densities and efficiencies to be realized in SiC devices. Figures 6.2 and 6.3 demonstrate the theoretical advantage of SiC's high breakdown field compared to silicon in shrinking the drift-region and associated parasitic on-state resistance of a 3000 V rated unipolar power MOSFET device.[8] The high breakdown field of SiC relative to silicon enables the blocking voltage region to be roughly 10X thinner and 10X heavier-doped, permitting a roughly 100-fold decrease in the dominant blocking region (N-Drift Region) resistance R_D of Fig. 6.2 for the SiC device relative to an identically rated 3000-V silicon power MOSFET.

Significant energy losses in many silicon high-power system circuits, particularly hard-switching motor drive and power conversion circuits, arise from semiconductor switching energy loss.[1,19] While the physics of semiconductor device switching loss are discussed in detail elsewhere,[15–17] switching energy loss is often a function of the turn-off time of the semiconductor switching device, generally defined as the time lapse between when a turn-off bias is applied and the time that the device actually cuts off most current flow. The faster a device turns off, the smaller its energy loss in a switched power conversion circuit. For device-topology reasons discussed in Refs. 8, 20–22, SiC's high breakdown field and wide energy bandgap enable much faster power switching than is possible in comparably volt-amp rated silicon power-switching devices. Therefore, SiC-based power converters could operate at higher switching

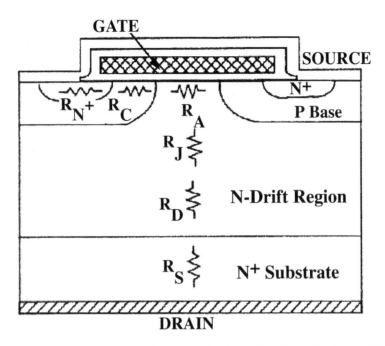

FIGURE 6.2 Cross-section of power MOSFET structure showing various internal resistances. The resistance R_D of the N-Drift Region is the dominant resistance in high-voltage power devices. (From Ref. 8. With permission.)

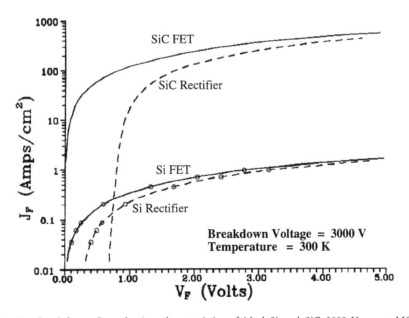

FIGURE 6.3 Simulated forward conduction characteristics of ideal Si and SiC 3000 V power MOSFETs and Schottky rectifiers. The high breakdown field of SiC relative to silicon (Table 6.1) enables the blocking voltage region (N-Drift Region in Fig. 6.2) to be roughly 10X thinner and 10X heavier-doped, permitting a roughly 100-fold increase in on-state current density for the 3000-V SiC devices relative to 3000-V silicon devices. (From Ref. 8. With permission.)

frequencies with much greater efficiency (i.e., less switching energy loss). Higher switching frequency in power converters is highly desirable because it permits use of proportionally smaller capacitors, inductors, and transformers, which in turn can greatly reduce overall system size and weight.

While SiC's smaller on-resistance and faster switching help minimize energy loss and heat generation, SiC's higher thermal conductivity enables more efficient removal of waste heat energy from the active device. Because heat energy radiation efficiency increases greatly with increasing temperature difference between the device and the cooling ambient, SiC's ability to operate at high junction temperatures permits much more efficient cooling to take place, so that heatsinks and other device-cooling hardware (i.e., fan cooling, liquid cooling, air conditioning, etc.) typically needed to keep high-power devices from over-heating can be made much smaller or even eliminated.

While the preceding discussion focused on high-power switching for power conversion, many of the same arguments can be applied to devices used to generate and amplify RF signals used in radar and communications applications. In particular, the high breakdown voltage and high thermal conductivity, coupled with high carrier saturation velocity, allow SiC microwave devices to handle much higher power densities than their silicon or GaAs RF counterparts, despite SiC's disadvantage in low-field carrier mobility (Section 6.6).[6,7,23]

System Benefits of High-Power, High-Temperature SiC Devices

Uncooled operation of high-temperature and/or high-power SiC electronics would enable revolutionary improvements to aerospace systems. Replacement of hydraulic controls and auxiliary power units with distributed "smart" electromechanical controls capable of harsh-ambient operation will enable substantial jet-aircraft weight savings, reduced maintenance, reduced pollution, higher fuel efficiency, and increased operational reliability.[24–26] SiC high-power solid-state switches will also enable large efficiency gains in electric power management and control. Performance gains from SiC electronics could enable the public power grid to provide increased consumer electricity demand without building additional generation plants, and improve power quality and operational reliability through "smart" power management. More efficient electric motor drives will benefit industrial production systems as well as transportation systems such as diesel-electric railroad locomotives, electric mass-transit systems, nuclear-powered ships, and electric automobiles and buses.

From the above discussions, it should be apparent that SiC high-power and/or high-temperature solid-state electronics promise tremendous advantages that could significantly impact transportation systems and power usage on a global scale. By improving the way in which electricity is distributed and used, improving electric vehicles so that they become more viable replacements for internal combustion-engine vehicles, and improving the fuel efficiency and reducing pollution of remaining fuel-burning engines and generation plants, SiC electronics promises the potential to better the daily lives of all citizens of planet Earth.

6.4   SiC Semiconductor Crystal Growth

As of this writing, much of the outstanding theoretical promise of SiC electronics highlighted in the previous section has largely gone unrealized. A brief historical examination quickly shows that serious shortcomings in SiC semiconductor material manufacturability and quality have greatly hindered the development of SiC semiconductor electronics. From a simple-minded point of view, SiC electronics development has very much followed the general rule of thumb that a solid-state electronic device can only be as good as the semiconductor material from which it is made.

Historical Lack of SiC Wafers

Most of silicon carbide's superior intrinsic electrical properties have been known for decades. At the genesis of the semiconductor electronics era, SiC was considered an early transistor material candidate, along with germanium and silicon. However, reproducible wafers of reasonable consistency, size, quality, and availability are a prerequisite for commercial mass-production of semiconductor electronics. Many semiconductor materials can be melted and reproducibly recrystallized into large, single crystals with the aid of a seed crystal, such as in the Czochralski method employed in the manufacture of almost all silicon wafers, enabling reasonably large wafers to be mass-produced. However, because SiC sublimes instead of melting at reasonably attainable pressures, SiC cannot be grown by conventional melt-growth techniques. This prevented the realization of SiC crystals suitable for mass-production until the late 1980s. Prior to 1980, experimental SiC electronic devices were confined to small (typically ~1 cm^2), irregularly shaped SiC crystal platelets (Fig. 6.4, right side) grown as a by-product of the Acheson process for manufacturing industrial abrasives (e.g., sandpaper)[27] or by the Lely process.[28] In the Lely process, SiC sublimed from polycrystalline SiC powder at temperatures near 2500°C are randomly condensed on the walls of a cavity forming small hexagonally shaped platelets. While these small, nonreproducible crystals permitted some basic SiC electronics research, they were clearly not suitable for semiconductor mass-production. As such, silicon became the dominant semiconductor fueling the solid-state technology revolution, while interest in SiC-based microelectronics was limited.

Growth of 3C-SiC on Large-Area (Silicon) Substrates

Despite the absence of SiC substrates, the potential benefits of SiC hostile-environment electronics nevertheless drove modest research efforts aimed at obtaining SiC in a manufacturable wafer form. Toward this end, the heteroepitaxial growth of single-crystal SiC layers on top of large-area silicon substrates was first carried out in 1983,[29] and subsequently followed by a great many others over the years using a variety of growth techniques. Primarily due to large differences in lattice constant (20% difference between SiC and Si) and thermal expansion coefficient (8% difference), heteroepitaxy of SiC using silicon as a substrate always results in growth of 3C-SiC with a very high density of crystallographic structural defects such as stacking faults, microtwins, and inversion domain boundaries.[30,31] Furthermore, the as-grown surface morphology of 3C-SiC grown on silicon is microscopically textured, making sub-micron lithography somewhat problematic. Other large-area wafer materials, such as sapphire, silicon-on-insulator, TiC, etc., have been employed as substrates for heteroepitaxial growth of SiC epilayers, but the resulting films have been of comparably poor quality with high crystallographic defect densities.

While some limited semiconductor electronic devices and circuits have been implemented in 3C-SiC grown on silicon,[32,33] the performance of these electronics can be summarized as severely limited by the high density of crystallographic defects to the degree that almost none of the operational benefits discussed in Section 6.3 have been viably realized. Among other problems, the crystal defects "leak" parasitic current across reverse-biased device junctions where current flow is not desired. Because excessive crystal defects lead to electrical device shortcomings, there are as yet no commercial electronics manufactured in 3C-SiC grown on large-area substrates.

Despite the lack of major technical progress, there is strong economic motivation to continue to pursue heteroepitaxial growth of SiC on large-area substrates, as this would provide cheap wafers for SiC electronics that would be immediately compatible with silicon integrated circuit manufacturing equipment. If ongoing work ever solves the extremely challenging crystallographic defect problems associated with the heteroepitaxial growth of SiC, it would likely become the material of choice for mass-production of SiC-based electronics. Given the present electrical deficiencies of heteroepitaxial SiC, 3C-SiC grown on silicon is more likely to be commercialized as a mechanical material in microelectromechanical systems (MEMS) applications (Section 6.6) instead of being used purely as a semiconductor in traditional solid-state electronics.

Sublimation Growth of SiC Wafers

In the late 1970s, Tairov and Tzvetkov established the basic principles of a modified seeded sublimation growth process for growth of 6H-SiC.[34,35] This process, also referred to as the modified Lely process, was a breakthrough for SiC in that it offered the first possibility of reproducibly growing acceptably large

FIGURE 6.4 Mass-produced 2.5-cm diameter 6H-SiC wafer manufactured circa 1990 via seeded sublimation by Cree Research (left), and 6H-SiC Lely and Acheson platelet crystals (right) representative of single-crystal SiC substrates available prior to 1989. 5.1-cm diameter seeded sublimation SiC wafers entered the commercial market in 1997.

single crystals of SiC that could be cut and polished into mass-produced SiC wafers. The basic growth process is based on heating polycrystalline SiC source material to ~2400°C under conditions where it sublimes into the vapor phase and subsequently condenses onto a cooler SiC seed crystal. This produces a somewhat cylindrical boule of single-crystal SiC that grows taller at a rate of a few millimeters per hour. To date, the preferred orientation of the growth in the sublimation process is such that vertical growth of a taller cylindrical boule proceeds along the [0001] crystallographic c-axis direction (i.e., vertical direction in Fig. 6.1). Circular "c-axis" wafers with surfaces that lie normal (perpendicular) to the c-axis can be sawed from the roughly cylindrical boule. While other growth orientations (such as growth along the a-axis) continue to be investigated, the electronic quality of this material has thus far proven inferior to c-axis-grown wafers.[36,37]

Commercially Available SiC Wafers

After years of further development of the sublimation growth process, Cree Research became the first company to sell 2.5-cm diameter semiconductor wafers of 6H-SiC (Fig. 6.4, left side) in 1989.[38] Only with the development of the modified Lely seeded sublimation growth technique have acceptably large and reproducible single-crystal SiC wafers of usable electrical quality become available. Correspondingly, the vast majority of silicon carbide semiconductor electronics development has taken place since 1990. Other companies have subsequently entered the SiC wafer market, and sublimation grown wafers of the 4H-SiC polytype have also been commercialized, as summarized in Table 6.2.

Commercially available 4H- and 6H-SiC wafer specifications are given in Table 6.2. N-type, p-type, and semi-insulating SiC wafers are commercially available at different prices. Wafer size, cost, and quality are all very critical to the manufacturability and process yield of mass-produced semiconductor micro-electronics. Compared to commonplace silicon and GaAs wafer standards, present-day 4H- and 6H-SiC wafers are small, expensive, and generally of inferior quality. In addition to high densities of crystalline defects such as micropipes and closed-core screw dislocations discussed in the next subsection,

TABLE 6.2   Commercial Vendors and Specifications of Selected Sublimation-Grown SiC Single-Crystal Wafers

Vendor [Ref.]	Year	Product	Wafer Diameter	Micropipes (#/cm²)	Price (U.S.$)
Cree	1993	6H n-type, Si-face, R-Grade	3.0 cm	200–1000	1000
[38]		6H n-type, Si-face, P-Grade	3.0 cm	200–1000	2900
		6H n-type, C-face, P-Grade	3.0 cm	200–1000	3000
		6H p-type, Si-face, P-Grade	3.0 cm	200–1000	3300
		4H n-type, Si-face, R-Grade	3.0 cm	200–1000	3800
Cree	1997	4H n-type, Si-face, R-Grade	3.5 cm	100–200	750
[38]		4H n-type, Si-face, P-Grade	3.5 cm	100–200	1300
		4H n-type, Si-face, P-Grade	3.5 cm	<30	2300
	1998	4H n-type, Si-face, R-Grade	5.1 cm	<200	2100
		4H n-type, Si-face, P-Grade	5.1 cm	<200	3100
	1997	4H p-type, Si-face, R-Grade	3.5 cm	<200	1900
		4H Semi-Insulating, R-Grade	3.5 cm	<200	4800
		6H n-type, Si-face, P-Grade	3.5 cm	<200	1000
		6H p-type, Si-face, P-Grade	3.5 cm	<200	2200
Nippon Steel [142]	1997	4H n-type	2.5 cm	NA	NA
SiCrystal [143]	1997	4H n-type, Quality I	3.5 cm	<200	1200
		4H n-type, Quality III	3.5 cm	400–1000	900
		4H n-type, Quality I	2.5 cm	<200	600
		6H n-type, Quality I	3.5 cm	<200	1200
Sterling and ATMI/Epitronics [144][145]	1998	6H n-type	3.5 cm	<100	800
		4H n-type	3.5 cm	<100	800

commercial SiC wafers also exhibit significantly rougher surfaces, and larger warpage and bow than is typical for silicon and GaAs wafers.[39] This disparity is not surprising considering that silicon and GaAs wafers have undergone several decades of commercial process refinement, and that SiC is an extraordinarily hard material, making it very difficult to properly saw and polish. Nevertheless, ongoing wafer sawing and polishing process improvements should eventually alleviate wafer surface quality deficiencies.

SiC Wafer Crystal Defects

While the specific electrical effects of SiC crystal defects are discussed later in Section 6.6, the micropipe defect (Table 6.2) is regarded as the most damaging defect that is limiting upscaling of SiC electronics capabilities.[9,40] A micropipe is a screw dislocation with a hollow core and a larger Burgers vector, which becomes a tubular void (with a hollow diameter on the order of micrometers) in the SiC wafer that extends roughly parallel to the crystallographic c-axis normal to the polished c-axis wafer surface.[41–44] Sublimation-grown 4H- and 6H-SiC wafers also contain high densities of closed-core screw dislocation defects that, like micropipes, cause a considerable amount of localized strain and SiC lattice deformation.[42,43,45,46] Similar to horizontal branches on a tree with its trunk running up the c-axis, dislocation loops emanate out along the basal plane from screw dislocations.[41,47] As shown in Table 6.2, micropipe densities in commercial SiC wafers have shown steady improvement over a 5-year period, leading to wafers with less than 30 micropipes per square centimeter of wafer area. However, as discussed in Section 6.6, SiC wafer improvement trends will have to accelerate if some of SiC's most beneficial high-power applications are going to reach timely commercial fruition.

SiC Epilayers

Most SiC electronic devices are not fabricated directly in sublimation-grown wafers, but are instead fabricated in much higher quality epitaxial SiC layers that are grown on top of the initial sublimation-grown wafer. Well-grown SiC epilayers have superior electrical properties and are more controllable and reproducible than bulk sublimation-grown SiC wafer material. Therefore, the controlled growth of high-quality epilayers is highly important in the realization of useful SiC electronics.

SiC Epitaxial Growth Processes

An interesting variety of SiC epitaxial growth methodologies — ranging from liquid-phase epitaxy, molecular beam epitaxy, and chemical vapor deposition — has been investigated.[11,12,32] The chemical vapor deposition (CVD) growth technique is generally accepted as the most promising method for attaining epilayer reproducibility, quality, and throughputs required for mass-production. In simplest terms, variations of SiC CVD are carried out by heating SiC substrates in a chamber with flowing silicon and carbon containing gases that decompose and deposit Si and C onto the wafer, allowing an epilayer to grow in a well-ordered single-crystal fashion under well-controlled conditions. Conventional SiC CVD epitaxial growth processes are carried out at substrate growth temperatures between 1400 and 1600°C at pressures from 0.1 to 1 atm, resulting in growth rates on the order of micrometers per hour.[39,48–50] Higher-temperature (up to 2000°C) SiC CVD growth processes are also being pioneered to obtain higher SiC epilayer growth rates — on the order of hundreds of micrometers per hour.[51]

SiC Homoepitaxial Growth

Homoepitaxial growth, whereby the polytype of the SiC epilayer matches the polytype of the SiC substrate, is accomplished by step-controlled epitaxy.[39,49,52] Step-controlled epitaxy is based on growing epilayers on an SiC wafer polished at an angle (called the "tilt-angle" or "off-axis angle") of typically 3° to 8° off the (0001) basal plane, resulting in a surface with atomic steps and flat terraces between steps, as schematically depicted in Fig. 6.5. When growth conditions are properly controlled and there is a sufficiently short distance between steps, Si and C atoms impinging onto the growth surface find their way to steps where they bond and incorporate into the crystal. Thus, ordered lateral "step flow" growth takes place, which enables the polytypic stacking sequence of the substrate to be exactly mirrored in the growing epilayer. When growth conditions are not properly controlled or when steps are too far apart (as can occur with SiC substrate surfaces that are polished to within less than 1° of the basal plane), growth adatoms can nucleate and bond in the middle of terraces instead of at the steps; this leads to heteroepitaxial growth of poor-quality 3C-SiC.[39,49] To help prevent spurious nucleation of 3C-SiC "triangular inclusions" during epitaxial growth, most commercial 4H- and 6H-SiC substrates are polished to tilt angles of 8° and 3.5° off the (0001) basal plane, respectively.

It is important to note that most present-day, as-grown SiC epilayers contain varying densities of undesirable surface morphological features that could affect SiC device processing and performance.[39,48] In addition to "triangular inclusions," these include "growth pits" as well as large macrosteps formed by coalescence of multiple SiC growth steps (i.e., "step bunching") during epitaxy. Pre-growth wafer polishing as well as growth initiation procedures have been shown to strongly impact the formation of undesirable epitaxial growth features.[39,48] Further optimization of pre-growth treatments and epitaxial growth initiation processes are expected to reduce undesired morphological growth features.

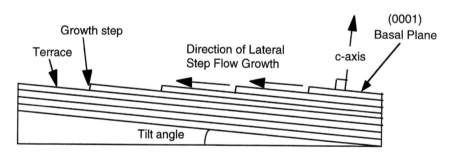

FIGURE 6.5 Cross-sectional schematic representation of "off-axis" polished SiC surface used for homoepitaxial growth. When growth conditions are properly controlled and there is a sufficiently short distance between steps, Si and C atoms impinging onto the growth surface find their way to steps where they bond and incorporate into the crystal. Thus, ordered lateral "step-flow" growth takes place, which enables the polytypic stacking sequence of the substrate to be exactly mirrored in the growing epilayer. (Modified from Ref. 10. With permission.)

SiC Epilayer Doping

In situ doping during CVD epitaxial growth is primarily accomplished through the introduction of nitrogen (usually N_2) for n-type and aluminum (usually trimethyl- or triethylaluminum) for p-type epilayers.[12] Some alternative dopants such as phosphorous, boron, and vanadium have also been investigated for n-type, p-type, and semi-insulating epilayers, respectively. While some variation in epilayer doping can be carried out strictly by varying the flow of dopant gasses, the site-competition doping methodology[53,54] has enabled a much broader range of SiC doping to be accomplished. In addition, site-competition epitaxy has also made moderate epilayer dopings more reliable and repeatable. The site-competition dopant-control technique is based on the fact that many dopants of SiC preferentially incorporate into either Si lattice sites or C lattice sites. As an example, nitrogen preferentially incorporates into lattice sites normally occupied by carbon atoms. By epitaxially growing SiC under carbon-rich conditions, most of the nitrogen present in the CVD system (whether it is a residual contaminant or intentionally introduced) can be excluded from incorporating into the growing SiC crystal. Conversely, by growing in a carbon-deficient environment, the incorporation of nitrogen can be enhanced to form very heavily doped epilayers for ohmic contacts. Aluminum, which is opposite to nitrogen, prefers the Si-site of SiC, and other dopants have also been controlled through site competition by properly varying the Si/C ratio during crystal growth. SiC epilayer dopings ranging from 9×10^{14} to 1×10^{19} cm^{-3} are commercially available, and researchers have reported obtaining dopings nearly a factor of 10 larger and smaller than this range for n-type and p-type dopings. Commercial epilayer thickness and doping tolerances are presently specified at 25% and 100%, respectively,[38] while doping uniformities of 7% and thickness uniformities of 4% over a 30-mm wafer have been reported in developmental research.[48]

SiC Epilayer Crystal Defects

Improvements in epilayer quality are needed as SiC electronics upscale toward production integrated circuits, as there are presently many observable defects present in state-of-the-art SiC homoepilayers. Non-ideal surface morphological features, such as "growth pits," 3C-SiC triangular inclusions ("triangle defects") introduced in Section 6.4, are generally more prevalent in 4H-SiC epilayers than 6H-SiC epilayers. Most of these features appear to be manifestations of non-optimal "step flow" during epilayer growth arising from substrate defects, non-ideal substrate surface finish, contamination, and/or unoptimized epitaxial growth conditions. While by no means trivial, it is anticipated that SiC epilayer surface morphology will greatly improve as refined substrate preparation and epilayer growth processes are developed.

Many impurities and crystallographic defects found in sublimation-grown SiC wafers do not propagate into SiC homoepitaxial layers. For example, basal-plane dislocation loops emanating from micropipes and screw dislocations in sublimation-grown SiC wafers (Section 6.4) are not generally observed in SiC epilayers.[47] Unfortunately, however, screw dislocations (both micropipes and closed-core screw dislocations) present in commercial *c*-axis wafers do replicate themselves up the crystallographic *c*-axis into SiC homoepilayers grown on commercial wafers. Therefore, as discussed later in Section 6.6, devices fabricated in commercial epilayers are still subject to electrical performance and yield limitations imposed by commercial substrate screw-dislocation defect densities.

Alternative Growth Methods to Reduce SiC Epilayer Dislocations

As of this writing, there is no known practical method of realizing screw-dislocation-free 4H- or 6H-SiC homoepilayers on conventional sublimation-grown substrates. Some non-conventional epitaxial growth techniques have been attempted in an effort to prevent the propagation of micropipes into an epilayer.[55,56] While these approaches have scored modest success in closing and covering up micropipes, to date there has been little, if any, improvement demonstrated in electrical devices fabricated in the resulting material. This is perhaps due to the fact that screw dislocations and associated harmful stresses may still be present in the epilayer, despite the fact that some open cores may have been converted to closed cores.

Because screw dislocations propagate up the *c*-axis, one could conceivably alleviate screw dislocations by growing epilayers on SiC wafers with their surface parallel to the *c*-axis using "a-axis" wafers. Unfortunately, efforts directed at realizing a-axis wafers and epilayers have to date been much less successful than *c*-axis wafers and epilayers, primarily because defects that form and propagate up the basal plane (the vertical wafer and epilayer growth direction in a-axis-oriented wafers) have proven more harmful and difficult to eliminate than screw dislocations in conventional *c*-axis wafers and epilayers.[36,37]

Selected-area epitaxial growth techniques have recently led to startling reductions in GaN epilayer defect densities.[57] While selective-area epitaxial growth of 3C-SiC has been demonstrated, the applicability of similar techniques to realizing superior electrical-quality SiC will be much more difficult due to the step-flow homoepitaxial growth mechanism of α-SiC as well as high growth temperatures (>1400°C), which are incompatible with conventional growth-masking materials like SiO_2.

6.5   SiC Device Fundamentals

In order to minimize the development and production costs of SiC electronics, it is essential that SiC device fabrication take advantage of the existing silicon and GaAs wafer processing infrastructure as much as possible. As will be discussed in this section, most of the steps necessary to fabricate SiC electronics starting from SiC wafers can be accomplished using somewhat modified commercial silicon electronics processes and fabrication tools.

Choice of Polytype for Devices

As discussed in Section 6.4, 4H-SiC and 6H-SiC are the far superior forms of semiconductor device quality SiC commercially available in mass-produced wafer form. Therefore, only 4H-SiC and 6H-SiC device processing methods will be explicitly considered in the rest of this section. It should be noted, however, that most of the processing methods discussed in this section are applicable to other polytypes of SiC, except for the case of 3C-SiC grown on silicon where all processing temperatures need to be kept well below the melting temperature of silicon (~1400°C).

It is generally accepted that 4H-SiC's substantially higher carrier mobility and shallower dopant ionization energies compared to 6H-SiC (Table 6.1) should make it the polytype of choice for most SiC electronic devices, provided that all other device processing, performance, and cost-related issues play out as being roughly equal between the two polytypes. Furthermore, the inherent mobility anisotropy that degrades conduction parallel to the crystallographic *c*-axis in 6H-SiC[58] will particularly favor 4H-SiC for vertical power device configurations (Section 6.6).

SiC Selective Doping: Ion Implantation

The fact that diffusion coefficients of most SiC dopants are negligibly small below ~1800°C is excellent for maintaining device junction stability, because dopants do not undesirably diffuse as the device is operated long-term at high-temperatures. Unfortunately, however, this characteristic also precludes the use of conventional dopant diffusion, a highly useful technique widely employed in silicon microelectronics manufacturing for patterned doping of SiC.

Laterally patterned doping of SiC is carried out by ion implantation. This somewhat restricts the depth to which most dopants can be conventionally implanted to less than 1 μm using conventional dopants and implantation equipment. Compared to silicon processes, SiC ion-implantation requires a much higher thermal budget to achieve acceptable dopant implant electrical activation. Summaries of ion-implantation processes for various dopants can be found in Refs. 11, 59, and 60. Most of these processes are based on carrying out implantation at elevated temperatures (~500 to 800°C) using a patterned high-temperature masking material. The elevated temperature during implantation promotes some lattice self-healing during the implant, so that damage and segregation of displaced silicon and carbon atoms does not become excessive, especially in high-dose implants often employed for ohmic contact formation.[59,60] Co-implantation of carbon with p-type dopants has recently been investigated as a means to improve the electrical conductivity of implanted p-type contact layers.[61]

Following implantation, the patterning mask is stripped and a much higher temperature (~1200 to 1800°C) anneal is carried out to achieve maximum electrical activation of dopant donor or acceptor ions. The final annealing conditions are crucial for obtaining desired electrical properties from ion-implanted layers. At higher implant anneal temperatures, the SiC surface morphology can seriously degrade as damage-assisted sublimation etching of the SiC surface begins to take place.[62] Because sublimation etching is driven primarily by loss of silicon from the crystal surface, annealing in silicon overpressures can be used to prevent surface degradation during high-temperature anneals. Such overpressure can be achieved by close-proximity solid sources, such as using an enclosed SiC crucible with SiC lid and/or SiC powder near the wafer, or by annealing in a silane-containing atmosphere.

SiC Contacts and Interconnects

All useful semiconductor electronics require conductive signal paths in and out of each device, as well as conductive interconnects to carry signals between devices on the same chip and to external circuit elements that reside off-chip. While SiC itself is theoretically capable of fantastic operation under extreme conditions (Section 6.3), such functionality is useless without contacts and interconnects that are also capable of operation under the same conditions to enable complete extreme-condition circuit functionality. Previously developed conventional contact and interconnect technologies will likely not be sufficient for reliable operation in extreme conditions that SiC enables. The durability and reliability of metal–semiconductor contacts and interconnects are two of the main factors limiting the operational high-temperature limits of SiC electronics. Similarly, SiC high-power device contacts and metallizations will have to withstand both high-temperature and high current density stress never before encountered in silicon power electronics experience.

The subject of metal–semiconductor contact formation is a very important technical field, too broad to be discussed in detail here. For general background discussions on metal–semiconductor contact physics and formation, the reader should consult narratives presented in Refs. 15 and 63. These references primarily discuss ohmic contacts to conventional narrow-bandgap semiconductors such as silicon and GaAs. Specific overviews of SiC metal–semiconductor contact technology can be found in Refs. 64 to 67.

As discussed in Refs. 64 to 67, there are both similarities and a few differences between SiC ohmic contacts and ohmic contacts to conventional narrow-bandgap semiconductors (e.g., silicon, GaAs). The same basic physics and current transport mechanisms that are present in narrow-bandgap contacts — such as surface states, Fermi-pinning, thermionic emission, and tunneling — also apply to SiC contacts. A natural consequence of the wider bandgap of SiC is higher effective Schottky barrier heights. Analogous with narrow-bandgap ohmic contact physics, the microstructural and chemical state of the SiC–metal interface is crucial to contact electrical properties. Therefore, pre-metal-deposition surface preparation, metal deposition process, choice of metal, and post-deposition annealing can all greatly impact the resulting performance of metal–SiC contacts. Because the chemical nature of the starting SiC surface is strongly dependent on surface polarity, it is not uncommon to obtain significantly different results when the same contact process is applied to the silicon face surface vs. the carbon face surface.

SiC Ohmic Contacts

Ohmic contacts serve the purpose of carrying electrical current into and out of the semiconductor, ideally with no parasitic resistance. The properties of various ohmic contacts to SiC reported to date are summarized in Refs. 66 and 67. While SiC specific ohmic contact resistances at room temperature are generally higher than in contacts to narrow-bandgap semiconductors, they are nevertheless sufficiently low for most envisioned SiC applications. Lower specific contact resistances are usually obtained to n-type 4H- and 6H-SiC (~10^{-4} to 10^{-6} ohm-cm^2) than to p-type 4H- and 6H-SiC (~10^{-3} to 10^{-5} ohm-cm^2). Consistent with narrow-bandgap ohmic contact technology, it is easier to make low-resistance ohmic contacts to heavily doped SiC. While it is possible to achieve ohmic contacts to lighter-doped SiC using high-temperature annealing, the lowest-resistance ohmic contacts are most easily implemented on SiC degenerately doped by site competition (Section 6.4) or high-dose ion implantation (Section 6.5). If the SiC doping is sufficiently degenerate, many metals deposited on a relatively clean SiC surface are ohmic in the "as deposited" state.[68] Regardless of doping, it is common practice in SiC to thermally anneal contacts to obtain the minimum possible ohmic contact resistance. Most SiC ohmic contact anneals are performed at temperatures around 1000°C in non-oxidizing environments. Depending on the contact metallization employed, this anneal generally causes limited interfacial reactions (usually metal-carbide or metal-silicide formation) that broaden and/or roughen the metal–semiconductor interface, resulting in enhanced conductivity through the contact.

Truly enabling harsh-environment SiC electronics will require ohmic contacts that can reliably withstand prolonged harsh-environment operation. Most reported SiC ohmic metallizations appear sufficient for long-term device operation up to 300°C. SiC ohmic contacts that withstand heat soaking under no electrical bias at 500 to 600°C for hundreds or thousands of hours in non-oxidizing gas or vacuum environments have also been demonstrated. In air, however, there has only been demonstration to date of a contact that can withstand heat soaking (no electrical bias) for 60 hours at 650°C.[69] Some very beneficial aerospace systems will require simultaneous high-temperature (T > 300°C) and high current density operation in oxidizing air environments. Electromigration, oxidation, and other electrochemical reactions driven by high-temperature electrical bias in a reactive oxidizing environment are likely to limit SiC ohmic contact reliability for the most demanding applications. The durability and reliability of SiC ohmic contacts is one of the critical factors limiting the practical high-temperature limits of SiC electronics.

SiC Schottky Contacts

Rectifying metal–semiconductor Schottky barrier contacts to SiC is useful for a number of devices, including metal–semiconductor field-effect transistors (MESFETs) and fast-switching rectifiers. References 64, 65, 67, and 70 summarize electrical results obtained in a variety of SiC Schottky studies to date. Due to the wide bandgap of SiC, almost all unannealed metal contacts to lightly doped 4H- and 6H-SiC are rectifying. Rectifying contacts permit extraction of Schottky barrier heights and diode ideality factors by well-known current-voltage (I-V) and capacitance-voltage (C-V) electrical measurement techniques.[63] While these measurements show a general trend that Schottky junction barrier height does somewhat depend on metal–semiconductor workfunction difference, the dependence is weak enough to suggest that surface state charge also plays a significant role in determining the effective barrier height of SiC Schottky junctions. At least some experimental scatter exhibited for identical metals can be attributed to cleaning and metal deposition process differences, as well as different barrier height measurement procedures. The work by Teraji et al.,[71] in which two different surface cleaning procedures prior to titanium deposition lead to ohmic behavior in one case and rectifying behavior in the other, clearly shows the important role that process recipe can play in determining SiC Schottky contact electrical properties.

It is worth noting that barrier heights calculated from C-V data are often somewhat higher than barrier heights extracted from I-V data taken from the same diode. Furthermore, the reverse current drawn in experimental SiC diodes, while small, is nevertheless larger than expected based on theoretical substitution of SiC parameters into well-known Schottky diode reverse leakage current equations developed for narrow-bandgap semiconductors. Bhatnagar et al.[72] proposed a model to explain these behaviors, in which localized surface defects, perhaps elementary screw dislocations where they intersect the SiC-metal interface, cause locally reduced junction barriers in the immediate vicinity of defects. Because current is exponentially dependent on Schottky barrier height, this results in the majority of measured current flowing at local defect sites instead of evenly distributed over the entire Schottky diode area. In addition to local defects, electric field crowding along the edge of the SiC Schottky barrier can also lead to increased reverse-bias leakage current and reduced reverse breakdown voltage.[15,16,63] Schottky diode edge termination techniques to relieve electric field edge crowding and improve Schottky rectifier reverse properties are discussed later in Section 6.6. Quantum mechanical tunneling of carriers through the barrier may also account for some excess reverse leakage current in SiC Schottky diodes.[73]

The high-temperature operation of rectifying SiC Schottky diodes is primarily limited by reverse-bias thermionic leakage of carriers over the junction barrier. Depending on the specific application and the barrier height of the particular device, SiC Schottky diode reverse leakage currents generally grow to excessive levels at around 300 to 400°C. As with ohmic contacts, electrochemical interfacial reactions must also be considered for long-term Schottky diode operation at the highest temperatures.

Patterned Etching of SiC for Device Fabrication

At room temperature, no known wet chemical etches single-crystal SiC. Therefore, most patterned etching of SiC for electronic devices and circuits is accomplished using dry etching techniques. The reader should consult Ref. 74; it contains an excellent summary of dry SiC etching results obtained to date. The most commonly employed process involves reactive ion etching (RIE) of SiC in fluorinated plasmas. Sacrificial etch masks (often aluminum metal) are deposited and photolithographically patterned to protect desired areas from being etched. The SiC RIE process can be implemented using standard silicon RIE hardware, and typical 4H- and 6H-SiC RIE etch rates are on the order of hundreds of angstroms per minute. Well-optimized SiC RIE processes are typically highly anisotropic with little undercutting of the etch mask, leaving smooth surfaces. One of the keys to achieving smooth surfaces is preventing "micromasking," wherein masking material is slightly etched and randomly redeposited onto the sample, effectively masking very small areas on the sample that were intended for uniform etching. This can result in "grass"-like etch-residue features being formed in the unmasked regions, which is undesirable in most cases. In special cases, RIE etching under conditions promoting micromasking is useful in greatly roughening the SiC surface to reduce the contact resistance of subsequently deposited ohmic metallizations.

While RIE etch rates are sufficient for many electronic applications, much higher SiC etch rates are necessary to carve features on the order of tens to hundreds of micrometers deep that are needed to realize advanced sensors, microelectromechanical systems (MEMS), and some very high-voltage power device structures. High-density plasma dry etching techniques, such as electron cyclotron resonance (ECR) and inductively coupled plasma (ICP), have been developed to meet the need for deep-etching of SiC. Residue-free patterned etch rates exceeding a thousand angstroms a minute have been demonstrated.[74–76]

Patterned etching of SiC at very high etch rates has also been demonstrated using photo-assisted and dark electrochemical wet etching.[77,78] By choosing proper etching conditions, this technique has demonstrated a very useful dopant-selective etch-stop capability. However, there are major incompatibilities of the electrochemical process that make it undesirable for VLSI mass-production, including extensive pre-etching and post-etching sample preparation, etch isotropy and mask undercutting, and somewhat non-uniform etching across the sample.

SiC Insulators: Thermal Oxides and MOS Technology

The vast majority of semiconductor integrated circuit chips in use today rely on silicon metal-oxide-semiconductor field effect transistors (MOSFETs), whose electronic advantages and operational device physics are summarized in Choma's chapter on devices and their models and elsewhere.[15,16,79] Given the extreme usefulness and success of MOSFET-based electronics in VLSI silicon, it is naturally desirable to implement high-performance inversion channel MOSFETs in SiC. Like silicon, SiC forms a thermal SiO_2 oxide when it is sufficiently heated in an oxygen environment. While this enables SiC MOS technology to somewhat follow the highly successful path of silicon MOS technology, there are nevertheless important differences in insulator quality and device processing that are presently preventing SiC MOSFETs from realizing their full beneficial potential. While the following discourse attempts to quickly highlight key issues facing SiC MOSFET development, more detailed insights can be found in Refs. 80 to 83. In highlighting the difficulties facing SiC MOSFET development, it is important to keep in mind that early silicon MOSFETs faced similar developmental challenges that took many years of dedicated research effort to successfully overcome.

From a purely electrical point of view, there are two prime operational deficiencies of SiC oxides and MOSFETs compared to silicon MOSFETs. First, effective inversion channel mobilities in most SiC MOS-FETs are much lower (typically well under 100 cm²/V-s for inversion electrons) than one would expect based on silicon inversion channel MOSFET carrier mobilities. This seriously reduces the transistor gain and current-carrying capability of SiC MOSFETs, so that SiC MOSFETs are not nearly as advantageous as theoretically predicted. Second, SiC oxides have not proven as reliable and immutable as well-developed silicon oxides, in that SiC MOSFETs are more prone to threshold voltage shifts, gate leakage, and oxide failures than comparably biased silicon MOSFETs. The excellent works by Cooper[80] and Brown et al.[83] discuss noteworthy differences between the basic electrical properties of n-type versus p-type SiC MOS devices. SiC MOSFET oxide electrical performance deficiencies appear mostly attributable to differences between silicon and SiC thermal oxide quality and interface structure that cause the SiC oxide to exhibit undesirably higher levels of interface state densities ($\sim 10^{11}$ to 10^{13} eV^{-1}cm^{-2}), fixed oxide charges ($\sim 10^{11}$ to 10^{12} cm^{-2}), charge trapping, carrier oxide tunneling, and roughness-related scattering of inversion channel carriers.

One of the most obvious differences between thermal oxidation of silicon and SiC to form SiO_2 is the presence of C in SiC. While most of the C in SiC converts to gaseous CO and CO_2 and escapes the oxide layer during thermal oxidation, leftover C species residing near the SiC–SiO_2 interface nevertheless appear to have a detrimental impact on SiO_2 electrical quality.[80,81] Cleaning treatments and oxidation/anneal recipes aimed at reducing interfacial C appear to improve SiC oxide quality. Another procedure employed to minimize detrimental carbon effects has been to form gate oxides by thermally oxidizing layers of silicon deposited on top of SiC.[84] Likewise, deposited insulators also show promise toward improving SiC MOSFET characteristics, as Sridevan et al.[85] have recently reported greatly improved SiC inversion channel carrier mobilities (>100 cm²/V-s) using thick deposited gate insulators.

SiC surfaces are well known to be much rougher than silicon surfaces, due to off-angle polishing needed to support SiC homoepitaxy (Fig. 6.5) as well as step-bunching (particularly pronounced in 4H-SiC) that occurs during SiC homoepilayer growth (Section 6.4).[39,86] The impact of surface morphology on inversion channel mobility is highlighted by the recent work of Scharnholz et al.,[87] in which improved mobility (>100 cm²/V-s) was obtained by specifically orienting SiC MOSFETs in a direction such that current flowed parallel to surface step texture. The interface roughness of SiC may also be a factor in poor oxide reliability by assisting unwanted injection of carriers that damage and degrade the oxide.

As Agarwal et al.[88] have pointed out, the wide bandgap of SiC reduces the potential barrier impeding tunneling of damaging carriers through SiC thermal oxides, so that perfectly grown oxides on atomically smooth SiC would not be as reliable as silicon thermal oxides. Therefore, it is highly probable that alternative gate insulators will have to be developed for optimized implementation of inversion-channel SiC FETs for the most demanding high-power and/or high-temperature electronic applications.

SiC Device Packaging and System Considerations

Hostile-environment SiC semiconductor devices and ICs are of little advantage if they cannot be reliably packaged and connected to form a complete system capable of hostile-environment operation. With proper materials selection, modifications of existing IC packaging technologies appear feasible for non-power SiC circuit packaging up to 300°C.[89,90] Prototype electronic packages that can withstand over a thousand hours of heat soaking without electrical bias at 500°C have been demonstrated.[91] Much work remains before electronics system packaging can meet the needs of the most demanding aerospace electronic applications, whose requirements include high-power operation in high-vibration, 500 to 600°C, oxidizing-ambient environments. Similarly, harsh-environment passive components, such as inductors, capacitors, and transformers, must also be developed for operation in demanding conditions before the full system-level benefits of SiC electronics discussed in Section 6.3 can be successfully realized.

6.6 SiC Electronic Devices and Circuits

This section briefly summarizes a variety of SiC electronic device designs broken down by major application areas. The operational performance of experimental SiC devices is compared to theoretically predicted SiC performance as well as the capabilities of existing silicon and GaAs devices. SiC process and materials technology issues limiting the capabilities of various SiC device topologies are highlighted as key issues to be addressed in further SiC technology maturation.

SiC Optoelectronic Devices

The wide bandgap of SiC is useful for realizing short wavelength blue and ultraviolet (UV) optoelectronics. 6H-SiC-based blue pn junction light-emitting diodes (LEDs) were the first silicon carbide-based devices to reach high-volume commercial sales. These epitaxially grown, dry-etch, mesa-isolated pn junction diodes were the first mass-produced LEDs to cover the blue (~250 to 280 nm peak wavelength) portion of the visible color spectrum, which in turn enabled the realization of the first viable full-color LED-based displays.[92] Because the SiC bandgap is indirect (i.e., the conduction minimum and valence band maximum do not coincide in crystal momentum space), luminescent recombination in the LEDs is governed by inherently inefficient indirect transitions mediated by impurities and phonons.[93] Therefore, the external quantum efficiency of SiC blue LEDs (i.e., percentage of light energy output obtained vs. electrical energy input) was limited to well below 1%. While commercially successful during the 1989 to 1995 timeframe, SiC-based blue LEDs have now been totally obsoleted by the emergence of much brighter, much more efficient, direct-bandgap GaN blue LEDs.

SiC has proven much more efficient at absorbing short-wavelength light, which has enabled the realization of SiC UV-sensitive photodiodes that serve as excellent flame sensors in turbine-engine combustion monitoring and control.[92,94] The wide bandgap of 6H-SiC is useful for realizing low photodiode dark currents, as well as sensors that are blind to undesired near-infrared wavelengths produced by heat and solar radiation. Commercial SiC-based UV flame sensors, again based on epitaxially grown, dry-etch, mesa-isolated 6H-SiC pn junction diodes, have successfully reduced harmful pollution emissions from gas-fired, ground-based turbines used in electrical power generation systems. Prototype SiC photodiodes are also being developed to improve combustion control in jet-aircraft engines.[95]

SiC RF Devices

The main use of SiC RF devices appears to lie in high-frequency solid-state high-power amplification at frequencies from around 600 MHz (UHF-band) to perhaps around 10 GHz (X-band). As discussed in better detail in Refs. 6, 7, 23, 96, and 97, the high breakdown voltage and high thermal conductivity, coupled with high carrier saturation velocity, allow SiC RF transistors to handle much higher power densities than their silicon or GaAs RF counterparts, despite SiC's disadvantage in low-field carrier mobility (Section 6.2). This power output advantage of SiC is briefly illustrated in Fig. 6.6 for the specific case of a Class A MESFET-based RF amplifier. The maximum theoretical RF power of a Class A MESFET operating along the DC load line shown in Fig. 6.6 is approximated by[7]:

$$P_{max} = \frac{I_{dson}(V_b - V_{knee})}{8} \tag{6.1}$$

The higher breakdown field of SiC permits higher drain breakdown voltage (V_b), permitting RF operation at higher drain biases. Given that there is little degradation in I_{dson} and V_{knee} for SiC vs. GaAs and silicon, the increased drain voltage directly leads to higher SiC MESFET output power densities. The higher thermal conductivity of SiC is also crucial in minimizing channel self-heating so that phonon scattering does not seriously degrade channel carrier velocity and I_{dson}. As discussed in Refs. 7 and 97, similar RF output power arguments can be made for SiC-based static induction transistors (SITs).

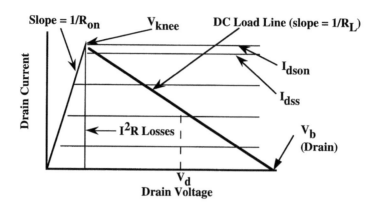

FIGURE 6.6   Piecewise linear MESFET drain characteristic showing DC load line used in Class A RF amplifier operation. The higher breakdown voltage V_b enabled by SiC's higher breakdown field enables operation at higher drain biases, leading to higher RF power densities. (From Ref. 7. With permission.)

The high power density of high-frequency SiC transistors could prove very useful in realizing solid-state transmitters for cell phone base stations, high-definition television (HDTV) transmitters, and radar transmitters, because it reduces the number of devices needed to generate sufficient RF power for these applications. Fewer transistors capable of operating at higher temperatures reduces matching and cooling requirements, leading to reduced overall size and cost of these systems. While excellent for fixed-base high-power RF transmission systems, SiC RF transistors are not well suited for portable handheld RF transceivers where drain voltage and power are restricted to function within the operational limitations of small-sized battery packs.

Because rapid progress is being made toward improving the capabilities of SiC RF power transistors, the reader should consult the latest electron device literature for up-to-date SiC RF transistor capabilities. A late-1997 summary of solid-state high-power RF amplification transistor results, including 4H-SiC, 6H-SiC, silicon, GaAs, and GaN device results, is given in Fig. 6.7.[7] Despite the fact that SiC RF transistors are not nearly as optimized, they have still demonstrated higher power densities than silicon and GaAs RF power transistors. The commercial availability of semi-insulating SiC substrates to minimize parasitic capacitances is crucial to the high-frequency performance of SiC RF MESFETs. MESFET devices fabricated on semi-insulating substrates are conceivably less susceptible to adverse yield consequences arising from micropipes than vertical high-power switching devices, primarily because a c-axis micropipe can no longer short together two conducting sides of a high field junction in most areas of the lateral channel MESFET structure. In addition to micropipes, other non-idealities, such as variations in epilayer doping and thickness, surface morphological defects, and slow charge trapping/detrapping phenomena causing unwanted device I-V drift,[98] also limit the yield, size, and manufacturability of SiC RF transistors. However, increasingly beneficial SiC RF transistors should continue to evolve as SiC crystal quality and device processing technology continues to improve.

In addition to high-power RF transistors, SiC mixer diodes show excellent promise for reducing undesired intermodulation interference in RF receivers.[99] More than 20-dB dynamic range improvement was demonstrated using non-optimized SiC Schottky diode mixers. Following further development and optimization, SiC-based mixers should improve the interference immunity of a number of RF systems where receivers and high-power transmitters are closely located, as well as improve the reliability and safety of flight RF-based avionics instruments used to guide aircraft in low-visibility weather conditions.

High-Temperature Signal-Level Devices

Most analog signal conditioning and digital logic circuits are considered "signal level" in that individual transistors in these circuits do not require any more than a few milliamperes of current and less than 20 V to function properly. Commercially available silicon-on-insulator circuits can perform complex

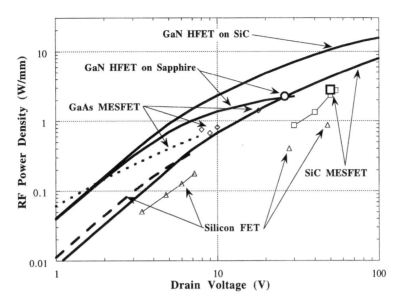

FIGURE 6.7 Theoretical (lines) and experimental (symbols) RF power densities of RF transistors fabricated in silicon, GaAs, SiC, and GaN as of late 1997. (From Ref. 96. With permission.)

digital and analog signal-level functions up to 300°C when high-power output is not required.[100,101] Aside from ICs where it is advantageous to combine signal-level functions with high-power or unique SiC sensors/MEMS onto a single chip, more expensive SiC circuits solely performing low-power signal-level functions appear largely unjustifiable for low-radiation applications at temperatures below 250 to 300°C.

Achieving long-term operational reliability is one of the primary challenges of realizing 300 to 600°C devices and circuits. Circuit technologies that have been used to successfully implement VLSI circuits in silicon and GaAs, such as CMOS, ECL, BiCMOS, DCFL, etc., are to varying degrees candidates for T > 300°C SiC integrated circuits. High temperature gate-insulator reliability (Section 6.5) is critical to the successful realization of MOSFET-based integrated circuits. Gate-to-channel Schottky diode leakage limits the peak operating temperature of SiC MESFET circuits to around 400°C (Section 6.5). Prototype bipolar SiC transistors have exhibited poor gains,[102] but improvements in SiC crystal growth and surface passivation should improve SiC BJT gains.[103] As discussed in Section 6.5, a common obstacle to all technologies is reliable long-term operation of contacts, interconnect, passivation, and packaging at T > 300°C. Because signal-level circuits are operated at relatively low electric fields well below the electrical failure voltage of most micropipes, micropipes affect signal-level circuit process yields to a much lesser degree than they affect high-field power device yields. Non-idealities in SiC epilayers, such as variations in epilayer doping and thickness, surface morphological defects, and slow charge trapping/detrapping phenomena causing unwanted device I–V drift, presently limit the yield, size, and manufacturability of SiC high-temperature integrated circuits.[83] However, continued progress in maturing SiC crystal growth and device fabrication technology should eventually enable the realization of SiC VLSI circuits.

Robust circuit designs that accommodate large changes in device operating parameters with temperature will be necessary for circuits to function successfully over the very wide temperature ranges (as large as 650°C spread) enabled by SiC. While there are similarities to silicon device behavior as a function of temperature, there are also significant differences that will present challenges to SiC integrated circuit designers. For example, in silicon devices, dopant atoms are fully ionized at standard operating temperatures of interest, so that free carrier concentrations correspond with dopant impurity concentrations.[14,15] Therefore, the resistivity of silicon increases with increasing temperature as phonon scattering reduces carrier mobility. SiC device layers, on the other hand, are significantly "frozen-out" due to deeper donor and acceptor dopant ionization energies, so that non-trivial percentages of dopants are not ionized to produce free carriers that carry current at or near room temperature. Thus, the resistivity of SiC layers

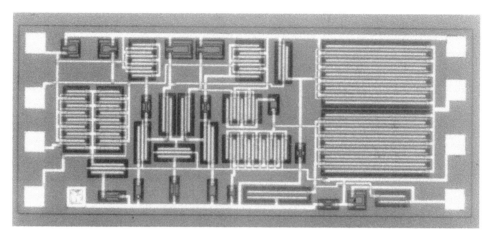

FIGURE 6.8   Optical micrograph of 1×2 mm² 300°C 6H-SiC operational amplifier integrated circuit. The chip contains 14 depletion-mode N-channel MOSFETs integrated with 19 resistors. (From Ref. 105. With permission.)

can sometimes initially decrease with increasing temperature as dopant atoms ionize to contribute more current-conducting free carriers, then decrease similar to silicon after most dopant atoms have ionized and increased phonon scattering degrades free carrier mobility. Thus, SiC transistor parameters can exhibit temperature variations not found in silicon devices, so that new device-behavior models are sometimes necessary to carry out proper design of wider-temperature range SiC integrated circuits. Because of carrier freeze-out effects, it will be difficult to realize SiC-based ICs operational at temperatures much lower than –55°C (the lower end of the U.S. Mil-Spec. temperature range).

Small-scale prototype logic and analog amplifier SiC-based ICs (one of which is shown in Fig. 6.8) have been demonstrated using SiC variations of NMOS, CMOS, JFET, and MESFET device topologies.[33,83,104–108] These prototypes are not commercially viable as of this writing, largely due to their high cost, unproven reliability, and limited temperature range that is mostly covered by silicon-on-insulator-based circuitry. However, increasingly capable and economical SiC integrated circuits will continue to evolve as SiC crystal growth and device fabrication technology continues to improve.

SiC High-Power Switching Devices

Operational Limitations Imposed by SiC Material Quality

As discussed in Section 6.3, the most lucrative system benefits of SiC electronics arguably lie in high-power devices. Unfortunately, these devices are also the most susceptible to present-day deficiencies in SiC material quality and process variations, mostly because they operate at high electric fields and high current densities that place the greatest electrical stresses on the semiconductor and surrounding device materials. Prototype SiC devices have demonstrated excellent area-normalized performance, often well beyond (>10X) the theoretical power density of silicon power electronics (Fig. 6.9). However, the presence of micropipe crystal defects has thus far prevented scale-up of small-area prototypes to large areas that can reliably deliver high total operating currents in large-scale power systems, as discussed in Section 6.4 and Refs. 9 and 40. Rectifying power device junctions responsible for OFF-state blocking fail at micropipe defects, leading to undesired (often damaging) localized current flow through micropipes at unacceptably low electric fields well below the critical reverse-breakdown field of defect-free SiC. Over the last decade, SiC micropipe densities have dropped from several hundred per square centimeter of wafer area to tens per square centimeter of wafer area (Section 6.4, Table 6.2), resulting in corresponding improvements in peak SiC device operating currents from less than 1 A to 10s of A. However, further defect reductions of at least an additional order of magnitude will be necessary before reasonably good power device yields and currents will be obtained.

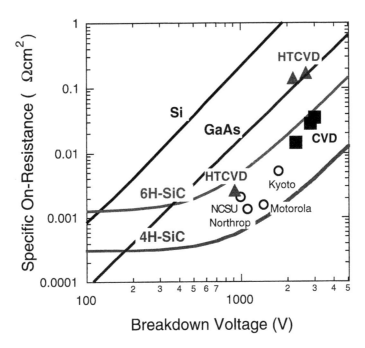

FIGURE 6.9 Experimental SiC (symbols) and theoretical SiC and silicon (lines) Schottky diode specific on resistance plotted as a function of off-state blocking voltage. While the graph clearly shows that the area-normalized performance of small-area SiC devices is orders of magnitude better than silicon, these results have not been successfully upscaled to realize large-area high-current SiC devices due to high densities of device-degrading defects present in commercial SiC wafers and epilayers. (From Ref. 117. With permission.)

In addition to micropipe defects, the density of non-hollow core (elementary) screw dislocation defects in SiC wafers and epilayers has been measured on the order of several thousands per square centimeter of wafer area (Section 6.4). While these defects are not nearly as detrimental to device performance as micropipes, recent experiments have shown that they degrade the leakage and breakdown characteristics of pn junctions.[109,110] Less direct experimental evidence exists to suggest that elementary screw dislocations may also cause localized reductions in minority carrier diffusion lengths[111,112] and non-uniformities and catastrophic localized failure to high-voltage Schottky rectifiers under reverse bias.[72,113] While localized breakdown is well known to adversely degrade silicon device reliability in high-power switching applications, the exact impact of localized breakdown in SiC devices has yet to be quantified. If it turns out that SiC power devices roughly adhere to the same reliability physics well known for silicon power devices, it is possible that SiC devices containing non-hollow core screw dislocations could prove unacceptably unreliable for use in the most demanding high-power conversion applications, such as large-motor control and public power distribution. Thus, these applications might require much larger (i.e., much longer-term) improvements in SiC material quality so as to eliminate all screw dislocations (both hollow core and non-hollow core) from any given device.

SiC High-Voltage Edge Termination

For SiC power devices to successfully function at high voltages, peripheral breakdown due to edge-related electric field crowding[15,16,63] must be avoided through careful device design and proper choice of insulating/passivating dielectric materials. The peak voltage of most prototype high-voltage SiC devices has been limited by often destructive edge-related breakdown, especially in SiC devices capable of blocking multiple kilovolts.[114,115] In addition, most testing of multi-kilovolt SiC devices has required the device to be immersed in specialized high-dielectric strength fluids or gas atmospheres to minimize damaging electrical arcing and surface flashover at device peripheries.[114,116,117]

A variety of edge termination methodologies, many of which were originally pioneered in silicon high-voltage devices, have been applied to prototype SiC power devices with varying degrees of success. Some of these approaches include tailored dopant guard rings,[122–124] tailored etches,[118–121] neutral ion implant damage rings,[125,126] metal guard rings,[127] and anode contact-insulator overlap.[128,129] The higher voltages and higher local electric fields of SiC power devices will place larger stresses on packaging and on wafer insulating materials, so it is unclear that traditional materials used to insulate/passivate silicon high-voltage devices will prove sufficient for reliable use in SiC high-voltage devices, especially if those devices are to be operated at high temperatures.

SiC High-Power Rectifiers

The high-power diode rectifier is a critical building block of power conversion circuits. A good review of experimental SiC rectifier results is given in Ref. 67. As discussed in Refs. 8 and 20 to 22, the most important SiC diode rectifier device design tradeoffs roughly parallel well-known silicon rectifier tradeoffs, except for the fact that numbers for current densities, voltages, power densities, and switching speeds are much higher in SiC. SiC's high breakdown field and wide energy bandgap permit operation of SiC metal–semiconductor Schottky diodes at much higher voltages (i.e., kilovolts) and current densities (kA/cm^2) than is practical with silicon-based Schottky diodes. A drawback of the wide bandgap of SiC is that it requires larger forward bias voltages (~1 V) to reach the turn-on "knee" where significant ON-state current begins flowing, and this can lead to an undesirable increase in ON-state power dissipation. However, the benefits of 100X decreased drift region resistance and much faster dynamic switching should greatly overcome SiC ON-state knee voltage disadvantages in most high-power systems. Figure 6.9 summarizes experimental Schottky diode specific on-resistance versus breakdown voltage results published up through 1997.[117] For blocking voltages up to 3 kV, unipolar SiC Schottky rectifiers offer lower turn-on voltages (~1 to 2 V vs. ~3 V) and faster switching speeds (due to no appreciable minority carrier injection/charge storage) than SiC pn junctions.[21,22]

In rectifiers that block over ~3 kV, bipolar minority carrier charge injection (i.e., conductivity modulation) should enable SiC pn diodes to carry higher current densities than unipolar Schottky diodes whose drift regions conduct solely using dopant-atom majority carriers.[20–22] SiC pn junction blocking voltages as high as 5.5 kV have been realized as of this writing,[130] and further blocking voltage improvements are expected as SiC materials growth and processing further improve. Consistent with silicon rectifier experience, SiC pn junction generation-related reverse leakage is usually smaller than thermionic-assisted Schottky diode reverse leakage. While it has not yet been experimentally verified for SiC, silicon power device experience[131] strongly suggests that SiC pn junction rectifiers should offer significantly better reverse breakdown immunity to overvoltage/overcurrent faults that can occur in high-power switching circuits with large inductors than SiC Schottky rectifiers. As with silicon bipolar devices, reproducible localized control of minority carrier lifetime will be essential in optimizing the switching-speed versus ON-state current density performance tradeoffs of SiC bipolar devices for specific applications. SiC minority carrier lifetimes on the order of several microseconds have been obtained in high-quality epilayers,[132] and lifetime reduction via intentional impurity incorporation and introduction of radiation-induced structural defects appears feasible.

Hybrid Schottky/pn rectifier structures first developed in silicon that combine pn junction reverse blocking with low Schottky forward turn-on should prove extremely useful to realizing application-optimized SiC rectifiers.[133,134] Similarly, combinations of dual Schottky metal structures and trench pinch rectifier structures can also be used to optimize SiC rectifier forward turn-on and reverse leakage properties.[135]

SiC High-Power Switching Transistors

Three terminal power switches that use small drive signals to control large voltages and currents are also critical building blocks of high-power conversion circuits. As well summarized in Ref. 21, a variety of prototype three-terminal SiC power switches have been demonstrated in recent years. For the most part,

SiC solid-state switches are based on well-known silicon device topologies, like the thyristor, vertical MOSFETs, IGBT, GTO, etc., that try to maximize power density via vertical current flow using the substrate as one of the device terminals. Because these switches all contain high-field junctions responsible for blocking current flow in the OFF-state, their maximum operating currents are primarily restricted by the material quality deficiencies discussed in Section 6.6. Therefore, while blocking voltages over 2 kV have been demonstrated in low-current devices,[116] experimental SiC power switches have only realized modest current ratings (under 1 A in most devices).

Silicon power MOSFETs and IGBTs are extremely popular in power circuits, largely because their MOS gate drives are well insulated and require little drive signal power, and the devices are "normally off" in that there is no current flow when the gate is unbiased at 0 V. However, as discussed in Section 6.5, the performance and reliability of SiC power device structures with inversion channel MOS field-effect gates (i.e., MOSFETs, IGBTs, etc.) are limited by poor inversion channel mobilities and questionable oxide reliability at high temperatures. Thus, SiC device structures that do not rely on high-quality gate oxides, such as the thyristor, appear more favorable for more immediate realization, despite some non-trivial drawbacks in operational circuit design and switching speed.

Recently, some non-traditional power switch topologies have been proposed to somewhat alleviate SiC oxide and material quality deficiencies while maintaining normally off insulated gate operation. Shenoy et al.[136] and Hara,[137] respectively, have implemented lateral and vertical doped-channel depletion/accumulation mode power SiC MOSFETs that can be completely depleted by built-in potentials at zero gate bias so that they are "normally off." Spitz et al.[116] recently demonstrated high-voltage SiC lateral MOSFETs implemented on semi-insulating substrates. These devices could conceivably reduce the adverse yield consequences of micropipes, because a *c*-axis micropipe can no longer short together two conducting sides of a high-field junction in most regions of the device. With the assistance of lateral surface electric field tailoring techniques, Baliga[138] has suggested that lateral-conduction SiC power devices could deliver better power densities than traditional vertical SiC power device structures. Baliga has also proposed the advantageous high-voltage switching by pairing a high-voltage SiC MESFET or JFET with a lower-voltage silicon power MOSFET.[138]

SiC for Sensors and Microelectromechanical Systems (MEMS)

Silicon carbide's high-temperature capabilities have enabled the realization of catalytic metal–SiC and metal–insulator–SiC (MIS) prototype gas sensor structures with great promise for emissions monitoring applications.[139,140] High-temperature operation of these structures, not possible with silicon, enables rapid detection of changes in hydrogen and hydrocarbon content to sensitivities of parts-per-million in very small-sized sensors that could easily be placed unobtrusively anywhere on an engine. Once they have been more fully developed, these sensors could assist in active combustion control to reduce harmful pollution emissions from automobile and aircraft engines.

Hesketh's chapter on micromachining describes conventional silicon-based microelectromechanical systems (MEMS). While the previous sections in this chapter have centered on the use of SiC for traditional semiconductor electronic devices, SiC is also likely to play a significant role in emerging MEMS applications.[141] In addition to high-temperature electrical operation, SiC has excellent mechanical properties vital to durable operation of microsystems that address some shortcomings of silicon-based MEMS, such as extreme hardness and low friction, reducing mechanical wear-out as well as excellent chemical inertness to corrosive ambients. Unfortunately, the same properties that make SiC more durable than silicon also make SiC more difficult to process into MEMS structures than silicon. Nevertheless, SiC-based pressure sensors, accelerometers, resonators (Fig. 6.10), and other MEMS systems are being developed for use in harsh-environment applications beyond the reach of silicon-based microsystems. The general approaches to fabricating harsh-environment MEMS structures in SiC and prototype SiC-MEMS results obtained to date are discussed in Ref. 141.

FIGURE 6.10 Micromachined SiC-based lateral resonator device. The excellent mechanical and electrical properties of SiC are enabling the development of harsh-environment microelectromechanical systems (MEMS) for operation beyond the limits of conventional silicon-based MEMS. (From Prof. M. Mehregany, Case Western Reserve University. With permission.)

6.7   Further Recommended Reading

This chapter has presented a brief summary overview of evolving SiC semiconductor device technology. The following publications, which were heavily referenced in this chapter, are highly recommended as advanced, supplemental reading that more completely covers the work being done to develop SiC electronics in much greater technical detail than possible within this short chapter. Reference 12 is a two-volume collection of invited in-depth papers from recognized leaders in SiC technology development that first appeared in special issues of the journal *Physica Status Solidi* (a, 162, No. 1) and (b, 202, No. 1) in 1997. Reference 11 is a two-volume collection of papers from the *7th International Conference on Silicon Carbide, III-Nitrides, and Related Materials* held in Stockholm, Sweden, in September 1997. As SiC electronics is evolving rapidly to fulfill the needs of a steadily increasing array of applications, the reader should consult the current literature for updates on SiC device capabilities. One of the best ongoing sources of SiC electronics information is the *International Conference on Silicon Carbide and Related Materials*, which is held every two years. A meeting was scheduled for October 1999 in Research Triangle Park, North Carolina (internet Web site: www.ISCRM99.ncsu.edu). In addition, a variety of Internet Web sites contain useful SiC information and links, including www.grc.nasa.gov/WWW/SiC/SiC.html, www.hiten.com, www.cree.com, www.ecn.purdue.edu/WBG/, www.sterling-semiconductor.com/, www.imc.kth.se/sic/, and www.ifm.liu.se/Matephys/new_page/research/sic/index.html, among others that can be easily located using widely available World Wide Web Internet search engine services.

References

1. Baliga, B. J., Power Semiconductor Devices for Variable-Frequency Drives, *Proceedings of the IEEE*, 82, 1112, 1994.
2. Baliga, B. J., Power ICs in the Saddle, *IEEE Spectrum*, 32, 34, 1995.

3. Baliga, B. J., Trends in Power Semiconductor Devices, *IEEE Transactions on Electron Devices*, 43, 1717, 1996.

4. Heydt, G. T. and Skromme, B. J., Applications of High Power Electronic Switches in the Electric Power Utility Industry and the Needs for High Power Switching Devices, Power Semiconductor Materials and Devices, *Materials Research Society Symposia Proceedings*, 483, Pearton, S. J., Shul, R. J., Wolfgang, E., Ren, F. and Tenconi, S., Eds., Materials Research Society, Warrendale, PA, 1998, 3.

5. Trew, R. J., Yan, J.-B., and Mock, P. M., The Potential of Diamond and SiC Electronic Devices for Microwave and Millimeter-Wave Power Applications, *Proceedings of the IEEE*, 79, 598, 1991.

6. Weitzel, C. E., Palmour, J. W., Carter, C. H., Jr., Moore, K., Nordquist, K. J., Allen, S., Thero, C., and Bhatnagar, M., Silicon Carbide High Power Devices, *IEEE Transactions on Electron Devices*, 43, 1732, 1996.

7. Weitzel, C. E. and Moore, K. E., Silicon Carbide and Gallium Nitride RF Power Devices, Power Semiconductor Materials and Devices, *Materials Research Society Symposia Proceedings*, 483, Pearton, S. J., Shul, R. J., Wolfgang, E., Ren, F., and Tenconi, S., Eds., Materials Research Society, Warrendale, PA, 1998, 111.

8. Bhatnagar, M. and Baliga, B. J., Comparison of 6H-SiC, 3C-SiC, and Si for Power Devices, *IEEE Transactions on Electron Devices*, 40, 645, 1993.

9. Neudeck, P. G., Progress in Silicon Carbide Semiconductor Electronics Technology, *Journal of Electronic Materials*, 24, 283, 1995.

10. Powell, J. A., Pirouz, P., and Choyke, W. J., Growth and Characterization of Silicon Carbide Polytypes for Electronic Applications, *Semiconductor Interfaces, Microstructures, and Devices: Properties and Applications*, Feng, Z. C., Eds., Institute of Physics Publishing, Bristol, United Kingdom, 1993, 257.

11. Pensl, G., Morkoc, H., Monemar, B., and Janzen, E., *Silicon Carbide, III-Nitrides, and Related Materials, Materials Science Forum*, 264-268, Trans Tech Publications, Switzerland, 1998.

12. Choyke, W. J., Matsunami, H., and Pensl, G., Silicon Carbide — *A Review of Fundamental Questions and Applications to Current Device Technology*, Wiley-VCH, Berlin, 1997.

13. Harris, G. L., *Properties of SiC*, EMIS Datareviews Series, 13, The Institute of Electrical Engineers, London, 1995.

14. Pierret, R. F., *Advanced Semiconductor Fundamentals*, Modular Series on Solid State Devices, Vol. 6, Addison-Wesley, Reading, MA, 1987.

15. Sze, S. M., *Physics of Semiconductor Devices, 2nd ed.*, Wiley-Interscience, New York, 1981.

16. Baliga, B. J., *Modern Power Devices*, 1st ed., John Wiley & Sons, New York, 1987.

17. Neudeck, G. W., *The PN Junction Diode, Modular Series on Solid State Devices*, 2, 2nd ed., Addison-Wesley, Reading, MA, 1989.

18. Neudeck, G. W., *The Bipolar Junction Transistor*, Modular Series on Solid State Devices, 3, 2nd ed., Addison-Wesley, Reading, MA, 1989.

19. Divan, D., Low-Stress Switching for Efficiency, *IEEE Spectrum*, 33, 33, 1996.

20. Ruff, M., Mitlehner, H. and Helbig, R., SiC Devices: Physics and Numerical Simulation, *IEEE Transactions on Electron Devices*, 41, 1040, 1994.

21. Chow, T. P., Ramungul, N., and Ghezzo, M., Wide Bandgap Semiconductor Power Devices, Power Semiconductor Materials and Devices, *Materials Research Society Symposia Proceedings*, 483, Pearton, S. J., Shul, R. J., Wolfgang, E., Ren, F., and Tenconi, S., Eds., Materials Research Society, Warrendale, PA, 1998, 89.

22. Bakowski, M., Gustafsson, U., and Lindefelt, U., Simulation of SiC High Power Devices, *Physica Status Solidi (a)*, 162, 421, 1997.

23. Trew, R. J., Experimental and Simulated Results of SiC Microwave Power MESFETs, *Physica Status Solidi (a)*, 162, 409, 1997.

24. Nieberding, W. C. and Powell, J. A., High Temperature Electronic Requirements in Aeropropulsion Systems, *IEEE Transactions on Industrial Electronics*, 29, 103, 1982.

25. Carlin, C. M. and Ray, J. K., The Requirements for High Temperature Electronics in a Future High Speed Civil Transport (HSCT), *Second International High Temperature Electronics Conference*, Charlotte, NC, One, King, D. B. and Thome, F. V., Eds., Sandia National Laboratories, Albuquerque, NM, 1994, I-19.

26. Reinhardt, K. C. and Marciniak, M. A., Wide-Bandgap Power Electronics for the More Electric Aircraft, *Transactions 3rd International High Temperature Electronics Conference*, Albuquerque, NM, 1, Sandia National Laboratories, Albuquerque, NM, 1996, I-9.

27. Acheson, A. G., England Patent 17911, 1892.

28. Lely, J. A., Darstellung von Einkristallen von Silicium carbid und Beherrschung von Art und Menge der eingebautem Verunreinigungen, *Ber. Deut. Keram. Ges.*, 32, 229, 1955.

29. Nishino, S., Powell, J. A., and Will, H. A., Production of Large-Area Single-Crystal Wafers of Cubic SiC for Semiconductor Devices, *Applied Physics Letters*, 42, 460, 1983.

30. Pirouz, P., Chorey, C. M., and Powell, J. A., Antiphase Boundaries in Epitaxially Grown Beta-SiC, *Applied Physics Letters*, 50, 221, 1987.

31. Pirouz, P., Chorey, C. M., Cheng, T. T., and Powell, J. A., Lattice Defects in β-SiC Grown Epitaxially on Silicon Substrates, Heteroepitaxy on Silicon II, *Materials Research Society Symposia Proceedings*, 91, Fan, J. C., Phillips, J. M., and Tsaur, B.-Y., Eds., Materials Research Society, Pittsburgh, PA, 1987, 399.

32. Davis, R. F., Kelner, G., Shur, M., Palmour, J. W., and Edmond, J. A., Thin Film Deposition and Microelectronic and Optoelectronic Device Fabrication and Characterization in Monocrystalline Alpha and Beta Silicon Carbide, *Proceedings of the IEEE*, 79, 677, 1991.

33. Harris, G. L., Wongchotigul, K., Henry, H., Diogu, K., Taylor, C., and Spencer, M. G., Beta SiC Schottky Diode FET Inverters Grown on Silicon, Silicon Carbide and Related Materials: *Proceedings of the Fifth International Conference, Institute of Physics Conference Series*, 137, Spencer, M. G., Devaty, R. P., Edmond, J. A., Kahn, M. A., Kaplan, R., and Rahman, M., Eds., IOP Publishing, Bristol, United Kingdom, 1994, 715.

34. Tairov, Y. M. and Tsvetkov, V. F., Investigation of Growth Processes of Ingots of Silicon Carbide Single Crystals, *Journal of Crystal Growth*, 43, 209, 1978.

35. Tairov, Y. M. and Tsvetkov, V. F., General Principles of Growing Large-Size Single Crystals of Various Silicon Carbide Polytypes, *Journal of Crystal Growth*, 52, 146, 1981.

36. Eldridge, G. W., Barrett, D. L., Burk, A. A., Hobgood, H. M., Siergiej, R. R., Brandt, C. D., Tischler, M. A., Bilbro, G. L., Trew, R. J., Clark, W. H., and Gedridge, R. W., Jr., High Power Silicon Carbide IMPATT Diode Development, *2nd Annual AIAA SDIO Interceptor Technology Conference*, Albuquerque, NM, American Institute of Aeronautics and Astronautics, Report 93-2703, Washington D.C., 1993.

37. Takahashi, J. and Ohtani, N., Modified-Lely SiC Crystals Grown in [1100] and [1120] Directions, *Physica Status Solidi (b)*, 202, 163, 1997.

38. Cree Research, Inc., 4600 Silicon Drive, Durham, NC 27703, http://www.cree.com.

39. Powell, J. A. and Larkin, D. J., Processed-Induced Morphological Defects in Epitaxial CVD Silicon Carbide, *Physica Status Solidi (b)*, 202, 529, 1997.

40. Neudeck, P. G. and Powell, J. A., Performance Limiting Micropipe Defects in Silicon Carbide Wafers, *IEEE Electron Device Letters*, 15, 63, 1994.

41. Yang, J.-W., SiC: Problems in Crystal Growth and Polytypic Transformation, Ph. D. dissertation, Case Western Reserve University, Cleveland, OH, 1993.

42. Si, W., Dudley, M., Glass, R., Tsvetkov, V., and Carter, C. H., Jr., Hollow-Core Screw Dislocations in 6H-SiC Single Crystals: A Test of Frank's Theory, *Journal of Electronic Materials*, 26, 128, 1997.

43. Si, W. and Dudley, M., Study of Hollow-Core Screw Dislocations in 6H-SiC and 4H-SiC Single Crystals, Silicon Carbide, III-Nitrides, and Related Materials, *Materials Science Forum*, 264-268, Pensl, G., Morkoc, H., Monemar, B., and Janzen, E., Eds., Trans Tech Publications, Switzerland, 1998, 429.

44. Heindl, J., Strunk, H. P., Heydemann, V. D., and Pensl, G., Micropipes: Hollow Tubes in Silicon Carbide, *Physica Status Solidi (a)*, 162, 251, 1997.

45. Wang, S., Dudley, M., Carter, C. H., Jr., Tsvetkov, V. F. and Fazi, C., Synchrotron White Beam Topography Studies of Screw Dislocations in 6H-SiC Single Crystals, Applications of Synchrotron Radiation Techniques to Materials Science, *Material Research Society Symposium Proceedings*, 375, Terminello, L., Shinn, N., Ice, G., D'Amico, K., and Perry, D., Eds., Materials Research Society, Warrendale, PA, 1995, 281.

46. Dudley, M., Wang, S., Huang, W., Carter, C. H., Jr., and Fazi, C., White Beam Synchrotron Topographic Studies of Defects in 6H-SiC Single Crystals, *Journal of Physics D: Applied Physics*, 28, A63, 1995.

47. Wang, S., Dudley, M., Carter, C. H., Jr., and Kong, H. S., X-Ray Topographic Studies of Defects in PVT 6II-SiC Substrates and Epitaxial 6H-SiC Thin Films, Diamond, SiC and Nitride Wide Bandgap Semiconductors, *Materials Research Society Symposium Proceedings*, 339, Carter, C. H., Jr., Gildenblat, G., Nakamura, S., and Nemanich, R. J., Eds., Materials Research Society, Pittsburgh, PA, 1994, 735.

48. Burk, A. A., Jr. and Rowland, L. B., Homoepitaxial VPE Growth of SiC Active Layers, *Physica Status Solidi (b)*, 202, 263, 1997.

49. Kimoto, T., Itoh, A., and Matsunami, H., Step-Controlled Epitaxial Growth of High-Quality SiC Layers, *Physica Status Solidi (b)*, 202, 247, 1997.

50. Rupp, R., Makarov, Y. N., Behner, H., and Wiedenhofer, A., Silicon Carbide Epitaxy in a Vertical CVD Reactor: Experimental Results and Numerical Process Simulation, *Physica Status Solidi, (b)*, 202, 281, 1997.

51. Kordina, O., Hallin, C., Henry, A., Bergman, J. P., Ivanov, I., Ellison, A., Son, N. T., and Janzen, E., Growth of SiC by "Hot-Wall" CVD and HTCVD, *Physica Status Solidi (b)*, 202, 321, 1997.

52. Kong, H. S., Glass, J. T., and Davis, R. F., Chemical Vapor Deposition and Characterization of 6H-SiC Thin Films on Off-Axis 6H-SiC Substrates, *Journal of Applied Physics*, 64, 2672, 1988.

53. Larkin, D. J., Neudeck, P. G., Powell, J. A., and Matus, L. G., Site-Competition Epitaxy for Superior Silicon Carbide Electronics, *Applied Physics Letters*, 65, 1659, 1994.

54. Larkin, D. J., SiC Dopant Incorporation Control Using Site-Competition CVD, *Physica Status Solidi, (b)*, 202, 305, 1997.

55. Rendakova, S. V., Nikitina, I. P., Tregubova, A. S., and Dmitriev, V. A., Micropipe and Dislocation Density Reduction in 6H-SiC and 4H-SiC Structures Grown by Liquid Phase Epitaxy, *Journal of Electronic Materials*, 27, 292, 1998.

56. Khlebnikov, I., Sudarshan, T. S., Madangarli, V., and Capano, M. A., A Technique for Rapid Thick Film SiC Epitaxial Growth, Power Semiconductor Materials and Devices, *Materials Research Society Symposia Proceedings*, 483, Pearton, S. J., Shul, R. J., Wolfgang, E., Ren, F., and Tenconi, S., Eds., Materials Research Society, Warrendale, PA, 1998, 123.

57. Nam, O. H., Zheleva, T. S., Bremser, M. D., and Davis, R. F., Lateral Epitaxial Overgrowth of GaN Films on SiO_2 Areas Via Metalorganic Vapor Phase Epitaxy, *Journal of Electronic Materials*, 27, 233, 1998.

58. Schaffer, W. J., Negley, G. H., Irvine, K. G., and Palmour, J. W., Conductivity Anisotropy in Epitaxial 6H and 4H SiC, Diamond, SiC, and Nitride Wide-Bandgap Semiconductors, *Materials Research Society Symposia Proceedings*, 339, Carter, C. H., Jr., Gildenblatt, G., Nakamura, S., and Nemanich, R. J., Eds., Materials Research Society, Pittsburgh, PA, 1994, 595.

59. Troffer, T., Schadt, M., Frank, T., Itoh, H., Pensl, G., Heindl, J., Strunk, H. P., and Maier, M., Doping of SiC by Implantation of Boron and Aluminum, *Physica Status Solidi (a)*, 162, 277, 1997.

60. Kimoto, T., Itoh, A., Inoue, N., Takemura, O., Yamamoto, T., Nakajima, T., and Matsunami, H., Conductivity Control of SiC by In-Situ Doping and Ion Implantation, Silicon Carbide, III-Nitrides, and Related Materials, *Materials Science Forum*, 264-268, Pensl, G., Morkoc, H., Monemar, B., and Janzen, E., Eds., Trans Tech Publications, Switzerland, 1998, 675.

61. Zhao, J. H., Tone, K., Weiner, S. R., Caleca, M. A., Du, H., and Withrow, S. P., Evaluation of Ohmic Contacts to P-Type 6H-SiC Created by C and Al Coimplantation, *IEEE Electron Device Letters*, 18, 375, 1997.

62. Capano, M. A., Ryu, S., Melloch, M. R., Cooper, J. A., Jr., and Buss, M. R., Dopant Activation and Surface Morphology of Ion Implanted 4H- and 6H-Silicon Carbide, *Journal of Electronic Materials*, 27, 370, 1998.

63. Rhoderick, E. H. and Williams, R. H., Metal-Semiconductor Contacts, *Monographs in Electrical and Electronic Engineering*, 19, Clarendon Press, Oxford, UK, 1988.

64. Porter, L. M. and Davis, R. F., A Critical Review of Ohmic and Rectifying Contacts for Silicon Carbide, *Materials Science and Engineering B*, B34, 83, 1995.

65. Bozack, M. J., Surface Studies on SiC as Related to Contacts, *Physica Status Solidi (b)*, 202, 549, 1997.

66. Crofton, J., Porter, L. M., and Williams, J. R., The Physics of Ohmic Contacts to SiC, *Physica Status Solidi (b)*, 202, 581, 1997.

67. Saxena, V. and Steckl, A. J., Building Blocks for SiC Devices: Ohmic Contacts, Schottky Contacts, and p-n Junctions, *Semiconductors and Semimetals*, 52, Academic Press, New York, 1998, 77.

68. Petit, J. B., Neudeck, P. G., Salupo, C. S., Larkin, D. J., and Powell, J. A., Electrical Characteristics and High Temperature Stability of Contacts to N- and P-Type 6H-SiC, Silicon Carbide and Related Materials, *Institute of Physics Conference Series*, 137, Spencer, M. G., Devaty, R. P., Edmond, J. A., Kahn, M. A., Kaplan, R., and Rahman, M., Eds., IOP Publishing, Bristol, 1994, 679.

69. Okojie, R. S., Ned, A. A., Provost, G., and Kurtz, A. D., Characterization of Ti/TiN/Pt Contacts on N-Type 6H-SiC Epilayer at 650°C, *1998 4th International High Temperature Electronics Conference*, Albuquerque, NM, IEEE, Piscataway, NJ, 1998, 79.

70. Itoh, A. and Matsunami, H., Analysis of Schottky Barrier Heights of Metal/SiC Contacts and Its Possible Application to High-Voltage Rectifying Devices, *Physica Status Solidi (a)*, 162, 389, 1997.

71. Teraji, T., Hara, S., Okushi, H., and Kajimura, K., Ideal Ohmic Contact to N-Type 6H-SiC by Reduction of Schottky Barrier Height, *Applied Physics Letters*, 71, 689, 1997.

72. Bhatnagar, M., Baliga, B. J., Kirk, H. R., and Rozgonyi, G. A., Effect of Surface Inhomogeneities on the Electrical Characteristics of SiC Schottky Contacts, *IEEE Transactions on Electron Devices*, 43, 150, 1996.

73. Crofton, J. and Sriram, S., Reverse Leakage Current Calculations for SiC Schottky Contacts, *IEEE Transactions on Electron Devices*, 43, 2305, 1996.

74. Yih, P. H., Saxena, V., and Steckl, A. J., A Review of SiC Reactive Ion Etching in Fluorinated Plasmas, *Physica Status Solidi (b)*, 202, 605, 1997.

75. Cao, L., Li, B., and Zhao, J. H., Inductively Coupled Plasma Etching of SiC for Power Switching Device Fabrication, Silicon Carbide, III-Nitrides, and Related Materials, *Materials Science Forum*, 264-268, Pensl, G., Morkoc, H., Monemar, B., and Janzen, E., Eds., Trans Tech Publications, Switzerland, 1998, 833.

76. McLane, G. F. and Flemish, J. R., High Etch Rates of SiC in Magnetron Enhanced SF_6 Plasmas, *Applied Physics Letters*, 68, 3755, 1996.

77. Shor, J. S. and Kurtz, A. D., Photoelectrochemical Etching of 6H-SiC, *Journal of the Electrochemical Society*, 141, 778, 1994.

78. Shor, J. S., Kurtz, A. D., Grimberg, I., Weiss, B. Z., and Osgood, R. M., Dopant-Selective Etch Stops in 6H and 3C SiC, *Journal of Applied Physics*, 81, 1546, 1997.

79. Pierret, R. F., Field Effect Devices, *Modular Series on Solid State Devices*, 4, Addison-Wesley, Reading, MA, 1983.

80. Cooper, J. A., Jr., Advances in SiC MOS Technology, *Physica Status Solidi (a)*, 162, 305, 1997.

81. Afanasev, V. V., Bassler, M., Pensl, G., and Schulz, M., Intrinsic SiC/SiO_2 Interface States, *Physica Status Solidi (a)*, 162, 321, 1997.

82. Ouisse, T., Electron Transport at the SiC/SiO_2 Interface, *Physica Status Solidi (a)*, 162, 339, 1997.

83. Brown, D. M., Downey, E., Ghezzo, M., Kretchmer, J., Krishnamurthy, V., Hennessy, W., and Michon, G., Silicon Carbide MOSFET Integrated Circuit Technology, *Physica Status Solidi (a)*, 162, 459, 1997.

84. Tan, J., Das, M. K., Cooper, J. A., Jr., and Melloch, M. R., Metal-Oxide-Semiconductor Capacitors Formed by Oxidation of Polycrystalline Silicon on SiC, *Applied Physics Letters*, 70, 2280, 1997.

85. Sridevan, S. and Baliga, B. J., Inversion Layer Mobility in SiC MOSFETs, Silicon Carbide, III-Nitrides, and Related Materials, *Materials Science Forum*, 264-268, Pensl, G., Morkoc, H., Monemar, B., and Janzen, E., Eds., Trans Tech Publications, Switzerland, 1998, 997.

86. Powell, J. A., Larkin, D. J., and Abel, P. B., Surface Morphology of Silicon Carbide Epitaxial Films, *Journal of Electronic Materials*, 24, 295, 1995.

87. Scharnholz, S., Stein von Kamienski, E., Golz, A., Leonhard, C., and Kurz, H., Dependence of Channel Mobility on the Surface Step Orientation in Planar 6H-SiC MOSFETs, Silicon Carbide, III-Nitrides, and Related Materials, *Materials Science Forum*, 264-268, Pensl, G., Morkoc, H., Monemar, B., and Janzen, E., Eds., Trans Tech Publications, Switzerland, 1998, 1001.

88. Agarwal, A. K., Seshadri, S., and Rowland, L. B., Temperature Dependence of Fowler-Nordheim Current in 6H- and 4H-SiC MOS Capacitors, *IEEE Electron Device Letters*, 18, 592, 1997.

89. Grzybowski, R. R. and Gericke, M., 500°C Electronics Packaging and Test Fixturing, *Second International High Temperature Electronic Conference*, Charlotte, NC, 1, King, D. B., and Thome, F. V., Eds., Sandia National Laboratories, Albuquerque, NM, 1994, IX-41.

90. Bratcher, M., Yoon, R. J., and Whitworth, B., Aluminum Nitride Package for High Temperature Applications, *Transactions 3rd International High Temperature Electronics Conference*, Albuquerque, NM, 2, Sandia National Laboratories, Albuquerque, NM, 1996, P-21.

91. Salmon, J. S., Johnson, R. W., and Palmer, M., Thick Film Hybrid Packaging Techniques for 500°C Operation, *1998 4th International High Temperature Electronics Conference*, Albuquerque, NM, IEEE, Piscataway, NJ, 1998, 103.

92. Edmond, J., Kong, H., Suvorov, A., Waltz, D., and Carter, C., Jr., 6H-Silicon Carbide Light Emitting Diodes and UV Photodiodes, *Physica Status Solidi (a)*, 162, 481, 1997.

93. Bergh, A. A. and Dean, P. J., *Light-Emitting Diodes*, Clarendon Press, Oxford, 1976.

94. Brown, D. M., Downey, E., Kretchmer, J., Michon, G., Shu, E., and Schneider, D., SiC Flame Senors for Gas Turbine Control Systems, *Solid-State Electronics*, 42, 755, 1998.

95. Przybylko, S. J., Developments in Silicon Carbide for Aircraft Propulsion System Applications, *AIAA/SAE/ASME/ASEE 29th Joint Propulsion Conference and Exhibit*, American Institute of Aeronautics and Astronautics, Report 93-2581, Washington, D.C., 1993.

96. Weitzel, C., Pond, L., Moore, K., and Bhatnagar, M., Effect of Device Temperature on RF FET Power Density, Silicon Carbide, III-Nitrides, and Related Materials, *Materials Science Forum*, 264-268, Pensl, G., Morkoc, H., Monemar, B., and Janzen, E., Eds., Trans Tech Publications, Switzerland, 1998, 969.

97. Sriram, S., Siergiej, R. R., Clarke, R. C., Agarwal, A. K., and Brandt, C. D., SiC for Microwave Power Transistors, *Physica Status Solidi (a)*, 162, 441, 1997.

98. Noblanc, O., Arnodo, C., Chartier, E., and Brylinski, C., Characterization of Power MESFETs on 4H-SiC Conductive and Semi-Insulating Wafers, Silicon Carbide, III-Nitrides, and Related Materials, *Materials Science Forum*, 264-268, Pensl, G., Morkoc, H., Monemar, B., and Janzen, E., Eds., Trans Tech Publications, Switzerland, 1998, 949.

99. Fazi, C. and Neudeck, P., Use of Wide-Bandgap Semiconductors to Improve Intermodulation Distortion in Electronic Systems, Silicon Carbide, III-Nitrides, and Related Materials, *Materials Science Forum*, 264-268, Pensl, G., Morkoc, H., Monemar, B., and Janzen, E., Eds., Trans Tech Publications, Switzerland, 1998, 913.

100. AlliedSignal Microelectronics and Technology Center, 9140 Old Annapolis Road, Columbia, MD 21045, http://www.mtcsemi.com.

101. Honeywell Solid State Electronics Center, 12001 State Highway 55, Plymouth, MN 55441, http://www.ssec.honeywell.com.

102. Wang, Y., Xie, W., Cooper, J. A., Jr., Melloch, M. R., and Palmour, J. W., Mechanisms Limiting Current Gain in SiC Bipolar Junction Transistors, Silicon Carbide and Related Materials 1995, *Institute of Physics Conference Series*, 142, Nakashima, S., Matsunami, H., Yoshida, S., and Harima, H., Eds., IOP Publishing, Bristol, U.K., 1996, 809.

103. Neudeck, P. G., Perimeter Governed Minority Carrier Lifetimes in 4H-SiC p⁺n Diodes Measured by Reverse Recovery Switching Transient Analysis, *Journal of Electronic Materials*, 27, 317, 1998.

104. Xie, W., Cooper, J. A., Jr., and Melloch, M. R., Monolithic NMOS Digital Integrated Circuits in 6H-SiC, *IEEE Electron Device Letters*, 15, 455, 1994.

105. Brown, D. M., Ghezzo, M., Kretchmer, J., Krishnamurthy, V., Michon, G., and Gati, G., High Temperature Silicon Carbide Planar IC Technology and First Monolithic SiC Operational Amplifier IC, *Second International High Temperature Electronics Conference*, Charlotte, NC, 1, Sandia National Laboratories, Albuquerque, NM, 1994, XI-17.

106. Ryu, S. H., Kornegay, K. T., Cooper, J. A., Jr., and Melloch, M. R., Digital CMOS IC's in 6H-SiC Operating on a 5-V Power Supply, *IEEE Transactions on Electron Devices*, 45, 45, 1998.

107. Diogu, K. K., Harris, G. L., Mahajan, A., Adesida, I., Moeller, D. F., and Bertram, R. A., Fabrication and Characterization of a 83 MHz High Temperature β-SiC MESFET Operational Amplifier with an AlN Isolation Layer on (100) 6H-SiC, *54th Annual IEEE Device Research Conference*, Santa Barbara, CA, IEEE, Piscataway, NJ, 1996, 160.

108. Neudeck, P. G., 600°C Digital Logic Gates, *NASA Lewis 1998 Research & Technology Report*, 1999.

109. Neudeck, P. G., Huang, W., and Dudley, M., Breakdown Degradation Associated with Elementary Screw Dislocations in 4H-SiC P⁺N Junction Rectifiers, Power Semiconductor Materials and Devices, *Materials Research Society Symposia Proceedings*, 483, Pearton, S. J., Shul, R. J., Wolfgang, E., Ren, F., and Tenconi, S., Eds., Materials Research Society, Warrendale, PA, 1998, 285.

110. Neudeck, P. G., Huang, W., Dudley, M., and Fazi, C., Non-Micropipe Dislocations in 4H-SiC Devices: Electrical Properties and Device Technology Implications, Wide-Bandgap Semiconductors for High Power, High Frequency and High Temperature, *Materials Research Society Symposia Proceedings*, 512, Denbaars, S., Shur, M. S., Palmour, J., and Spencer, M., Eds., Materials Research Society, Warrendale, PA, 1998, 107.

111. Doolittle, W. A., Rohatgi, A., Ahrenkiel, R., Levi, D., Augustine, G., and Hopkins, R. H., Understanding the Role of Defects in Limiting the Minority Carrier Lifetime in SiC, Power Semiconductor Materials and Devices, *Materials Research Society Symposia Proceedings*, 483, Pearton, S. J., Shul, R. J., Wolfgang, E., Ren, F., and Tenconi, S., Eds., Materials Research Society, Warrendale, PA, 1998, 197.

112. Hubbard, S. M., Effect of Crystal Defects on Minority Carrier Diffusion Length in 6H SiC Measured Using the Electron Beam Induced Current Method, Master of Science dissertation, Case Western Reserve University, Cleveland, OH, 1998.

113. Raghunathan, R. and Baliga, B. J., Role of Defects in Producing Negative Temperature Dependence of Breakdown Voltage in SiC, *Applied Physics Letters*, 72, 3196, 1998.

114. Neudeck, P. G., Larkin, D. J., Powell, J. A., Matus, L. G., and Salupo, C. S., 2000 V 6H-SiC p-n Junction Diodes Grown by Chemical Vapor Deposition, *Applied Physics Letters*, 64, 1386, 1994.

115. Domeij, M., Breitholtz, B., Linnros, J., and Ostling, M., Reverse Recovery and Avalanche Injection in High Voltage SiC PIN Diodes, Silicon Carbide, III-Nitrides, and Related Materials, *Materials Science Forum*, 264-268, Morkoc, H., Pensl, G., Monemar, B., and Janzen, E., Eds., Trans Tech Publications, Switzerland, 1998, 1041.

116. Spitz, J., Melloch, M. R., Cooper, J. A., Jr., and Capano, M. A., 2.6 kV 4H-SiC Lateral DMOSFET's, *IEEE Electron Device Letters*, 19, 100, 1998.

117. Kimoto, T., Wahab, Q., Ellison, A., Forsberg, U., Tuominen, M., Yakimova, R., Henry, A., and Janzen, E., High-Voltage (>2.5kV) 4H-SiC Schottky Rectifiers Processed on Hot-Wall CVD and High-Temperature CVD Layers, Silicon Carbide, III-Nitrides, and Related Materials, *Materials Science Forum*, 264-268, Pensl, G., Morkoc, H., Monemar, B., and Janzen, E., Eds., Trans Tech Publications, Switzerland, 1998, 921.

118. Peters, D., Schorner, R., Holzlein, K. H., and Friedrichs, P., Planar Aluminum-Implanted 1400 V 4H Silicon Carbide p-n Diodes with Low On Resistance, *Applied Physics Letters*, 71, 2996, 1997.

119. Ueno, K., Urushidani, T., Hashimoto, K., and Seki, Y., The Guard-Ring Termination for the High Voltage SiC Schottky Barrier Diodes, *IEEE Electron Device Letters*, 16, 331, 1995.

120. Itoh, A., Kimoto, T., and Matsunami, H., Excellent Reverse Blocking Characteristics of High-Voltage 4H-SiC Schottky Rectifiers with Boron-Implanted Edge Termination, *IEEE Electron Device Letters*, 17, 139, 1996.

121. Singh, R. and Palmour, J. W., Planar Terminations in 4H-SiC Schottky Diodes with Low Leakage and High Yields, *9th International Symposium on Power Semiconductor Devices and IC's*, IEEE, Piscataway, NJ, 1997, 157.

122. Ramungul, N., Khemka, V., Chow, T. P., Ghezzo, M., and Kretchmer, J., Carrier Lifetime Extraction from a 6H-SiC High-Voltage P-i-N Rectifier Reverse Recovery Waveform, Silicon Carbide, III-Nitrides, and Related Materials 1997, *Materials Science Forum*, 264-268, Pensl, G., Morkoc, H., Monemar, B., and Janzen, E., Eds., Trans Tech Publications, Switzerland, 1998, 1065.

123. Konstantinov, A. O., Wahab, Q., Nordell, N., and Lindefelt, U., Ionization Rates and Critical Fields in 4H Silicon Carbide, *Applied Physics Letters*, 71, 90, 1997.

124. Harris, C. I., Konstantinov, A. O., Hallin, C., and Janzen, E., SiC Power Device Passivation Using Porous SiC, *Applied Physics Letters*, 66, 1501, 1995.

125. Alok, D., Baliga, B. J., and McLarty, P. K., A Simple Edge Termination for Silicon Carbide Devices with Nearly Ideal Breakdown Voltage, *IEEE Electron Device Letters*, 15, 394, 1994.

126. Alok, D. and Baliga, B., SiC Device Edge Termination Using Finite Area Argon Implantation, *IEEE Transactions on Electron Devices*, 44, 1013, 1997.

127. Raghunathan, R. and Baligà, B. J., EBIC Measurements of Diffusion Lengths in Silicon Carbide, *1996 Electronic Materials Conference*, Santa Barbara, CA, TMS, Warrendale, PA, 1996, 18.

128. Su, J. N. and Steckl, A. J., Fabrication of High Voltage SiC Schottky Barrier Diodes by Ni Metallization, Silicon Carbide and Related Materials 1995, *Institute of Physics Conference Series*, 142, Nakashima, S., Matsunami, H., Yoshida, S., and Harima, H., Eds., IOP Publishing, Bristol, United Kingdom, 1996, 697.

129. Brezeanu, G., Fernandez, J., Millan, J., Badila, M., and Dilimot, G., MEDICI Simulation of 6H-SiC Oxide Ramp Profile Schottky Structure, Silicon Carbide, III-Nitrides, and Related Materials, *Materials Science Forum*, 264-268, Pensl, G., Morkoc, H., Monemar, B., and Janzen, E., Eds., Trans Tech Publications, Switzerland, 1998, 941.

130. Singh, R., Irvine, K. G., Kordina, O., Palmour, J. W., Levinshtein, M. E., and Rumyanetsev, S. L., 4H-SiC Bipolar P-i-N Diodes With 5.5 kV Blocking Voltage, *56th Annual Device Research Conference*, Charlottesville, VA, IEEE, Piscataway, NJ, 1998, 86.

131. Ghose, R. N., *EMP Environment and System Hardness Design*, D. White Consultants, Gainesville, VA, 1984, 4.1.

132. Kordina, O., Bergman, J. P., Hallin, C., and Janzen, E., The Minority Carrier Lifetime of N-Type 4H- and 6H-SiC Epitaxial Layers, *Applied Physics Letters*, 69, 679, 1996.

133. Held, R., Kaminski, N., and Niemann, E., SiC Merged P-N/Schottky Rectifiers for High Voltage Applications, Silicon Carbide, III-Nitrides, and Related Materials, *Materials Science Forum*, 264-268, Pensl, G., Morkoc, H., Monemar, B., and Janzen, E., Eds., Trans Tech Publications, Switzerland, 1998, 1057.

134. Dahlquist, F., Zetterling, C. M., Ostling, M., and Rottner, K., Junction Barrier Schottky Diodes in 4H-SiC and 6H-SiC, Silicon Carbide, III-Nitrides, and Related Materials, *Materials Science Forum*, 264-268, Pensl, G., Morkoc, H., Monemar, B., and Janzen, E., Eds., Trans Tech Publications, Switzerland, 1998, 1061.

135. Schoen, K. J., Henning, J. P., Woodall, J. M., Cooper, J. A., Jr., and Melloch, M. R., A Dual-Metal-Trench Schottky Pinch-Rectifier in 4H-SiC, *IEEE Electron Device Letters*, 19, 97, 1998.

136. Shenoy, P. M. and Baliga, B. J., The Planar 6H-SiC ACCUFET: A New High-Voltage Power MOSFET Structure, *IEEE Electron Device Letters*, 18, 589, 1997.

137. Hara, K., Vital Issues for SiC Power Devices, Silicon Carbide, III-Nitrides, and Related Materials, *Materials Science Forum*, 264-268, Pensl, G., Morkoc, H., Monemar, B., and Janzen, E., Eds., Trans Tech Publications, Switzerland, 1998, 901.

138. Baliga, B. J., Prospects For Development of SiC Power Devices, Silicon Carbide and Related Materials 1995, *Institute of Physics Conference Series*, 142, Nakashima, S., Matsunami, H., Yoshida, S. and Harima, H., Eds., IOP Publishing, Bristol, United Kingdom, 1996, 1.

139. Hunter, G. W., Neudeck, P. G., Chen, L. Y., Knight, D., Liu, C. C., and Wu, Q. H., SiC-Based Schottky Diode Gas Sensors, Silicon Carbide, III-Nitrides, and Related Materials, *Materials Science Forum*, 264-268, Pensl, G., Morkoc, H., Monemar, B., and Janzen, E., Eds., Trans Tech Publications, Switzerland, 1998, 1093.

140. Lloyd Spetz, A., Baranzahi, A., Tobias, P., and Lundstrom, I., High Temperature Sensors Based on Metal-Insulator-Silicon Carbide Devices, *Physica Status Solidi (a)*, 162, 493, 1997.

141. Mehregany, M., Zorman, C., Narayanan, N., and Wu, C. H., Silicon Carbide MEMS for Harsh Environments, *Proceedings of the IEEE*, 14, 1998.

142. Nippon Steel Corporation, 5-10-1 Fuchinobe, Sagamihara, Kanagawa 229, Japan.

143. SiCrystal AG, Heinrich-Hertz-Platz 2, D-92275 Eschenfelden http://www.sicrystal.de.

144. Sterling Semiconductor, 22660 Executive Drive, Suite 101, Sterling, VA 20166, http://www.sterling-semiconductor.com/.

145. Epitronics Corporation, 550 West Juanita Ave., Mesa, AZ 85210, http://www.epitronics.com.

7

Passive Components

Ashraf Lotfi
Bell Laboratories
Lucent Technologies

7.1 Magnetic Components

Integration Issues

It is well known and recognized that magnetic components should be avoided when designing integrated circuits due to their lack of integrability. New developments in the field of magnetic component fabrication are promising devices that can be integrated and miniaturized using monolithic fabrication techniques as opposed to today's bulk methods. The driving forces for such developments rest in certain applications that benefit or rely on inductive or magnetically coupled devices using ferromagnetic media. Examples of such applications include tuned RF tanks, matching networks, dc-dc power conversion and regulation, network filters, and line isolators/couplers.

Emerging applications requiring more mobility, lower power dissipation, and smaller component and system sizes have been drivers for the development of highly integrated systems and/or subsystems. In order to match these trends, it has become necessary to be able to integrate high-quality magnetic devices (i.e., inductors and transformers) with the systems they operate in as opposed to being stand-alone discrete devices. Not only does their discrete nature prevent further miniaturization, but their very nature also hampers improved performance (e.g., speed).

The main features of a monolithic magnetic device include:

1. High values of inductance compared to air core spirals
2. Enhanced high-frequency performance
3. Energy storage, dc bias, and power handling capabilities
4. Use of ferromagnetic materials as a magnetic core
5. Photolithographic fabrication of windings and magnetic core
6. Multi-layer mask fabrication for complete magnetic device design
7. Standard or semi-standard IC processing techniques

Due to the use of standard photolithography, etching, and patterning methods for their fabrication, monolithic magnetic devices may appear compatible with IC processes. However, two main characteristics make these devices more suitably fabricated off-line from a mainstream IC process:

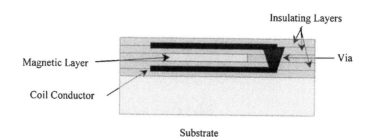

FIGURE 7.1   Cross-section of a monolithic micromagnetic device fabricated using IC methods.

- Coarser design rules. Usually, magnetic device designs do not require sub-micron geometries as demanded by semiconductor designs. This discrepancy means that an expensive sub-micron process for these components would unnecessarily raise the device cost.

- Use of ferromagnetic core materials. The use of iron, cobalt, nickel, and their alloys is at the heart of a high-quality magnetic device. Some of these materials are alien and contaminating to semiconductor cleanrooms. As a result, processing sequences and logistics for full integration with semiconductors is still in the development phase.

With these two major differences, integration of magnetics and semiconductors may require the use of multi-chip modules or single package multi-die cases. Full integration into a single monolithic die requires separate processing procedures using the same substrate.

The construction of a monolithic micromagnetic device fabricated on a substrate such as silicon or glass is shown in Fig. 7.1. In this diagram, the magnetic layer is sandwiched between upper and lower conductor layers that are connected together by means of an electrically conducting via. This structure is referred to as a *toroidal device* from its discrete counterpart.

Conversely, a dual structure can be made where two magnetic layers sandwich the conductor layer (or layers). This dual structure can be referred to as an *EE device* since it is derived from the standard discrete "EE" core type. In either case, as required by the operation of any magnetically coupled device, the magnetic flux path in the magnetic film and the current flow in the coil conductor are orthogonal, in accordance with Ampere's circuital law. Interlayer insulation between conductors and the magnetic layer(s) is necessary, both to reduce capacitive effects and to provide a degree of electrical voltage breakdown. These parameters are affected by the choice of insulator systems used in microfabricated circuits, due to the differing values of dielectric constants and breakdown voltages used. Some commonly used insulator systems include silicon dioxide and polyimide, each of which has distinctly different processing methods and physical characteristics. Conductor layers for the coil windings can be fabricated using standard aluminum metallization. In some cases, copper conductors are a better choice due to their higher conductivity and hence lower resistive losses. This is especially important if the device is to handle any significant power. The magnetic film layer is a thin film of chosen magnetic material typically between 1 and 10 μm in thickness. Such materials can be routinely deposited by standard techniques such as sputtering or electrodeposition. The specific method chosen must yield magnetic films with the desired properties, namely permeability, parallel loss resistance, and maximum flux density. These parameters vary with the deposition conditions and techniques, such that significant development and optimization has occurred to produce the desired results.

Since the design may call for energy storage and hence gaps in the core, the fabrication method can be modified to incorporate these features. Figure 7.2 shows the geometry of a planar magnetic core with a gap produced by photolithography. In this case, the gap is formed as a result of the artwork generated for the core design. Figure 7.3 shows the design of a gap using multi-layer magnetic films. The energy storage region exists in the edge insulation region between the two magnetic layers.

The fabrication and construction of conductors for the coil usually involve depositing standard interconnect metals (e.g., aluminum) by sputter deposition. The thicknesses are chosen based on the current-

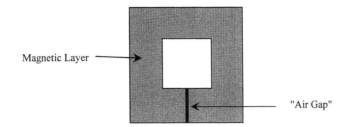

FIGURE 7.2 Planar, single-layer magnetic core configuration for energy storage (top view).

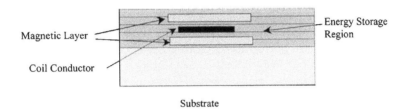

FIGURE 7.3 Multiple magnetic layers for energy storage micromagnetics (cross-sectional view).

carrying capability and the frequency of operation, as well as the desired configuration (inductor or transformer). The dc resistance of the conductors must be minimized to reduce dc losses, but the conductor thickness and arrangement must also result in minimal ac losses. This can be accomplished by reduced resistivity (i.e., copper vs. aluminum) and by multiple conductor layers to reduce skin and proximity effects at high frequencies.

Designs for Integrated Circuits

Unlike discrete magnetic components, monolithic micromagnetic devices are designed to operate at substantially higher frequencies. Due to their integrated nature and smaller physical size, interconnection and coupling parasitics are lower, thus enabling high-frequency response. However, the smaller physical size also places upper limits on such characteristics as inductance, current levels, power levels, and dissipation. With these limits, the maximum energy storage, *E*, is lower. In any inductor, the energy stored due to the current flowing (*I*) is related to the magnetic fields in the volume of the device by:

$$E = \frac{1}{2}LI^2 = \frac{1}{2}\iiint \bar{B} \cdot \bar{H} dv \tag{7.1}$$

where *L* is the transformer or inductor's inductance in Henries, and *I* (A) is the maximum current carried by the corresponding winding. This is related to the magnetic flux density, *B*, and magnetic field, *H*, present in the volume of the device. So, for a small physical volume, one can see from Eq. (7.1) that the energy stored is also small. This limited energy storage capability limits these devices to operation in low-power circuits. In order to obtain a high *B-H* product for more energy storage, a combination of high-permeability and low-permeability regions should be fabricated (i.e., a gap in the high permeability path is introduced). This gap region helps to maintain a high flux density as well as an appreciable field. The highly permeable region, on the other hand, while being able to maintain high flux density does not support large magnetic fields due to the fundamental relationship between magnetic field and flux:

$$\bar{B} = \mu_0 \mu_r \bar{H} \tag{7.2}$$

In Eq. (7.2), μ_0 is the permeability of vacuum ($4\pi \times 10^{-7}$ H/m) and μ_r is the relative permeability of the medium in which the magnetic field produces the corresponding magnetic flux density. The size of this gap determines both the energy storage levels and the inductance attainable (which is lower than the inductance attainable without a gap). In micromagnetic fabrication, two approaches may be taken to create this "air gap" region. One is to introduce a planar lithographical feature into the core structure (Fig. 7.2), and the other is to rely on multiple magnetic core layers separated by insulating layers (Fig. 7.3). The drawback of the lithographical gap is the limits imposed by the design rules. In this case, the gap may not be any smaller than the minimum design rule, which can be quite coarse. Excessive gap sizes result in very low inductance, requiring an increase in number of turns to compensate for this drop. Consequently, electrical losses in these windings increase and also the fabrication becomes more complicated. The drawback of multiple magnetic core layers is the need to add another level of processing to obtain at least a second (or more) magnetic layer(s). The stack-up of these layers and the edge terminations determine the amount of energy storage possible in the device. Unlike the lithographically produced gap, the energy storage in this case is much more difficult to estimate due to the two-dimensional nature of the edge termination fields in the gap region surrounding the multi-layer magnetic cores. In uniform field cases, the energy stored in volume of the gap can be obtained from Eqs. (7.1) and (7.2) due to the continuity and uniformity of the flux density vector in both the core and gap regions, giving:

$$E = \frac{1}{2}LI^2 \approx \frac{1}{2}\frac{B^2 V_{gap}}{\mu_0} \qquad (7.3)$$

where V_{gap} is the volume of the gap region in cubic meters (m³). The approximation is valid as long as the gap region carries a uniform flux density and is "magnetically long" compared to the length of the highly permeable core region (i.e., gap length/$\mu_{r\,gap}$ >> core length/$\mu_{r\,mag}$). Usually, this condition can be satisfied with most ferromagnetic materials of choice, but some ferromagnetic materials may have low enough permeabilities to render this approximation invalid. In this event, some energy is stored within the ferromagnetic material and Eq. (7.3) should be modified. Eq. (7.3) is very useful in determining the size of gap needed to support the desired inductance and current levels for the device. For example, if a 250-nH inductor operating at 250 mA of current bias were needed, the gap volume necessary to support these specifications would be about 2×10^{-5} mm³, assuming a material with a maximum flux density of 1.0 T. In the planar device of Fig. 7.2 with nominal magnetic film dimensions of 2 μm in the normal direction and 200 μm in the planar direction, the required gap width would be about 5 μm. Since the gap in this case is obtained by photolithography, the minimum feature size for this process would need to be 5 μm. If a different material of lower maximum flux density capability of 0.5 T were used instead, the rated current level of 250 mA would have to be downgraded to 62 mA to prevent saturation of the magnetic material. Conversely, the gap length of 5 μm could be increased to 20 μm while maintaining the same current level, assuming adjustments are made to the turns to maintain the desired inductance. Such tradeoffs are common, but are more involved due to the interaction of gap size with inductance level and number of turns.

Another aspect of the design is the conductor for coil windings for an inductor or for primary and secondary windings in the case of a transformer. The number of turns is usually selected based on the desired inductance and turns ratio (for a transformer), which are typically circuit design parameters. As is well known, the number of turns around a magnetic core gives rise to an inductance, L, given by:

$$L = \frac{\mu_0 \mu_r N^2 A}{l} H \qquad (7.4)$$

In this relation, N is the number of turns around a magnetic core of cross-sectional area A (m²) and magnetic path length l (m). The inductance is reduced by the presence of a gap since this will serve to increase the path length. The choice of conductor thickness is always made in light of the ac losses occurring when conductors carry high-frequency currents. The conductors will experience various current redistribution effects due to the presence of eddy currents induced by the high-frequency magnetic fields surrounding the conductors. The well-known *skin effect* is one of such effects. Current will crowd toward the surface of the conductor and flow mainly in a thickness related to the skin depth, δ,

$$\delta = \frac{1}{\sqrt{\pi f \mu_0 \sigma}} \ \text{m} \tag{7.5}$$

For a copper conductor, $\delta = 66/\sqrt{f(\text{MHz})}$ μm. At 10 MHz, the skin depth in copper is 20 μm, placing an upper limit on conductor thickness. When the interconnect metallization is aluminum, $\delta = 81/\sqrt{f(\text{MHz})}$ μm, so the upper limit at 10 MHz becomes 25 μm of metal thickness. Usually, the proximity of conductors to one another forces further optimization due to the introduction of losses due to eddy currents induced by neighboring conductors. In this case, the conductor thickness should be further adjusted with respect to the skin depth to reduce the induced eddy currents. The increase in conductor resistance due to the combined skin and proximity effects in a simple primary-secondary winding metallization scheme (shown in Fig. 7.4) can be calculated as an increase over the dc resistance of the conductor from:

$$Rac = Rdc \cdot \frac{1}{2}\frac{h}{\delta}\left\{ \frac{\sinh h/\delta + \sin h/\delta}{\cosh h/\delta - \cos h/\delta} + \frac{\sinh h/\delta - \sin h/\delta}{\cosh h/\delta + \cos h/\delta} \right\} \tag{7.6}$$

In this relationship, h is the thickness of the metallization being used and *Rac* is obtained once the dc resistance (*Rdc*, also a function of h) is known. A distinct minimum for *Rac* can be obtained and yields the lowest possible ac resistance when:

$$\frac{h}{\delta} = \frac{\pi}{2} \tag{7.7}$$

with a corresponding minimum value of ac resistance of:

$$Rac = \frac{\pi}{2}Rdc \ \tanh\frac{\pi}{2} = 1.44Rdc \tag{7.8}$$

When the geometry differs from the simple primary-to-secondary interface of Fig. 7.4 to more turns, layers, shapes, etc., a more complicated analysis is necessary.[4] The simple relation of Eq. (7.7) is nolonger valid. Nevertheless, this relation provides a very good starting point for many designs. A qualitative

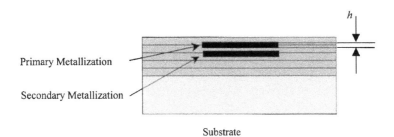

FIGURE 7.4 Configuration of primary and secondary transformer metallization for ac resistance calculation.

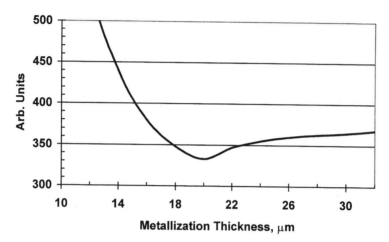

FIGURE 7.5 The variation of high-frequency metallization resistance with metal thickness for the configuration in Fig. 7.4. A distinct minimum is observed due to skin and proximity effects.

explanation of this behavior stems from the fact that a thicker metallization will produce less dc resistance, but provides poor ac utilization due to current crowding near the surface. On the other hand, a thinner metallization increases the dc resistance, while providing better ac conductor utilization. The optimum situation is somewhere in between these two extreme cases, as can be seen from Fig. 7.5.

The principles presented in this section regarding design issues are at the core of every magnetic component design for integrated circuits. However, many design details — especially at elevated frequencies — are beyond the scope of this text. It is important to note that many of the limitations on high-frequency designs (100 MHz and higher) are imposed by the properties of the magnetic materials used in the cores of these devices.

Magnetic Core Materials

The most common magnetic materials used for discrete magnetic components operating at higher frequencies are ferrites. This is mainly due to their high resistivity (1 to 10 Ωm). Despite a low saturation flux density of about 0.3T, such a high resistivity makes ferrites suitable for applications up to 1 MHz, where hysteresis core losses are still limited. When the frequency is raised over 1 MHz, the core losses become excessive, thus degrading the quality factor and efficiency of the circuit. Moreover, the permeability of all magnetic materials experiences a roll-off beyond a maximum upper frequency. Commonly used ferrites (e.g., MnZn ferrites) operate up to 1 to 2 MHz before permeability roll-off occurs. Higher roll-off frequencies are available, but with higher loss factors (e.g., NiZn ferrites). Ferrites, however, are not amenable to integrated circuit fabrication since they are produced by a high-temperature sintering process. In addition, their low flux saturation levels would not result in the smallest possible device per unit area. A set of more suitable materials for integrated circuit fabrication are the magnetic metal alloys usually derived from iron, cobalt, or nickel. These alloys can be deposited as thin films using IC fabrication techniques such as sputtering or electrodeposition and possess saturation flux levels of 0.8T to as high as 2.0T. Their main drawback due to their metallic nature is a much lower resistivity. Permalloy, a common magnetic alloy (80% nickel and 20% iron), has a resistivity of 20×10^{-8} Ωm, with a saturation flux density of 0.8T. Other materials such as sendust (iron-aluminum-silicon) have improved resistivity of 120×10^{-8} Ωm and saturation flux density of 0.95T.

To overcome the problem of low resistivity, the magnetic layers must be deposited in thin films with limited thickness. Since eddy currents flow in the metallic films at high frequencies, their effect can be greatly reduced by making the film thickness less than a skin depth. The skin depth in the magnetic film, δ_m, is given by:

$$\delta_m = \frac{1}{\sqrt{\pi f \mu_o \mu_r \sigma}} \text{ m} \tag{7.9}$$

In a thin film of permalloy ($\mu_r = 2000$), the skin depth at 10 MHz is 3 μm. In order to limit eddy current losses in the film, its thickness must be chosen to be less than 3 μm. This limitation will conflict with the inductance requirement, since a larger inductance requires a thicker magnetic film (see Eq. 7.4). Such difficulties can be overcome by depositing the magnetic film in multiple layers insulated from one another to restrict eddy current circulation. Such a structure would still provide the overall thickness needed to achieve the specified inductance while limiting the eddy current loss factor. The use of multi-layers also allows the reduction of die size due to the build-up of magnetic core cross-section (A in Eq. 7.4) in vertical layers, rather than by increasing the planar dimensions. As a result, it can be seen that a tradeoff exists between number of layers and die size to yield the most economical die cost.

In addition to eddy current losses due to the low magnetic metal resistivity, hysteresis losses occur in any magnetic material due to the traversing of the non-linear *B-H* loop at the frequency of operation. This is due to the loss of energy needed to rotate magnetic domains within the material. This loss is given by

$$P_{hys} = f \cdot \oint \bar{H} \cdot d\bar{B} \tag{7.10}$$

which is the area enclosed by the particular *B-H* loop demanded by the circuit operation and f is the frequency of operation. Total loss is expressed in many forms, depending on the application. In many cases, it is given in the form of a "parallel" or "shunt resistance" (Fig. 7.6) and it therefore presents a reduction in impedance to the source as well as a reduction in the overall quality factor of the inductor or transformer. It also represents a finite power loss since this loss is simply V^2/R_p watts, where V is the applied voltage.

Notice that R_p is a non-linear resistance with both frequency and flux level dependencies. It can be specified at a given frequency and flux level and is usually experimentally measured. It can also be extracted from core loss data usually available in the form

$$P = kf^\alpha B^\beta = \frac{V^2}{R_p} \tag{7.11}$$

In this relation, k, α, and β are constants for the material at hand. This model is useful for circuit simulation purposes, thereby avoiding the non-linear properties of the magnetic material. Care, however, should be exercised in using such models since with a large enough excitation, the value of the shunt resistor changes.

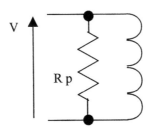

FIGURE 7.6 Magnetic material losses are represented by a parallel shunt resistance.

7.2   Air Core Inductors

Air core inductors do not use a magnetic core to concentrate the lines of magnetic flux. Instead, the flux lines exist in the immediate neighborhood of the coils without tight confinement. As a result, the inductance of an air core coil is considerably lower than one with a magnetic core. In fact, at low frequencies, the quality factor is reduced by a factor of μ_r when the loss factor is small. At high frequencies, however, the loss factor may be so high that the addition of a magnetic core and increased inductance actually ends up degrading the quality factor below the air core value. Discrete air core inductors have been used in RF applications and have been wound using discrete magnet or Litz wire onto forming cylinders.

For integrated circuits, especially RF and microwave circuits, spiral metallization deposited on a substrate is a common means to obtain small amounts of inductance with relatively high quality factors. Inductance values that can be obtained by these techniques are usually in the low nH range (1 to 20 nH). Estimating inductance values using air cores is much more complicated than in the case of highly permeable cores due to the lack of flux concentration. Formulas have been derived for different spiral air coil shapes (*assuming perfectly insulating substrates*) and are tabulated in several handbooks for inductance calculations.[6–9] An example of a useful inductance formula[9] is:

$$ L = 4\pi \times 10^{-7} N^2 r_{avg} \left\{ \ln \frac{8 r_{avg}}{(r_{out} - r_{in})} + \frac{1}{24} \left(\frac{(r_{out} - r_{in})}{r_{avg}} \right)^2 \left(\ln \frac{8 r_{avg}}{(r_{out} - r_{in})} + 3.583 \right) - 0.5 \right\} \quad (7.12) $$

In this formula, r_{out} and r_{in} are the outer and inner radii of the spiral, respectively, and the average radius r_{avg} is

$$ r_{avg} = \frac{(r_{out} + r_{in})}{2} \quad (7.13) $$

The formula is an approximation that loses accuracy as the device size becomes large (i.e., large r_{out} with respect to r_{in}).

The loss factors of such devices are strongly influenced by the non-idealities of the substrates and insulators on which they are deposited. For example, aluminum spiral inductors fabricated on silicon with highly doped substrates and epitaxial layers can have significant reductions in quality factor due to the conductivity of the underlying layers. These layers act as ground planes, producing the effect of an image of the spiral underneath. This in turn causes a loss in inductance. This can be as much as 30 to 60% when compared to a spiral over a perfect insulator. In addition, an increase in the loss factor occurs due to circulating eddy currents in the conductive under-layers. Increases in the effective resistance of 5 to 10 times the perfect insulator case are possible, increasing with increased frequency. All these effects can be seen to degrade the performance of these inductors, thus requiring design optimization.[10–12]

These substrate effects appear in the form of coupling capacitances from the spiral metal to the substrates, as well as spreading resistances in the substrate itself. The spreading resistance is frequency dependent, increasing with higher frequency. The amount of coupling to the substrate depends on the coupling capacitances and hence the separation of the spiral from the substrate. This distance is the dielectric thickness used in the IC process. Only with very large dielectric thicknesses are the substrate effects negligible. In practical cases where it is relatively thin and limited to a few microns, the effects are very large, giving an overall quality factor, Q, which is significantly lower than the Q of the spiral without the substrate. Fig. 7.7 shows a typical degradation curve of Q on a resistive substrate for "thick" and "thin" separations or dielectric thicknesses. The trends of this curve are also similar if the dielectric thickness variable is replaced by the substrate resistivity as a variable. The exact amount of degradation depends on the separation involved, the dielectric constant, and the resistivity of the substrate. With these quantities known, it is possible to construct a circuit model to include these effects and hence solve for the overall quality factor, including the substrate effects.

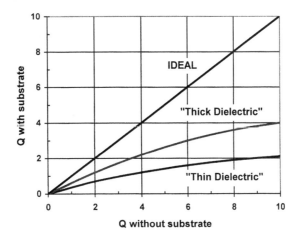

FIGURE 7.7 Degradation of inductor quality factor by placement on a resistive substrate.

In order to improve the inductor quality factor on a resistive substrate, some design solutions are possible. One solution to this problem is to increase the substrate resistivity. Another is to design a spiral with a small footprint to reduce coupling to the substrate. In order to offset the increased resistance (which also reduces Q), thicker metallization would be necessary and clearly a tradeoff situation arises requiring some design optimization by circuit modeling or, more accurately, by electromagnetic finite-element analysis.

7.3 Resistors

Resistors have been available for use in integrated circuits for many years.[13-16] Some of these are made in silicon and thus are directly integrated with the rest of the IC process. Others, similar to the magnetic device case, are thin-film resistors fabricated in an off-line process that is not necessarily compatible with silicon IC processing. Integrated silicon resistors offer simplicity in fabrication but have less than ideal characteristics with loose tolerances. For this reason, many circuits rely on the ratio of resistor values rather than on their absolute values. Thin-film resistors, on the other hand, are far superior, offering tight tolerances and the ability to trim their absolute value down to very precise values. They also display more stability in terms of temperature and frequency dependence.[17]

Usually, resistors in integrated circuits are characterized in terms of their sheet resistance rather than their absolute resistance value. Sheet resistance, R_{sheet}, is defined as the resistance of a resistive strip with equal length and width so that

$$R_{sheet} = \frac{\rho}{t} (\Omega/\Box) \tag{7.14}$$

where ρ is the material resistivity ($\Omega \cdot m$) and t is its thickness (m). Once R_{sheet} is given, the resulting resistor value is obtained by multiplying by its length-to-width aspect ratio. In order to avoid very high aspect ratios, an appropriate sheet resistivity should be used. For example, with $R_{sheet} = 10\ \Omega/\Box$ a 10:1 length-to-width ratio would give a 100-Ω resistor. However, to obtain a 1-kΩ resistor, it would be better to use a different material with, for example, $R_{sheet} = 100\ \Omega/\Box$ with the same 10:1 ratio instead of using a 100:1 ratio with the low-resistivity material.

Integrated Semiconductor Resistors

In this category, the existing semiconductor is used as the resistive material. The resistor may be fabricated at a number of stages during the IC process, giving rise to different resistors with different characteristics. Some of the most common include:

Diffused Resistors

This can be formed during either the base or emitter diffusion of a bipolar process. For an npn process, the base diffusion resistor is a p-type of moderate sheet resistivity, typically in the range of 100 to 200 $\Omega/\square$. This can provide resistors in the 50 to 10 kΩ range. The heavily doped n$^+$ emitter diffusion will produce an n$^+$-type resistor with low sheet resistivity of 2 to 10 $\Omega/\square$. This can provide resistors with low values in the 1 to 100 Ω range. Due to tolerances in the photolithographic and etching processes, the tolerance in the absolute resistance can be as high as range. Due to tolerances on the photolithographic and etching processes, the tolerance on the absolute resistance can be as high as $\pm$30%. However, resistor pairs can be matched closely in temperature coefficients and doping profiles, especially when placed side-by-side on the chip, so that the resultant tolerance of the resistor ratio can be made to be less than $\pm$1%. Since a diffusion resistor is based on a p-type base over an n-type epitaxy, or an n$^+$-type emitter over a p-type base, it is essential that the formed p-n junctions are always reverse-biased to ensure that current flows in the intended portion of the resistor. The presence of such a reverse-biased p-n junction also introduces a distributed capacitance from the resistor body to the substrate. This will cause high-frequency degradation, whereby the resistor value drops from its nominal design value to a lower impedance value due to the shunting capacitance.

Pinched Resistors

A variation to the diffused resistor that is used to increase the sheet resistivity of base region is to use the n$^+$-type emitter as a means to reduce the cross-sectional area of the base region, thereby increasing the sheet resistivity. This can increase the sheet resistance to about 1 k$\Omega/\square$. In this case, one end of the n$^+$-type emitter must be tied to one end of the resistor to contain all current flow to the pinched base region.

Epitaxial Resistors

High resistor values can be formed using the epitaxial layer since it has higher resistivity than other regions. Epitaxial resistors can have sheet resistances around 5 k$\Omega/\square$. However, epitaxial resistors have even looser tolerances due to the wide tolerances on both epitaxial resistivity and epitaxial layer thickness.

MOS Resistors

A MOSFET can be biased to provide a non-linear resistor. Such a resistor provides much greater values than diffused ones while occupying a much smaller area. When the gate is shorted to the drain in a MOSFET, a quadratic relation between current and voltage exists and the device conducts current only when the voltage exceeds the threshold voltage. Under these circumstances, the current flowing in this resistor (i.e., the MOSFET drain current) depends on the ratio of channel width-to-length. Hence, to increase the resistor value, the aspect ratio of the MOSFET should be reduced to give longer channel length and narrower channel width.

Thin-Film Resistors

As mentioned before in the magnetic core case, a resistive thin-film layer can be deposited (e.g., by sputtering) on the substrate to provide a resistor with very tight absolute-value tolerance. In addition, given a large variety of resistor materials, a wide range of resistor values can be obtained in small footprints, thereby providing very small parasitic capacitances and small temperature coefficients. Some common thin-film resistor materials include tantalum, tantalum nitride, and nickel-chromium. Unlike semiconductor resistors, thin-film resistors can be laser trimmed to adjust their values to very high accuracies of up to 0.01%. Laser trimming can only increase the resistor value since the fine beam evaporates a portion of the thin-film material. By its nature, laser trimming is a slow and costly operation that is only justified when very high accuracy on absolute values is necessary.

7.4   Capacitors

As in the inductor case, the limitation on integrated capacitors is die size, due to the limited capacitance/unit area available on a die. These limitations are imposed by the dielectrics used with their dielectric constants and breakdown voltages. Most integrated capacitors are either junction capacitors or MOS capacitors.

Junction Capacitors

A *junction capacitor* is formed when a p-n junction is reversed-biased. This can be formed using the base–emitter, base–collector, or collector–substrate junctions of an npn structure in bipolar ICs. Of course, the particular junction must be maintained in reverse-bias to provide the desired capacitance. Since the capacitance arises from the parallel plate effect across the depletion region, whose thickness in turn is voltage dependent, the capacitance is also voltage dependent, decreasing with increased reverse-bias. The capacitance depends on the reverse-voltage in the following form:

$$C(V) \;=\; \frac{C_0}{\left(1 + V/\psi_0\right)^n} \tag{7.15}$$

The built-in potential, Ψ_0, depends on the impurity concentrations of the junction being used. For example, $\Psi_0 = 0.7$ V for a typical bipolar base–emitter junction. The exponent n depends on the doping profile of the junction. The approximations $n = 1/2$ for a step junction and $n = 1/3$ for a linearly graded junction are commonly used. The resultant capacitance depends on C_0, the capacitance per unit area with zero bias applied. This depends on the doping level and profile. The base–emitter junction provides the highest capacitance per unit area, around 1000 pF/mm^2 with a low breakdown voltage (~5 V). The base–collector junction provides about 100 pF/mm^2 with a higher breakdown voltage (~40 V).

MOS Capacitors

MOS capacitors are usually formed as parallel plate devices with a top metallization and a high conductivity n^+ emitter diffusion as the two plates, with a thin oxide dielectric sandwiched in between. The oxide is usually a thin layer of SiO$_2$ with a relative dielectric constant ε_r of 3 to 4, or Si$_3$N$_4$ with ε_r of 5 to 8. Since the capacitance obtained is $\varepsilon_0\varepsilon_r A/t_{oxide}$, the oxide thickness, t_{oxide}, is critical. The lower limit on the oxide thickness depends on the process yields and tolerances, as well as the desired breakdown voltage and reliability. MOS capacitors can provide around 1000 pF/mm^2, with breakdown voltages up to 100 V. Unlike junction capacitors, MOS capacitors are voltage independent and can be biased either positively or negatively. Their breakdown, however, is destructive since the oxide fails permanently. Care should be taken to prevent overvoltage conditions.

References

1. Saleh, N. and Qureshi, A., "Permalloy thin-film inductors," *Electronics Letters*, vol. 6, no. 26, pp. 850-852, 1970.
2. Soohoo, R., "Magnetic film inductors for integrated circuit applications," *IEEE Trans. Magn.*, vol. MAG-15, pp. 1803, 1979.

3. Mino, M. et al., "A new planar microtransformer for use in micro-switching converters," *IEEE Trans. Magn.*, vol. 28, pp. 1969, 1992.

4. Vandelac, J. and Ziogas, P., "A novel approach for minimizing high frequency transformer copper loss," *IEEE Trans. Power Elec.*, vol. 3, no. 3, pp. 266-76, 1988.

5. Sato, T., Tomita, H., Sawabe, A., Inoue, T., Mizoguchi, T., and Sahashi, M., "A magnetic thin film inductor and its application to a MHz Switching dc-dc converter," *IEEE Trans. Magn.*, vol. 30, no. 2, pp. 217-223, 1994.

6. Grover, F., *Inductance Calculations*, Dover Publishing, New York, 1946.

7. Welsby, V., *Theory and Design of Inductance Coils*, MacDonald & Co., London, 2nd ed., 1960.

8. Walker, C., *Capacitance, Inductance, and Crosstalk Analysis*, Artech House, Boston, 1990.

9. Gupta, K. C., Garg, R., and Chadha, R., *Computer-Aided Design of Microwave Circuits*, Artech House, Dedham, MA, 1981.

10. Remke, R. and Burdick, G.,"Spiral inductors for hybrid and microwave applications," *Proc. 24th Electron Components Conf.*, May 1974, pp. 152-161.

11. Arnold, R. and Pedder, J.,"Microwave characterization of microstrip lines and spiral inductors in MCM-D technology," *IEEE Trans. Comp., Hybrids, and Manuf. Tech.*, vol. 15, pp. 1038-45, 1992.

12. Nguyen, N. M. and Meyer, R. G.,"Si IC-compatible inductors and LC passive filters," *IEEE J. Solid-State Circuits*, vol. 25, pp. 1028-1031, Aug. 1990.

13. Glaser, A. and Subak-Sharpe, G., *Integrated Circuit Engineering*, Addison-Wesley, Reading, MA, 1977.

14. Goodge, M. E., *Semiconductor Device Technology*, Howard Sams & Co., Inc., Indiana, 1983.

15. Grebene, A. B., *Bipolar and MOS Analog Integrated Circuit Design*, John Wiley, New York, 1984.

16. Gray, P. R. and Meyer, R. G., *Analysis and Design of Analog Integrated Circuits*, John Wiley, New York, 1993.

17. Sergent, J. E. and Harper, C. A., *Hybrid Microelectronics Handbook*, McGraw-Hill, New York, 1995.

18. Levy, R. A., *Microelectronic Materials and Processes*, Kluwer Academic Publishers, Dordrecht, Netherlands, 1989.

8

Power IC Technologies

Akio Nakagawa
Toshiba Corporation

8.1 Introduction

VLSI technology has advanced so greatly that Gigabit DRAMs have become a reality, and the technology faces an optical lithography limit. Microelectronics mostly advances signal processing LSIs such as memories and microprocessors. Power systems and the related circuits cannot be outside the influence of VLSI technology.[1] It would be quite strange for power systems alone to still continue to consume a large space while brains become smaller and smaller. On the other hand, almost all of the systems require actuators or power devices to control motors, displays, and multimedia equipment. The advances in microelectronics have made it possible to integrate large-scale circuits in a small silicon chip, ending up in high system performance and resultant system miniaturization. The system miniaturization inevitably necessitated power IC development. Typical early power ICs were audio power amplifiers, which used bipolar transistors as output devices. The pn junction isolation method was well suited to integrate bipolar transistors with control circuits.

Real advancements in intelligent power ICs were triggered by the invention of power DMOSFETs[2] in the 1970s. DMOS transistors have ideal features for output devices of power ICs. No driving dc current is necessary, and large currents can be controlled simply by changing the gate voltage. In addition, DMOS switching speed is sufficiently fast.

The on-resistance of vertical DMOSFETs has been greatly reduced year by year with advances in fine lithography in LSI technology. In the mid-1980s, the new concept "Smart Power"[3] was introduced. Smart Power integrates bipolar and CMOS devices with vertical DMOS, using a process primarily optimized for poly-silicon gate self-aligned DMOS. The main objective is to integrate control and protection circuits with vertical power devices, not only to increase device reliability and performance, but also to realize easy use of power devices. The concept of Smart Power was applied to high-voltage vertical DMOS with drain contact on the back side of the chip because discrete DMOS technology was already well advanced in the early 1980s. The main application field was automotive, replacing mechanical relays and eliminating wire harnesses.

As the technology of microlithography has further advanced, the on-resistance of DMOS, especially low-voltage DMOS, has continuously decreased. In the early 1990s, the on-resistance of low-voltage lateral DMOS became lower than that of bipolar transistors.[4] It was even realized that low-voltage lateral DMOS is superior to vertical planar discrete DMOS since fine lithography does not contribute to a decrease in on-resistance of vertical DMOS because of JFET resistance. Recently, with the introduction of a 0.6-μm design rule, lateral DMOS has become predominant over the wide voltage range — from 20 V up to 150 V. Mixed technology, called BCD,[4] integrating BiCMOS and DMOS, is now widely accepted for low-voltage power ICs.

For high-voltage power ICs, DMOS is not suitable for output devices because of a high on-resistance. Thyristor-like devices, such as GTOs, have conventionally been used for high-voltage applications. Integration of thyristor-like devices needs a method of dielectric device isolation (DI). The conventional DI method, called EPIC,[5] has been used for high-voltage telecommunication ICs, called SLIC. However, it has problems of high cost and large wafer warpage. In 1985 and 1986, wafer direct-bonding technology was invented,[6,7] and low-cost DI wafers became available. Wafer warpage of directly bonded SOI wafers is very small. This made it possible not only to fabricate large-diameter (8-in.) SOI wafers, but also to apply advanced lithography to DI power ICs. The chip size of DI power ICs can be reduced by narrow-trench isolation and by the use of high-performance lateral IGBTs. The low-cost DI wafers and the chip size reduction have widened the application fields of DI power ICs, covering automotive, motor control, and PDP drivers.

8.2 Intelligent Power ICs

Figure 8.1 shows typical functions integrated into intelligent power ICs. The most important feature of the intelligent power IC is that a large power can be controlled by logic-level input signals and all the cumbersome circuits such as driving circuits, sense circuits, and protection circuits required for power device control are inside the power ICs.

Technologies for realizing such power ICs are classified into three categories, as shown in Figs. 8.2 to 8.4. These are self-isolation, junction isolation, and dielectric isolation. Self-isolation is a method that does not use any special means to isolate each device, and each device structure automatically isolates itself from the other.

Junction isolation (JI) is a method that uses reverse-biased junction depletion layers to isolate each device. JI is the most frequently used method for low-voltage power ICs, using bipolar transistor or DMOS outputs.

Junction isolation is not sufficient to isolate IGBTs or thyristors. Dielectric isolation is a method that uses silicon dioxide film to isolate devices and thus offers complete device isolation. Although the EPIC method has conventionally been used, the high cost of wafer fabrication has been a problem. Recently, wafer direct-bonding technology was invented and bonded SOI wafers are available at a lower price.

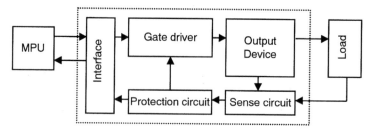

FIGURE 8.1 Typical integrated functions in intelligent power ICs.

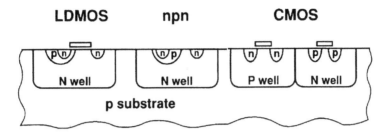

FIGURE 8.2   Self-isolation technology.

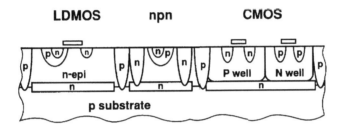

FIGURE 8.3   Junction isolation (JI) technology.

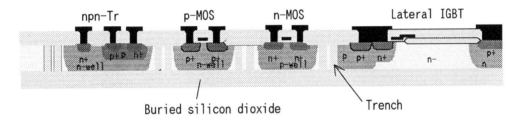

FIGURE 8.4   Dielectric isolation (DI) technology.

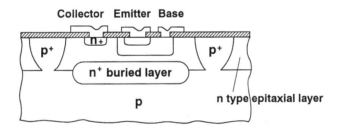

FIGURE 8.5   Junction isolation structure.

pn Junction Isolation

One of the fundamental issues in integrated circuits is how to electrically isolate each device from the others. pn junction isolation is the most familiar method and has been used since the beginning of bipolar IC history. Figure 8.5 shows the cross-section of a typical junction isolation structure. First, an n-type epitaxial layer is formed on p-type silicon substrate. p-type diffusion layers are then formed to reach the p-type substrate, resulting in isolated n-type islands surrounded by p-type regions. By keeping the substrate potential in the lowest level, the pn junctions, surrounding the islands, are reverse-biased and the depletion layers are formed to electrically isolate each island from the others.

If this method is applied to high-voltage power ICs, a thick n-type epitaxial layer is required and deep isolation diffusions are necessary. Deep diffusion accompanies large lateral diffusion, ending up in a large

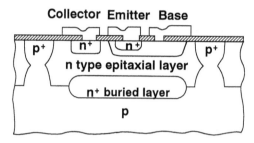

FIGURE 8.6   Junction isolation with upward isolation diffusions.

isolation area. One solution for this is to use buried p[+] diffusion layers for upward isolation diffusions, as shown in Fig. 8.6. However, 200 V is a practical limit for conventional pn junction isolation.

A variety of methods have been proposed to overcome this voltage limit. Figure 8.7 shows a typical example for this.[8] A shallow hole is formed where a high-voltage device is formed before the n-type epitaxial growth. This allows a locally thicker n-type epitaxial layer for high-voltage transistors.

Another distinguished example is shown in Fig. 8.8, where an n[+]-substrate is used in place of a p-type substrate. p-type and n-type epitaxial layers are subsequently formed. This example makes it possible to integrate a vertical DMOSFET with a backside drain contact with junction-isolated BiCMOS control circuits. This structure was proposed as "Smart Power" in the mid-1980s.

Impact of Dielectric Isolation

Dielectric isolation (DI) is a superior method for integrating many kinds of devices in a single chip. DI has many advantages[9–11] over junction-isolation techniques, including,

1. Virtually all integrated components can be treated as if they were discrete devices, so that circuit design becomes easy.
2. Bipolar devices, including thyristors, can be integrated without any difficulties.

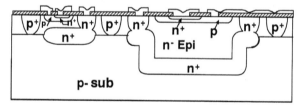

FIGURE 8.7   An example to overcome the junction isolation voltage limit.

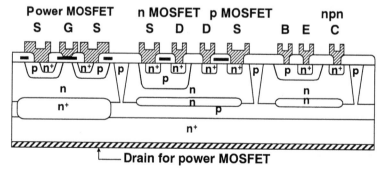

FIGURE 8.8   Junction isolation with VDMOSFET.

3. Coupling between two devices can be minimized, thus attaining better IC performances: no latch-up, high speed, large noise immunity, and ruggedness.
4. High-temperature operation is feasible because there are virtually no parasitics and leakage current is low.
5. Radiation hardness for space use.

Figure 8.9 shows a cross-section of the conventional DI, called EPIC. The crystalline silicon islands completely surrounded by silicon dioxide film are floating in the supporting substrate made of a thick polysilicon layer. The fabrication process of EPIC wafers is complicated and illustrated in Fig. 8.10. The problem with EPIC is the high cost of wafers and large wafer warpage. The development of the EPIC method was initiated by the early works of J.W. Lathlop et al.,[12] and J. Bouchard et al.,[13] in 1964. The EPIC method was first applied to high-speed bipolar ICs owing to its low parasitic capacitance.

Early work on high-voltage integrated circuits was triggered by the need for display drivers and high-voltage telecommunication circuits. Efforts to achieve high-voltage lateral MOSFETs started in the early 1970s, and the 800-V lateral MOSFET, using RESURF concept and DMOS (DSA[2]) technology, was developed for display drivers in 1976, before the RESURF concept was fully established.[14]

The need for high-voltage SLICs advanced the EPIC technology because it required electrically floating high-voltage bi-directional switches, which were realized only by the DI technique.

A variety of dielectric isolation methods, classified as silicon on insulator (SOI) technology, were invented in the 1970s. These are SOS (silicon on sapphire[15]), SIMOX,[16] and recrystallized poly-silicon such as ZMR.[17] And, silicon wafer direct-bonding (SDB)[6,7] was proposed in 1985.

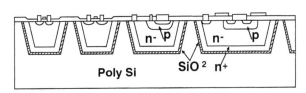

FIGURE 8.9   Dielectric isolation with EPIC technology.

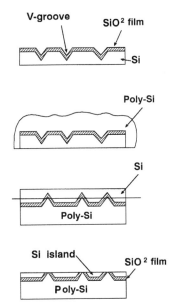

FIGURE 8.10   Fabrication process of EPIC wafers. A very thick poly-crystalline silicon layer is deposited on oxidized single-crystal silicon with a grooved surface. The crystalline silicon is grounded and polished so that the silicon layers isolate each other by the grooves.

The SOI wafer structure is simple. A single crystalline silicon layer is formed on the buried oxide layer or insulator substrate. Major methods are SIMOX and wafer bonding. SIMOX is a method that forms a buried oxide layer by a high dose of oxygen ion implantation and subsequent high-temperature annealing. Wafer bonding is a method that bonds an oxidized wafer and a substrate wafer at room temperature and strengthens the bond by annealing at high temperature. The thickness of the bonded SOI layer is adjusted by mechanical grinding and polishing.

In the late 1980s, MOS gate power device technology was greatly improved. In particular, the success of the MOS bipolar composite devices such as IGBTs[18,19] and MCTs[20] made it possible to control a large current by the MOS gate. The large current-handling capability of IGBTs has accelerated adopting DI with IGBT outputs.

In the early SLICs, double-injection devices with current control gates such as gated diodes and GTOs were used for such switches.[21] Recently developed SLICs (telecommunication ICs) have adopted lateral IGBTs or MOS gated thyristors because of the ease of gate drive. All the commercialized SLICs, so far, have adopted the conventional DI method. The success of SLIC was supported by the fact that monolithic integration and added function deserved expensive DIs for telecommunications application.

In the 1990s, wafer bonding technology was well established and low-cost SOI wafers were made available. A low-cost DI method realized by SOI technology, using several micron thick or less silicon layers, changed the situation of DI research and widened the application fields.

If the silicon layer is thin, devices in the SOI layer are isolated with narrow trenches. This makes SOI technology very attractive for high-voltage applications because chip size can be reduced and resultant chip cost is reduced. The SOI technology widened the application field of DI toward consumer use.

High-voltage SOI research work started in the early 1990s. Research efforts have been directed toward:

1. Monolithic device integration of multiple number of high-voltage, high-current devices with control circuits
2. ICs allowing high temperature operation and ruggedness
3. Low-cost DI power IC process development
4. High-current, high-speed, MOS-controlled lateral output devices with self-protection functions

8.3   High-Voltage Technology

It is quite important to realize a high breakdown voltage in an integrated device structure. There are two major techniques for high-voltage power ICs. These are the field plate and resurf techniques.

Field Plate

It is very important to realize a high breakdown voltage in a planar device structure. In other words, it is ideal if a one-dimensional pn junction breakdown voltage is realized in an actual pn junction, formed by thermal impurity diffusion. Actual pn junctions consist of cylindrical junctions and spherical junctions near the surface. Generally, the breakdown voltage of cylindrical or spherical junctions is significantly lower than that of an ideal 1-D planar junction, if junction curvature is small.

A *field plate* is a simple and frequently used technique to increase the breakdown voltage of an actual planar junction. Figure 8.11 shows an example. Field plates, placed on the thick-field oxide, induce depletion layers underneath themselves. The curvature of the formed depletion layers can be increased with the induced depletion layers, thereby relaxing the curvature effects of the field plate.

Resurf Technique

The *resurf* technique was originally proposed in 1979[14] as a method to obtain a high breakdown voltage in a conventional JI structure, where the breakdown voltage is limited by the thickness of the epitaxial layer. Figure 8.12 shows a high-voltage structure, where the depletion layer develops in the p-substrate and n-epitaxial-layer. If the epi-layer is thick or impurity doping is high (a), breakdown occurs before

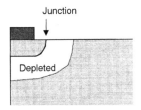

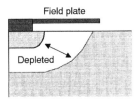

FIGURE 8.11 Field plate structure.

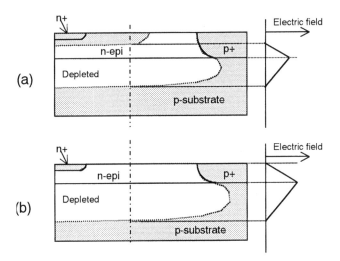

FIGURE 8.12 Resurf technique.

n-epi layer is completely depleted. If an appropriate epi-layer thickness is chosen (b), the epi-layer is completely depleted when breakdown occurs. The achieved breakdown voltage is very high because the depletion layer is sufficiently thick, both in lateral direction and vertical direction. The important point is that the total charge Q_c in the epi-layer is chosen so that the value satisfies the equation:

$$Q_c = \varepsilon\, E_c \qquad (8.1)$$

where E_c denotes critical electric field in silicon (3×10^5 V/cm). This charge can be depleted just when the electric field becomes E_c or breakdown occurs. In other words, the epi-layer is completely depleted just when breakdown occurs, if the total epi-layer dose is Q_c/q, which is approximately 2×10^{12}/cm².

8.4 High-Voltage Metal Interconnection

In high-voltage power ICs, there must be interconnection layers crossing high-voltage junctions. These high-voltage interconnection layers may cause degradation of the breakdown voltage of high-voltage

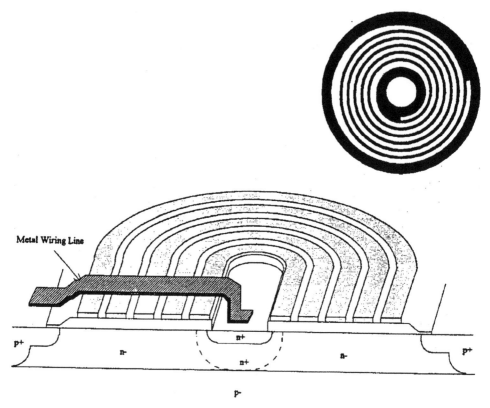

FIGURE 8.13 A method to shield the influence of the metal interconnection layer. (Copyright (1994) IEEE. With permission.)

devices. These problems are often solved with a thicker insulator layer under the interconnection layers. However, special means are required if the breakdown voltage is over 400 V.

Figure 8.13 shows one of the methods to shield the influence of metal interconnection layers on the underlying devices. A spiral-shaped high-resistance poly-silicon layer, connecting source and drain electrodes, effectively shields the influence of the interconnection layer on the depletion layer.[22] This is because the potential of the high-resistance poly-silicon layer is determined by small leakage current.

Another typical example is multiple floating field plates. The cross-section of the structure is similar to Fig. 8.13. The difference is that the poly-silicon forms multiple closed field rings, which are electrically floating each other. Multiple floating field plates also prevent breakdown voltage reduction due to metal interconnection.

8.5 High-Voltage SOI Technology

SOI power ICs are classified into two categories from the viewpoint of SOI wafer structure. The difference is whether there is a buried n+ layer on the buried oxide. Figure 8.14 shows a typical device structure, employing an n+ buried layer on the buried oxide. The breakdown voltage is determined with the thickness of the high-resistivity n-layer or the thickness of the depletion layer. The maximum breakdown voltage is limited to below 100 V because of SOI layer thickness or practically available trench depth. For this case, SOI wafers are used as a simple replacement for conventional DI wafers.

Figure 8.15 shows another typical SOI power IC structure, employing a high-voltage lateral IGBT. The n− drift layer is fully depleted by application of a high voltage. As the buried oxide and the depletion

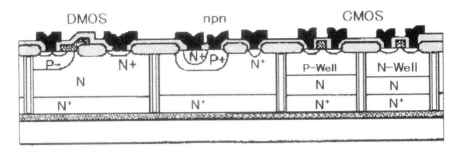

FIGURE 8.14 A high-voltage SOI device structure with n⁺ buried layer on the buried oxide.

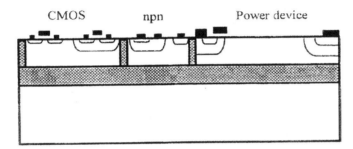

FIGURE 8.15 A high-voltage SOI device structure without n⁺ buried layer on the buried oxide.

layer both share the applied voltage, high breakdown voltage is realized in a relatively thin SOI. This type of power IC fully enjoys the features of SOI technology.

1. Complete device isolation by trench technique and small isolation region
2. Virtually no parasitic active component
3. A high breakdown voltage exceeding 500 V is realized by applying a large portion of the voltage across the thick buried oxide
4. Small wafer warpage and fine lithography is applicable
5. High-temperature operation is possible

   There are two big issues associated with high-voltage devices on SOI.[11] One is how to realize a high breakdown voltage under the influence of substrate ground potential. The other is how to attain a low on-resistance with a thin silicon layer. In the conventional DI, the wrap-around n⁺ region (see Fig. 8.14) is used in the DI island to prevent the influence of substrate potential on the device breakdown voltage. However, for thin silicon layers, this method cannot be used. The bottom silicon dioxide layer simply works as an undoped layer as far as the Poisson equation is concerned. Thus, a SOI layer on a grounded silicon substrate structure behaves in a way similar to the structure of a doped n-type thin silicon layer on undoped silicon layer (corresponding to silicon dioxide) on a grounded p silicon substrate. Thus, the SOI layer works in the same way as a resurf layer.

   A high breakdown voltage of a thin silicon layer device can be realized by sharing a large applied voltage with the buried dioxide film, whose breakdown field is far greater than that of silicon. The buried oxide film is able to sustain a large share of applied voltage, because the dielectric breakdown field is larger than that of silicon.

   Figure 8.16 shows a typical SOI diode structure and its potential distribution. It is seen that almost a half of the voltage is applied across the buried oxide. Figure 8.17 shows the electric field distribution along the symmetry axis of the diode of Fig. 8.16. The electric field in the oxide is larger than that in silicon because the following relation holds.

   The two electric field components $E_t(Si)$, $E_t(I)$, normal to the interface of the silicon and the bottom insulator layer, have the relation:

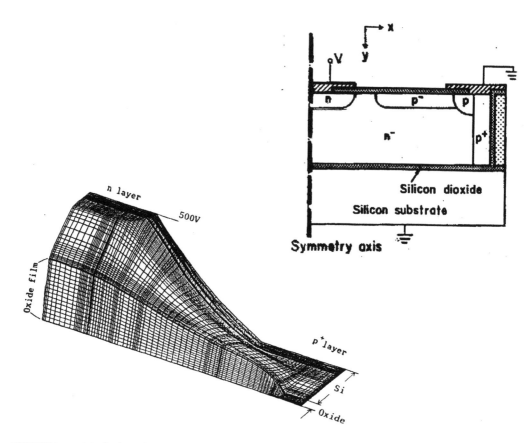

FIGURE 8.16   SOI diode and potential distribution in the diode.

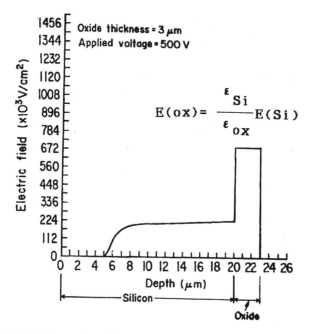

FIGURE 8.17   Electric field distribution along the symmetry axis of diode shown in Fig. 8.16.

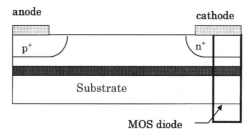

FIGURE 8.18 1-D MOS-diode structure in SOI diode.

$$\varepsilon(\text{Si})E_t(\text{Si}) = \varepsilon(\text{I})E_t(\text{I}) \qquad (8.2)$$

where $\varepsilon(\text{Si})$, $\varepsilon(\text{I})$ denote dielectric constants for silicon and silicon dioxide, respectively. Using an insulator film with a lower dielectric constant will increase the device breakdown voltage because the insulator layer sustains a larger share of the applied voltage.

For optimized SOI diodes, the breakdown voltage is substantially limited to the breakdown voltage of the 1-D MOS diode portion, as illustrated in Fig. 8.18, consisting of $n^+/n^-/\text{oxide/substrate}$. Figure 8.19 shows the measured SOI device breakdown voltage as a function of SOI layer thickness with buried oxide thickness as a parameter. The calculated breakdown voltage of 1-D MOS diodes are shown together. A 500-V breakdown voltage can be obtained with a 13-µm thick SOI with 3-µm thick buried oxide.

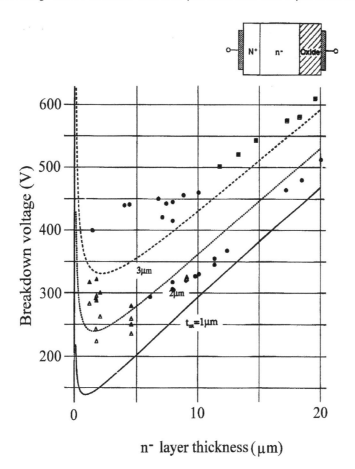

FIGURE 8.19 Measured and calculated SOI device breakdown voltage.

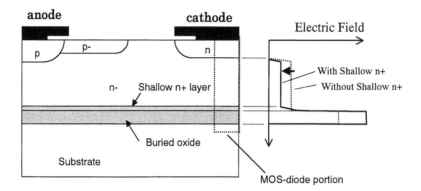

FIGURE 8.20   SOI diode with shallow n⁺ layer diffused from the bottom of the SOI layer.

It is very difficult to achieve a high breakdown voltage exceeding 600 V in simple SOI structures, because a thicker buried oxide layer of 4 µm or more is required. Maximum breakdown voltage is substantially limited with the breakdown voltage of the 1-D MOS diode and actually the realized breakdown voltage is lower than this limit. If the influence of the substrate potential can be shielded, it is possible to achieve a higher breakdown voltage in the SOI device.

A new high-voltage SOI device structure, free from the above constraints, was proposed in 1991,[1] that realizes a 1200-V breakdown voltage.[23]

To improve the breakdown voltage, an SOI structure with a shallow n⁺ layer diffused from the bottom of SOI layer was proposed.[24] Figure 8.20 shows the structure of an SOI diode with a shallow n⁺ layer and the electric field strength in the MOS diode portion compared to that without a shallow n⁺ layer. In general, if a larger portion of the applied voltage is carried with the bottom oxide layer, a higher breakdown voltage can be achieved. The problem is how to apply a higher electric field across the buried oxide without increasing the electric field strength in the SOI layer. This problem can be solved by placing a certain amount of positive charge on the SOI layer– buried oxide interface. The positive charge at the interface shields the high electric field in the buried oxide, so that a voltage across the oxide layer can be increased without applying a higher electric field in the SOI layer. The shallow n⁺ layer diffused from the bottom is a practical technique to place the positive charge on the SOI layer–buried oxide interface, as shown in Fig 8.20. The required dose of the shallow n⁺ layer is around $1 \times 10^{12} \text{cm}^{-2}$.

Very Thin SOI Case

Merchant et al.[25] showed that the SOI diode breakdown voltage is significantly enhanced if the SOI layer thickness is very thin, such as 0.1 µm. As shown in Fig. 8.19, reduction in the SOI layer thickness enhances the breakdown voltage if the thickness is less than 1 µm. This is because the carrier path along the vertical high electric field is as short as the SOI layer thickness, so that the carriers reach the top or bottom surface of the SOI layer before ionizing a sufficient amount of carriers for avalanche multiplication along the path. They proposed a combination of the very thin SOI layer and a linearly graded impurity profile of the n-type silicon layer for a high-voltage n⁺n⁻p⁺ lateral diode. A 700-V breakdown voltage was realized by this structure.

The exact 2-D simulations revealed that the ideal profile for a lateral diode is approximated by a function that is similar to a tangent function, as shown in Fig. 8.21. The important point is that the p-layer impurity profile should also be graded and that the linearly graded portion is terminated with the exponentially increasing ending portions. By using the proposed profile, a 5000-V lateral diode was predicted to be realized on 0.1-µm SOI on a 600-µm thick quartz substrate. A completely uniform lateral electric field is realized at 5000 V (see Fig. 8.21).

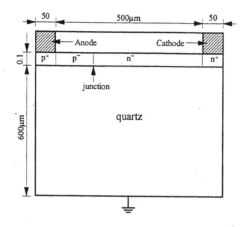

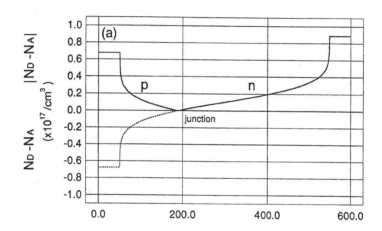

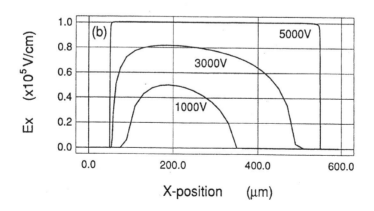

FIGURE 8.21 Calculated ideal profile for a lateral diode on thin SOI.

8.6   High-Voltage Output Devices

High-voltage output devices are the most important part of power ICs. An entire power system can be integrated on a single silicon chip if high-voltage power devices can be integrated with analog and digital circuits as well as MPUs. Recently, MOS gate power devices were adopted, primarily because of the low on-resistance and the ease of gate control. These are DMOSFETs and IGBTs.

Lateral Power MOSFET

pn junction-isolated power ICs are frequently used for low-voltage applications, where DMOS is the primary choice for output devices. Since the reliability of junction isolation is not sufficient, SOI DMOS power ICs will be used where high reliability is required. In this section, DMOS electrical characteristics are described, using mostly the junction-isolated DMOS data.

For above a 60-V breakdown voltage range, the vertical DMOS structure with upside surface drain contact (up-drain DMOS, see Fig. 8.22) has conventionally been used. However, recently, the lateral DMOS (LDMOS) structure (Fig. 8.23) tends to be used for the entire voltage range. This is because the LDMOS on-resistance can be directly improved by adopting finer lithography. On the other hand, up-drain vertical DMOS on-resistance includes the resistances of the buried n^+ layer and sinker plug diffusions, which are not improved by finer lithography.

Figure 8.24 shows state-of-the-art DMOS on-resistance as a function of breakdown voltage. The figure also shows state-of-the-art on-resistance for vertical discrete trench MOSFETs as a comparison. Black circles show lateral DMOS, and open squares show trench MOSFETs. Recently, battery-operated mobile equipment and computer peripherals have opened a large applications area, and lateral MOSFETs of the less than 60 V are the major output devices. It is astonishing that the state-of-the-art on-resistances of lateral DMOS and vertical trench MOSFETs are almost the same. This implies that power ICs with a

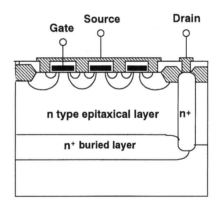

FIGURE 8.22   Vertical DMOS structure with upside surface drain contact.

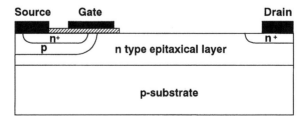

FIGURE 8.23   Lateral DMOS (LDMOS) structure.

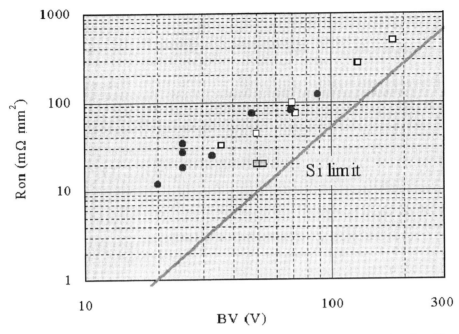

FIGURE 8.24 State-of-the-art lateral DMOS and vertical trench MOSFET on-resistance as a function of breakdown voltage (black circles show lateral DMOS, and open squares show trench MOSFETs).

vertical DMOS output will be replaced by power ICs with a lateral DMOS output, if current capacity is small — for example, less than 10 A.

High-side switching operation is an important function in automotive applications, especially in case of H-bridges for motor control. The on-resistance of conventional junction-isolated, high-voltage MOS-FETs, shown in Fig. 8.23, is significantly influenced by the source to substrate bias,[26] because the drift layer is depleted. However, in the SOI MOSFETs shown in Fig. 8.25, the drift layer is not depleted, but a hole inversion layer is formed. Thus, the substrate bias influence of SOI LDMOS is small.[26]

Figure 8.26 shows a 60-V DMOS in a 2-μm thick p-type SOI. The fabrication process is completely compatible with the CMOS process. The threshold voltage is controlled by channel implant. The experimentally obtained specific on-resistance of the 60-V LDMOS is 100 mΩ · mm.[2] The developed power MOSFET is completely free from substrate bias influence.[27] This is because the hole accumulation layer is induced on the buried oxide, leaving the n-drift layer unchanged. These results indicate that this device can be used for high-side switches without on-resistance increase.

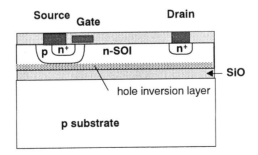

FIGURE 8.25 SOI MOSFET at high-side operation.

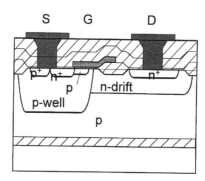

FIGURE 8.26 60-V DMOS in 5-μm thick p-type SOI.

Lateral IGBTs on SOI

IGBTs are suitable for high-voltage, medium current power ICs because of large current capability, as shown in Fig. 8.27 IGBT can be recognized as a pnp transistor driven by an n-channel MOSFET for a first order approximation. IGBTs should be fabricated by conventional CMOS-compatible processes so that conventional CMOS circuit libraries can be utilized without changes.

The switching speed of bipolar power devices is conventionally controlled by introduction of a lifetime killer. However, the lifetime control process is not compatible with a conventional CMOS process. There are two ways to control the switching speed of power devices. One way is to use thin SOI layers. The switching speed of IGBTs improves as the SOI thickness decreases,[28] because the carrier lifetime is effectively decreased by the influence of large carrier recombination at the silicon dioxide interfaces.[29] The other way is to reduce the emitter efficiency of the p$^+$ drain or collector. The effective methods are (1) emitter short, (2) low-dose emitter,[30] (3) high-dose n buffer, and (4) forming an n$^+$ layer in the p$^+$ emitter.[31]

Figure 8.28 shows a cross-section of large current lateral IGBTs. Large current capability has been realized by adopting multiple surface channels. Figure 8.29 shows typical current voltage curves of the

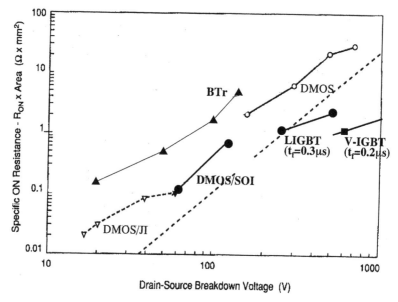

FIGURE 8.27 On-resistance versus breakdown voltage for bipolar transistor, DMOS, LIGBT, and VIGBT (discrete IGBTs).

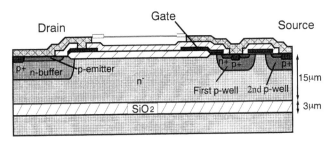

FIGURE 8.28 Cross-section of large current lateral IGBT.[30] (Copyright (1997) IEEE. With permission.)

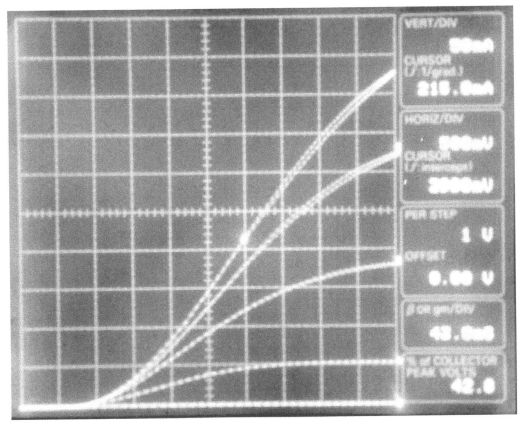

FIGURE 8.29 Typical current-voltage curves of a multi-channel LIGBT (Vertical scale: 50 mA/Div, Horizontal scale: 0.5 V/Div).

multi-channel LIGBT. The current is an exponential function of the drain bias (collector bias) for the low-voltage range. The current-voltage curves seem to have a 0.8-V offset voltage, just like a diode. The typical switching speed of the developed LIGBTs is 300 ns.

It is extremely important to increase the operating current density of LIGBTs in order to reduce chip size. This is because output devices occupy most of the chip area and the cost of the power ICs depends significantly on the size of the power devices. The current density of the developed LIGBT is 175 A/cm^2 for 3-V forward voltage.

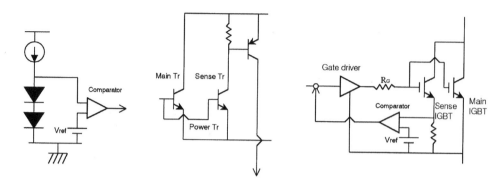

(a) Over temperature sense circuit   (b) Over current sense circuit for bipolar Tr.   (c) Over current sense circuit for IGBT(MOSFET)

FIGURE 8.30 Sense circuits for over-temperature and over-current.

8.7 Sense and Protection Circuit

Power ICs have a significant advantage over discrete devices in terms of power switch protection because the protection circuit can be integrated on the same chip. Figure 8.30(a) shows the over-temperature sense circuit. In this circuit, the junction temperature dependence of diode forward voltage drop is utilized to sense the temperature. The series diodes are located close to the power switches so that the junction temperature of the diodes responds to the temperature at the power switches. Comparing the forward voltage drop across the diodes with the reference voltage, the circuit senses the over-temperature at the power switches and the fault signal is fed back to the gate control logic.

Figure 8.30(b) shows the over-current sense circuit for bipolar power transistors. The magnitude of the current through the main transistor is reflected to the current through the sense transistor with the current mirror circuit configuration. Therefore, the voltage drop across the resistor in the collector of the sense transistor is proportional to the current through the main transistor. Comparing the voltage drop with the reference voltage, the circuit detects the over-current. Figure 8.30(c) shows the over-current sense circuit for IGBTs or MOSFETs, which has a similar configuration with Fig. 8.30(b). In this circuit, the over-current is detected by the voltage drop across the resistor in the sense IGBT (MOSFET) emitter (source).

In short-circuit protection, the collector current should be squeezed or terminated as soon as the short-circuit operation is detected. For this purpose, the protection circuit directly draws down the gate voltage with the short feedback loop. Figure 8.31 shows a typical short-circuit protection circuit for IGBTs. When high current flows through the IGBT, the voltage drop across the emitter resistor of the sense IGBT directly drives the npn transistor. The transistor draws down the gate voltage so that the collector current is reduced to the level of a safe turn-off operation.

8.8 Examples of High-Voltage SOI Power ICs with LIGBT Outputs

Current main applications of high voltage SOI power ICs are dc motor control and flat-panel display drivers. Recently, technologies for color plasma display panels have been greatly improved, and demands for PDP driver ICs have increased. There are several reports[32,33] on the development of such ICs using SOI wafers. Flat-panel display drivers have to integrate a large number of high-voltage devices. Trench isolation and LIGBTs are key techniques for reducing chip size and resultant cost.

Another large market is the motor control field. Home-use appliances use a number of small motors, which are directly controlled by a ac source line. Single-chip inverter ICs are able to reduce system size and increase system performance. Figure 8.32 shows a 500-V, 1-A, single-chip inverter IC[34] for dc brushless motors. It integrates six 500-V, 1-A LIGBTs, six 500-V diodes, control protection, and logic circuits.

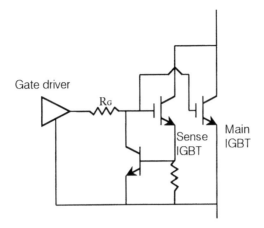

Short circuit protection

FIGURE 8.31 Typical short-circuit protection method for IGBTs.

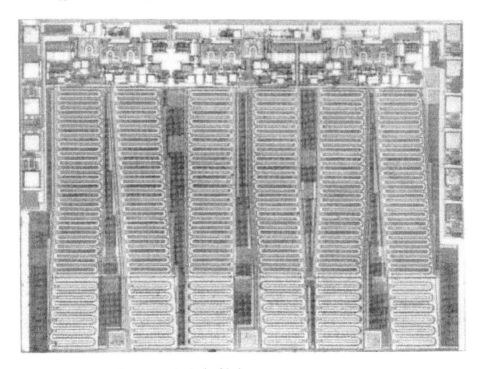

FIGURE 8.32 Photograph of a 500-V, 1-A, single-chip inverter.

8.9 SOI Power ICs for System Integration

Another prospective application of SOI technology is in the automotive field, which requires large current DMOS outputs. Conventional pn junction-isolated power ICs are frequently used for these applications; however, the reliability of junction isolation is not sufficient. SOI DMOS power ICs will be used where high reliability is required.

For less than 100-V applications, the required thickness of the buried oxide layer is less than 1 μm. The warpage of the SOI wafers is very small; thus, fine lithography can be applied. The same CMOS

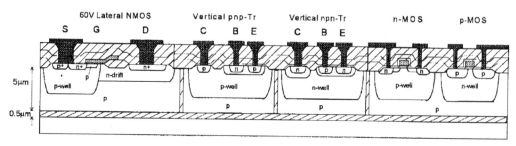

FIGURE 8.33 SOI power IC system integration.

circuit library can be used without changes because the same CMOS fabrication process can be applied without modification if a relatively thick SOI layer is used.

This section shows the possibility of integration of an MPU, together with BiCMOS analog circuits and 60-V power LDMOS. 4-bit MPUs, vertical npn, pnp, and 60-V power DMOS were fabricated on 2-μm SOI wafers by a conventional 0.8-μm BiCMOS process.[27] The 60-V DMOS used CMOS p-well without using self-alignment.

The fabricated 4-bit MPU, consisting of 30,000 FETs for the core, 6000 FETs for the cache, and 120,000 FETs for the ROM, operated at a 20% faster clock speed of 50 MHz at 25°C, as compared to 42 MHz of the bulk version MPU, and even operated at over 200°C. It was found that the clock speed could be improved and that a large latch-up immunity at high temperature was realized even if the MOSFETs were not isolated by trenches. The maximum operating temperature was more than 300°C. It was found that the yield of the MPU fabricated on SOI was the same as that on bulk wafers, verifying that the crystal quality of the currently available SOI wafers was sufficiently good. It was also found that both SOI and bulk MPUs could be operated at 300°C if MPUs consisted of pure CMOS, although the power consumption of the bulk MPU was larger than that of the SOI MPUs.

One of the characteristic features of the SOI power IC structure, shown in Fig. 8.33, is that there are no buried layers for bipolar transistors. It was found that vertical npn and pnp transistors fabricated on the n-well and p-well layers exhibited sufficiently good characteristics, and the typical current gains h_{FE} for the vertical npn and pnp transistors were 80 and 30, respectively.

All these results show that system integration including power LDMOS will be a reality in SOI wafers.

8.10   High-Temperature Operation of SOI Power ICs

The leakage current of SOI devices simply reduces as the SOI layer becomes thinner, as seen in Fig. 8.34.[35] Small leakage current enables high-temperature operation of SOI power ICs. It was experimentally shown that IGBTs can be operated at a switching frequency of 20 kHz at 200°C if they are fabricated in thin SOI of less than 5 μm. The maximum operating temperature of analog circuits in SOI increases as the SOI layer becomes thinner. Figure 8.35 shows the output voltage of bandgap reference circuits as a function of temperature with SOI thickness as a parameter. In the circuits, each device was not trench-isolated. If all the devices are trench-isolated, much higher temperature operation can be expected.

The 200°C operation of 250-V, 0.5-A, three-phase, 1-chip inverter ICs fabricated in a 5-μm thick SOI layer was demonstrated.[36] CMOS circuits on SOI were found to be capable of operating at 300°C. CMOS-based analog circuits with a minimum number of bipolar transistors were adopted. It was found that the bandgap reference circuit operated at 250°C, which was higher than that expected from Fig. 8.34. This was probably because each device was trench-isolated. A DC brushless motor was successfully operated by the single-chip inverter IC at 200°C. The carrier frequency was 20 kHz.

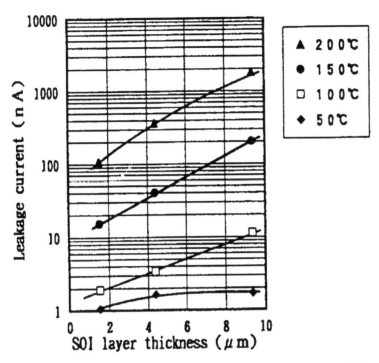

FIGURE 8.34 Leakage current vs. SOI layer thickness. (Copyright (1994) IEEE. With permission.)

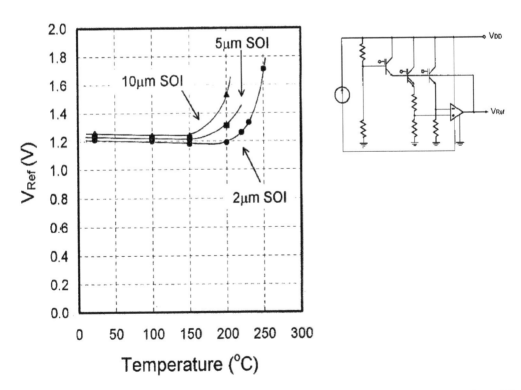

FIGURE 8.35 Output voltage of a bandgap reference circuit as a function of operation temperature. (Copyright (1995) IEEE. With permission.)

References

1. Nakagawa, A., "Impact of dielectric isolation technology on power ICs (Invited)," *Proc. of ISPSD*, 1991, 16.
2. Tarui, Y., Hayashi, Y., and Sekigawa, T., "Diffusion self-aligned MOST: a new approach for high speed device," *Proc. of the 1st Conference on Solid State Devices*, Tokyo, 1969, 105.
3. Wrathall, R. S., Tam, D. T., Terry, L., and Rob, S. P., "Integrated circuits for the control of high power," *IEDM Tech. Digest*, 1983, 408.
4. Murari, B., Bertotti, F., and Vignola, G. A., *Smart Power ICs*, Springer-Verlag, 1995.
5. Beasom, J. D., "A process for simultaneous fabrication of vertical npn and pnp's and p-ch MOS devices," *IEDM Tech. Digest*, 1973, 41.
6. Shimbo, M., Furukawa, K., Fukuda, K., and Tanzawa, K., "Silicon-to-silicon direct bonding method," *J. Appl. Phys.*, vol. 60, p. 2987, 1986.
7. Ohashi, H., Ohura, J., Tsukakoshi, T., and Shimbo, M., "Improved dielectrically isolated device intergation by silicon wafer-wafer direct bonding (SDB) technique," *IEDM Tech. Digest*, 1986, 210.
8. Okabe, T., Sakamoto, K., and Hoya, K., "Semi-well isolation-based intelligent power IC technology," *Proc. of ISPSD*, 1988, 96.
9. Becke, H. W., "Approaches to isolation in high voltage integrated circuits (invited paper)," *IEDM Tech. Digest*, 1985, 724.
10. Rumennik, V., "Power devices are in the chips," *IEEE SPECTRUM*, July 1985, 42.
11. Nakagawa, A., Yasuhara, N., and Baba, Y., "New 500V output device structure on silicon oxide film," *Proc. of ISPSD*, 1990, 97.
12. Lathrop, J. W., "Semiconductor-networking technology-1964," *Proc. IEEE*, vol. 52, p. 1430, 1964.
13. Bouchard, J. and Hammmond, F. W., Abstract No. 165, "The iso-layer process," *J. Electrochem. Soc.*, vol. 111, p. 197C, 1964.
14. Appels, J. A. and Vaes, H. M. J., "High voltage thin layer devices (RESUF DEVICES)," *IEDM Tech. Digest*, 1979, 238.
15. Rosen, R. S., Sprinter, M. R., and Tremain, R. E. Jr., "High voltage SOS/MOS devices and circuit elements: design, fabrication, and performance," *IEEE J. Solid State Circuits*, vol. SC-11, p. 431, 1976.
16. Izumi, Doken, M., and Ariyoshi, H., "C.M.O.S. devices fabricated on buried SiO2 layers formed by oxygen implantation into silicon," *Electron. Lett.*, vol. 14, p. 593, 1978.
17. Geis, M. W., Flanders, D. C., and Smith, H. I., "Crystallographic orientation of silicon on an amorphous substrate using an artificial surface-relief grating and laser crystallization," *Appl. Physics Lett.*, vol. 35, p. 71, 1970.
18. Baliga, B. J., Adler, M. S., Gray, P. V., and Love, R. P., "The insulated gate rectifier (IGR) a new switching device," *IEDM Tech. Digest*, 1982, 264.
19. Nakagawa, A., Ohashi, H., Kurata, M., Yamaguchi, H., and Watanabe, K., "Non-latch-up 75A bipolar-mode MOSFET with large ASO," *IEDM Tech. Digest*, 1984, 860.
20. Temple, V. A. K., "MOS controlled thyristor," *IEDM Tech. Digest*, 1984, 282.
21. Kamei, T., "High voltage integrated circuits for telecommunication," *IEDM Tech. Digest*, 1981, 254.
22. Endo, K., Baba, Y., Udo, Y. Yasui, M., and Sano, Y., "A 500A 1-chip inverter IC with new electric field reduction structure," *Proc. of ISPSD*, 1994, 379.
23. Funaki, H., Yamaguchi, Y., Hirayama, K., and Nakagawa, A., "New 1200V MOSFET structure on SOI with SIPOS shielding layer," *Proc. of ISPSD*, 1998, 25.
24. Yasuhara, N., Nakagawa, A., and Furukawa, K., "SOI device structure implementing 650V high voltage output devices on VLSIs," *IEDM Tech., Digest*, 1991, 141.
25. Merchant, S., Arnold, E., Baumgart, H., Mukherjee, S., Pein, H., and Pinker, R., "Realization of high breakdown voltage (>700V) in thin SOI devices," *Proc. of ISPSD*, 1991, 141.
26. Arnold, E., Merchant, S., Amato, M., Mukherjee, S., and Pein, H., "Comparison of junction-isolated and SOI high-voltage devices operating in the source-follower mode," *Proc. of ISPSD*, 1992, 242.

27. Funaki, H., Yamaguchi, Y., Kawaguchi, Y., Terazaki, Y., Mochizzuki, H., and Nakagawa, A., "High voltage BiCDMOS technology on bonded 2μm SOI integrating vertical npn, pnp, 60V-LDMOS and MPU, capable of 200°C operation," *IEDM Tech. Digest*, 1995, 967.

28. Yasuhara, N., Matsudai, T., and Nakagawa, A., "SOI thickness and buried oxide thickness dependencies of high voltage lateral IGBT switching characteristics," *Ext. Abstr. Int. Conf. SSDM*, 1993, 270.

29. Omura, I., Yasuhara, N., Nakagawa, A., and Suzuki, Y., "Numerical analysis of switching characteristics — switching speed enhancement by reducing the SOI thickness," *Proc. of ISPSD*, 1993, 248.

30. Funaki, H., Matsudai, T., Nakagawa, A., Yasuhara, N., and Yamaguchi, Y., "Multi-channel SOI lateral IGBTs with large SOA," *Proc. of ISPSD*, 1997, 33.

31. Yamaguchi, Y., Nakagawa, A., Yasuhara, N., Watanabe, K., and Ogura, T., "New anode structure for high voltage lateral IGBTs," *Ext. Abst. Int. Conf. SSDM*, 1990, 677.

32. Gonzalez, F., Shekhar, V., Chan, C., Choy, B., and Chen, N., "Fabrication of a 300V high current (300mA/output), smart-power IC using gate-controlled SCRs on bonded (BSOI) technology," *IEDM Tech. Digest*, 1995, 473.

33. Sumida, H., Hirabayashi, H., Shimabukuro, H., Takazawa, Y., and Shigeta, Y., "A high performance plasma display panel driver IC using SOI," *Proc. of ISPSD*, 1998, 137.

34. Nakagawa, A., Funaki, H., Yamaguchi, Y., and Suzuki, F., "Improvements in lateral IGBT design for 500V 3A one chip inverter ICs," *Proc. of ISPSD*, 1999, 321.

35. Matsudai, T., Yamaguchi, Y., Yasuara, N., Nakagawa, A., and Mochizki, H., "Thin SOI IGBT leakage current and a new device structure for high temperature operation," *Proc. of ISPSD*, 1994, 399.

36. Yamaguchi, Y., Yasuhara, N., Matsudai, T., and Nakagawa, A., "200°C High temperature operation of 250V 0.5A one chip inverter ICs in SOI," *Proc. of PCIM INTER'98*, Japan, 1998, 1.

9
Noise in VLSI Technologies

Samuel S. Martin
Thad Gabara
Kwok Ng
Bell Laboratories
Lucent Technologies

9.1 Introduction

The progress of VLSI technologies is a result of an intimate interaction between improvements in IC chip design and in device properties of the underlying process technologies. The semiconductor roadmap following Moore's law is responsible for an exponential decrease of minimum feature size of devices. The associated increase of device speed and decrease of supply voltage have strong implications for the available noise margin of a VLSI chip. This chapter addresses the main issues relevant for noise in VLSI technologies at various levels as indicated schematically in Fig. 9.1. At the microscopic level, the fundamental sources of noise associated with carrier transport are derived with emphasis on semiconductors. The next level deals with the noise properties of active devices and passive components. Three classes of active devices are chosen for illustration: bipolar junction transistors (BJTs), field effect transistors (FETs), and two terminal junction devices (diodes). Although the treatment of these device classes implies silicon-based technologies (Si-BJT, Si-MOSFET, Si-diode), it can generally be applied to other device classes as well (HBT, MESFET, etc.). Finally, the chip level noise is presented in terms of the major contributions being amplifier noise, oscillator noise, timing jitter, and interconnect noise. The evolution of VLSI technologies into the deep-sub-micron regime has significant implications for the treatment of fundamental noise mechanisms, which are briefly outlined in the last section of the chapter.

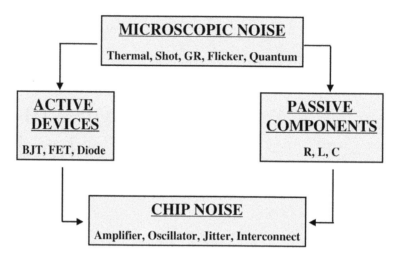

FIGURE 9.1   Schematic describing the various levels of noise studied in this chapter and their causal relationship.

9.2   Microscopic Noise

Thermal Noise

Thermal noise is, in general, associated with random motion of particles in a force-free environment. Since the mean available energy per degree of freedom is proportional to temperature, the resulting noise is referred to as thermal noise. Specifically, carrier transport in semiconductors is treated by considering energy distribution functions, such as the Fermi–Dirac distribution. The effect of local fields and scattering of carriers in position and momentum space is described by a change in the distribution function. This is expressed by the Boltzmann transport equation. Noise is a stochastic process and is characterized in terms of time-averaged quantities, as the instantaneous value cannot be predicted. The fundamental origins of thermal noise in semiconductors are microscopic velocity fluctuations. Velocity fluctuations occur due to various mechanisms, such as electron-phonon scattering, electron-electron scattering, etc. Since the same mechanisms give rise to macroscopic resistivity of a material, noise can be viewed as the microscopic property of transport. Consequently, noise is intimately related to bulk transport in a semiconductor. The voltage noise spectral density is given by:

$$S_{\Delta v}(\omega) = 2 \cdot \int_{-\infty}^{\infty} \langle \Delta v(t)\Delta v(t+\tau)\rangle\, e^{j\omega\tau} d\tau \tag{9.1}$$

In the above expression, $\Delta v(t)$ denotes the instantaneous velocity fluctuation, ω the frequency, and S the spectral density of noise. Velocity fluctuations can also be viewed as the driving force for diffusion, where the diffusion coefficient is given by:

$$D(\omega) = \int_{0}^{\infty} \langle \Delta v | \Delta v(t+\tau)\rangle\, e^{-j\omega\tau} d\tau \tag{9.2}$$

Therefore, the voltage noise spectral density can be directly related to the real part of the frequency-dependent diffusion function as:

$$S_{\Delta v}(\omega) = 4 \cdot Re[D(\omega)] \tag{9.3}$$

The current noise spectral density due to a charge q in a slab of material of length L can be derived from the carrier velocity fluctuation as:

$$S_{\Delta i}(\omega) = \frac{q^2}{L^2} \cdot S_{\Delta v}(\omega) \tag{9.4}$$

The current noise spectral density of the whole slab of material of area A is then obtained by integrating the carrier density n(x) and the diffusion coefficient over the length of the sample, as given by:

$$S_I(\omega) = \frac{4q^2 A}{L^2} \cdot \int_0^L n(x) Re[D(\omega, x)] dx \tag{9.5}$$

In the limit of uniform carrier density, thermal equilibrium, and neglecting quantum effects, the diffusion coefficient can be expressed in terms of the carrier mobility using the Einstein relation:

$$D = \frac{kT}{q}\mu \tag{9.6}$$

and the current noise spectral density is given by:

$$S_I(\omega) = 4kT \cdot \left(nq\mu \cdot \frac{A}{L}\right) \tag{9.7}$$

The above expression leads to the well-known Johnson thermal noise of a conductance $G = \sigma A/L$, where the conductivity is $\sigma = nq\mu$ and the carrier mobility is denoted by μ:

$$S_I(\omega) = 4kT \cdot G(\omega) \tag{9.8}$$

In general, the conductance is the real part of the frequency-dependent admittance of the material as indicated above. Note that the thermal noise expression given here is valid only when the mean scattering time of the carriers is negligible with respect to the inverse of the frequency at which the noise is measured. This assumption is true for most studies of semiconductors where the mean scattering times are on the order of 1 ps. The generalized expression for finite scattering times is discussed in a later section. The thermal noise expression also provides a direct relation between the microscopic noise and the bulk resistance (conductance), indicating the nature of noise as being due to the same type of scattering mechanisms as those causing resistance in a sample. A prominent feature of thermal noise is its direct dependence on temperature, giving rise to the possibility of measuring a thermodynamic quantity with noise.

Shot Noise

The discrete nature of particles undergoing an average flow in space, together with their velocity distribution, results in a random arrival at a fixed plane of incidence. The corresponding fluctuations of the flow are referred to as *shot noise*. Specifically, the discrete flow of charge in a field yields shot noise of current in semiconductors. Shot noise can be derived as being due to a random train of pulses corresponding to events, such as carrier emission from an electrode, carrier injection across a semiconductor junction, etc. The power spectral density of the resulting time-dependent waveform is given by:

$$S(\omega) = 2\nu \cdot \langle a^2 \rangle \cdot |F(\omega)|^2 \qquad (9.9)$$

Here, $F(\omega)$ is the Fourier transform of the impulse function, ν the mean rate of the events, and $\langle a^2 \rangle$ the mean square amplitude of the pulses. For a current pulse $\delta i(t)$ due to the transit of a charge q through a sample, the Fourier transform is given by:

$$F(\omega) = \frac{1}{q} \cdot \int_0^{\tau_T} \delta i(t) \cdot e^{-j\omega\tau} dt \qquad (9.10)$$

At sufficiently low frequencies, the transit time τ_T can be neglected with respect to $1/\omega$, and the current pulse can be considered a delta function. The Fourier transform function then becomes unity. The mean rate of current pulses is $\nu = I/q$, where I is the mean value of the current, and the mean pulse amplitude $\langle a \rangle = q$, so that the power spectral density of shot noise current is given by:

$$S_I(\omega) = 2qI \qquad (9.11)$$

Note that the above expression is valid only for negligible transmit times of the carriers through a sample region with respect to the inverse of the frequency at which the noise is measured. The shot noise for non-negligible transit times and scattering times is treated in more detail later. One of the most significant properties of shot noise is its dependence on the mean current flow or bias in a semiconductor, in contrast to thermal noise which is dependent only on the temperature and resistivity of the material. Shot noise can also be measured only when the energy distribution function of the carriers can be probed by a reference plane of incidence, such as a p-n junction in a semiconductor.

Generation-Recombination Noise

The generation and recombination of charge carriers due to traps in semiconductors results in fluctuations of the current flow through the semiconductor. The temporal change in the carrier number N from a rate of generation g(t) and a rate of recombination r(t) is given by:

$$\frac{d(\Delta N)}{dt} = -\frac{\Delta N}{\tau} + \Delta g(t) - \Delta r(t) \qquad (9.12)$$

Here, τ is the mean lifetime of the carriers. The noise spectral density of the carrier number fluctuations is found to be:

$$S_N(\omega) = 4 \langle \Delta N^2 \rangle \cdot \frac{\tau}{1 + \omega^2 \tau^2} \qquad (9.13)$$

A current fluctuation is related to a carrier number fluctuation through $\Delta I/I = \Delta N/N$. The current noise spectral density from generation and recombination of carriers is then given by:

$$S_I(\omega) = 4 \frac{I_0^2}{N_0^2} \langle \Delta N^2 \rangle \cdot \frac{\tau}{1 + \omega^2 \tau^2} \qquad (9.14)$$

Here, I_0 denotes the mean current and N_0 the mean carrier number. The form of the above expression implies a constant noise power below a characteristic frequency $\omega_c = 1/\tau$, and a $1/\omega^2$ dependence (Lorentzian) at higher frequencies. This type of noise is observed in systems with a single well-defined trapping time constant or energy level of traps. The most significant feature of GR noise is its explicit frequency dependence, in contrast to the "white" spectral densities of thermal and shot noise. This frequency dependence has strong implications for semiconductors, since the characteristic frequency is well within the frequency range of most applications.

Flicker Noise

Flicker noise is, in general, a phenomenon observed in many systems with an inverse frequency dependence of the noise spectral density over a wide frequency regime. In semiconductors, the presence of energy traps can lead to generation and recombination of carriers and a corresponding frequency-dependent noise of the flicker noise type. A phenomenological approach for deriving the frequency dependence of flicker noise is to consider the expression for generation-recombination noise given by:

$$S_I^\tau(\omega) = A(I) \cdot \frac{\tau}{1 + \omega^2 \tau^2} \tag{9.15}$$

Here, $A(I)$ summarizes the current-dependent pre-factor given earlier. A distribution of characteristic time constants given by a probability density $P(\tau)d\tau$ would result in a current noise spectral density of the form:

$$S_I(\omega) = A(I) \cdot \int_{\tau_0}^{\tau_1} \left(\frac{\tau}{1 + \omega^2 \tau^2} \cdot P(\tau)d\tau \right) \tag{9.16}$$

In the case of MOSFETs, the tunneling of carriers from the channel through an oxide to a trap at a distance x from the channel/oxide interface produces a probability distribution of time constants:

$$P(\tau) \propto \frac{1}{\tau} \tag{9.17}$$

Such a distribution of time constants yields the flicker noise expression for the current noise spectral density:

$$S_I(\omega) \propto \frac{A(I)}{\omega} \tag{9.18}$$

Note that the inverse frequency dependence of the noise is obtained from the integral above with the approximation: $1/\tau_1 < \omega < 1/\tau_0$. The most important characteristic of flicker noise is its explicit inverse frequency dependence, like GR noise. Another feature is the dependence of flicker noise magnitude on material properties such as trap densities, carrier mobility, etc. Consequently, no general form can be given for the flicker noise of semiconductors, but specific devices need to be considered.

Although the overall frequency dependence of flicker noise is universally found to be inversely proportional to frequency in semiconductors, the expression for the amplitude of flicker noise is strongly dependent on device technology. In general, for Si-BJTs and Si-MOSFETs, the flicker noise power is found to increase with bias current according to a power-law as given by:

$$S_I(\omega) = K \cdot \frac{I^\gamma}{\omega^\alpha} \tag{9.19}$$

The bias exponent γ has typical values between 1 and 2, and the frequency exponent α varies around unity by about 20%.

Quantum Limit

Quantum effects become relevant for noise processes for sufficiently high frequencies and low temperatures, such that a quantum of energy is comparable to or larger than the thermal energy:

$$\hbar\omega \geq kT \qquad (9.20)$$

Consequently, the treatment of thermal or shot noise power requires modifying the mean available energy for each degree of freedom, which results in the following generalized expression for the spectral density of current noise:

$$S_I^Q(\omega) = S_I^C(\omega) \cdot \left(\frac{1}{2} \frac{\hbar\omega}{kT} + \frac{\hbar\omega/kT}{[e^{\hbar\omega/kT} - 1]} \right) \qquad (9.21)$$

Here, the quantum correction (denoted by Q) to the classical noise power (denoted by C) is given by the expression in the parentheses, which includes both the Planck distribution function and the zero-point energy term. In the limit of low frequencies or high temperatures ($\hbar\omega \ll kT$), the expression in parentheses reduces to unity and the classical form for the noise is recovered.

A further implication of quantum effects is a fundamental lower limit on the noise power of a linear amplifier, which is given by:

$$\frac{\Delta P_{min}}{\Delta f} = \frac{(G-1)}{G} \cdot \hbar\omega_0 \qquad (9.22)$$

Here, P denotes the noise power G of the amplifier gain and Δf is a frequency interval centered around the signal frequency ω_0. This limit on the noise is a result of the uncertainty principle applied to the number of photons and the corresponding phase of an electromagnetic wave.

9.3 Device Noise

The microscopic noise mechanisms described in the previous section are responsible for producing macroscopic noise in active semiconductor devices as well as passive components. This section briefly outlines the major noise contributions in several device technologies relevant for IC design. The intrinsic noise sources are related to measurable noise parameters and to properties suitable for device modeling.

Passive Components

The most relevant contribution to noise from passive components is the thermal noise of a resistance R given by:

$$S_v^t = 4kTR \qquad (9.23)$$

The thermal noise is determined by the magnitude of the resistance and its equilibrium temperature only. It exhibits a "white" (constant) noise spectral density as a function of frequency and becomes frequency dependent at high frequencies, either due to the reactive components of a real resistor or fundamentally due to non-equilibrium effects.

Besides the thermal noise of resistors, the reactive components (inductors and capacitors) also play a major role in affecting noise in VLSI chips. The ideal reactive components are not associated with any noise sources. However, non-ideal properties such as series resistance in inductors and leakage currents in capacitors give rise to thermal and shot noise, respectively. Moreover, the reactive components introduce the frequency dependence of noise or cause phase shifts between voltage and current noise sources.

Finally, the thermal noise bandwidth of a resistance R or a conductance G is limited by the series inductance L or the shunt capacitance C, respectively. The resulting integrated thermal noise powers are given by:

$$\langle i_n^2 \rangle \; = \; \frac{kT}{L} \tag{9.24a}$$

$$\langle v_n^2 \rangle \; = \; \frac{kT}{C} \tag{9.24b}$$

Here, i_n and v_n denote the noise current and noise voltage, respectively, integrated over the whole frequency bandwidth.

Diodes

A p-n junction is characterized by a depletion region and an energy barrier. The depletion region can be modeled by a frequency-dependent admittance, comprised of a diffusion plus depletion conductance $G_J(\omega)$ and a diffusion plus depletion capacitance $C_J(\omega)$. In the presence of a bias across the junction, the current-voltage characteristic is given by:

$$I_D \; = \; I_S \cdot \left\{ \exp\left(\frac{qV}{kT} \right) - 1 \right\} \tag{9.25}$$

Here, I_D is the total current through the diode, I_S the reverse saturation current, and V the forward voltage across the diode. The junction noise is shown to be arising from the shot noise due to current across the junction, and the thermal noise from the conductance of the junction:

$$S_I(\omega) \; = \; 2q(I_D + 2I_S) + 4kT[G_J(\omega) - G_J(0)] \tag{9.26}$$

Here, the frequency-dependent junction conductance is denoted by $G_J(\omega)$ and the low-frequency junction conductance is given by:

$$G_J(0) \; = \; \frac{q}{kT} \cdot (I_D + I_S) \tag{9.27}$$

Note that the general expression for the diode noise (Eq. 9.26) reduces to the pure shot noise form 2qI at low frequencies and to the thermal noise form 4kTG at zero bias.

Bipolar Junction Transistors

Figure 9.2 shows a schematic of a bipolar junction transistor (BJT) in a conventional device layout. The macroscopic noise sources in such a BJT are thermal noise from all resistive regions (especially from the extrinsic base region), shot noise from the base–emitter and base–collector junctions, and flicker noise from base and collector currents. These noise sources can be summarized as:

$$S_{R_x}^t \; = \; 4kTR_x \tag{9.28a}$$

$$S_{I_B}^s \; = \; 2qI_B \tag{9.28b}$$

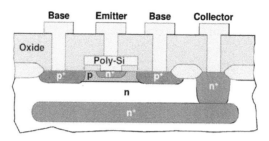

FIGURE 9.2   Schematic cross-section of an npn bipolar junction transistor in a conventional device layout.

$$S^s_{I_C} = 2qI_C \tag{9.28c}$$

$$S^f_{I_B} = K_B \cdot \frac{I^\gamma_B}{f^\alpha} \tag{9.28d}$$

$$S^f_{I_C} = K_C \cdot \frac{I^\gamma_C}{f^\alpha} \tag{9.28e}$$

The flicker noise magnitudes are strongly dependent on device processing and technology, and are summarized in the pre-factors K for each term. Note that, in general, the flicker noise exponents α and γ are different for base and collector currents. The resistor R_x denotes the total of base, emitter, and collector resistances, and the individual thermal noise terms need to be included appropriately within the device model. Since the noise sources are distributed in various regions of the device, it is useful to refer the noise either at the input or the output in order to estimate their contributions to the overall noise performance of the device. Here, we choose to refer the noise to the input of the device. The total input referred noise is given by taking into account the transconductance g_m:

$$S^{IN}_{tot} = S^t_{R_x} + (S^s_{I_B} + S^f_{I_B}) \cdot R^2_x + (S^s_{I_C} + S^f_{I_C}) \cdot \frac{(R_x + Z_\pi)^2}{g^2_m Z^2_\pi} \tag{9.29}$$

In the above expression, the input impedance relevant for noise is denoted as Z_π, indicating a π-model for the small-signal equivalent circuit of the device. For high gain BJTs, the general expression can be simplified and separated to obtain the voltage and current noise terms.

The input referred voltage noise is given by:

$$S^{IN}_v = 4kTR_B + \left(2qI_B + K_B \cdot \frac{I^\gamma_B}{f^\alpha}\right)R^2_B + 2qI_C \frac{[|R_B + Z_\pi|^2]}{g^2_m |Z_\pi|^2} \tag{9.30}$$

The input referred current noise is given by:

$$S^{IN}_i = 2qI_B + K_B \cdot \frac{I^\gamma_B}{f^\alpha} + \left(2qI_C + K_C \cdot \frac{I^\gamma_C}{f^\alpha}\right) \cdot \frac{1}{g^2_m |Z_\pi|^2} \tag{9.31}$$

The input referred voltage and current noise expressions can be used to completely characterize the noise properties of an individual device. The circuit noise performance is obtained by deriving the noise parameters from the above expressions, as described in the next section. Note that all types of bipolar devices (heterojunction bipolar transistors, etc.) can be treated in a manner similar to that given here.

The overall frequency dependence of noise in BJTs is discussed, together with that of MOSFETs, at the end of the next section.

Field Effect Transistors

Figure 9.3 shows a schematic of an n-channel MOSFET with a conventional device layout. The macroscopic noise sources in a MOSFET are thermal noise from resistive regions (especially from the gate resistance), shot noise from gate leakage current, dynamic channel thermal noise from drain current, and flicker noise from drain and gate currents. These noise sources can be summarized as:

$$S_{R_x}^t = 4kTR_x \tag{9.32a}$$

$$S_{I_G}^s = 2qI_G \tag{9.32b}$$

$$S_{I_D}^t = \beta \cdot 4kTg_\mu \tag{9.32c}$$

$$S_{I_G}^f = K_G \cdot \frac{I_G^\gamma}{f^\alpha} \tag{9.32d}$$

$$S_{I_D}^f = K_D \cdot \frac{I_D^\gamma}{f^\alpha} \tag{9.32e}$$

Note that, except for the channel thermal noise of the drain current, the intrinsic noise sources in a MOSFET are very similar to that of a BJT. The pre-factor β to the channel thermal noise was introduced to account for the specific operation of FETs in general. It was shown to have a value of 2/3 for MOSFETs and a value between 1/3 and 2/3 for JFETs. For sub-micron MOSFETs, high field effects are known to increase its value to well above 2/3. In a manner similar to that used for BJTs, one can express the noise at the input by taking into account the transconductance of the device, and then separate into voltage and current contributions.

The total input referred noise is given by:

$$S_{tot}^{IN} = S_{R_x}^t + (S_{I_G}^s + S_{I_G}^f) \cdot R_x^2 + (S_{I_D}^t + S_{I_D}^f) \cdot \frac{(R_x + Z_\pi)^2}{g_m^2 Z_\pi^2} \tag{9.33}$$

The input referred voltage noise is given by:

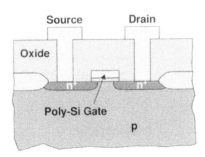

FIGURE 9.3 Schematic cross-section of an n-channel MOSFET in a conventional device layout.

$$S_v^{IN} = 4kTR_G + \left(2qI_G + K_G \cdot \frac{I_G^\gamma}{f^\alpha}\right) \cdot R_G^2 + \left(\beta \cdot 4kTg_m + K_D \cdot \frac{I_D^\gamma}{f^\alpha}\right)\frac{[|R_G + Z_\pi|^2]}{g_m^2|Z_\pi|^2} \quad (9.34)$$

The input referred current noise is given by:

$$S_i^{IN} = 2qI_G + K_G \cdot \frac{I_G^\gamma}{f^\alpha} + \left(\beta \cdot 4kTg_m + K_D \cdot \frac{I_D^\gamma}{f^\alpha}\right)\frac{1}{g_m^2|Z_\pi|^2} \quad (9.35)$$

The overall frequency behavior of the noise in FETs is similar to that of BJTs; however, the contributions from the device parameters are different in each case specific to the device properties. In the above expressions, the flicker noise becomes predominant below a "corner" frequency f_{c1}; whereas above a second "corner" frequency f_{c2}, the noise increases with frequency. Figure 9.4 illustrates this behavior schematically, where the high-frequency increase of noise is referred to as being due to impedance coupling. In this region, the thermal or shot noise from one region of the device is coupled into other regions due to the reactive components and gives rise to frequency dependence. Figure 9.5 also illustrates the overall behavior of noise figure vs. frequency measured on devices from different technologies. The bias-dependent behavior of noise in the low-frequency flicker noise regime is illustrated in Fig. 9.6, where measured output current noise power at 10 Hz is plotted as function of drain current for n-channel MOSFETs, with different gate lengths as given in the legend of the figure. Note that although all devices exhibit power law behavior, the exponent increases with decreasing gate length.

9.4   Chip Noise

The device noise mechanisms discussed in Section 9.3 give rise to noise at the chip level in a number of ways. This section outlines the principal noise mechanisms encountered in amplifiers, and oscillators, as well as timing jitter and interconnect noise in VLSI circuits.

Amplifier Noise

Amplifier noise is a small-signal phenomenon and can thus be treated by a linear approximation. It is directly determined by the noise and gain of the devices used in the design. The noise figure of an amplifier can be computed from the noise parameters of the active devices used in the circuit. The noise

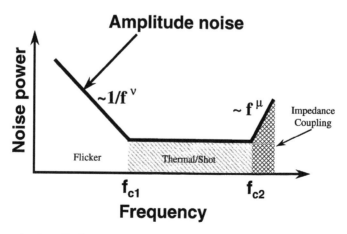

FIGURE 9.4   Schematic of amplitude noise as function of frequency, indicating the three distinct regions of noise (flicker, thermal/shot, and impedance coupling).

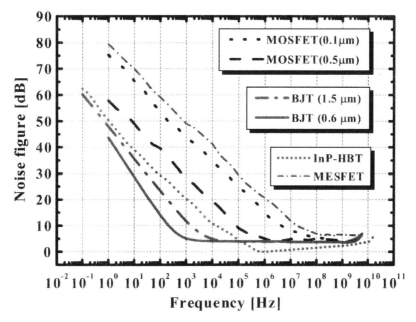

FIGURE 9.5 Measured noise figure vs. frequency for various device technologies. The plot indicates the relative magnitude of the three noise regions shown schematically in Fig. 9.4. The plot also presents the wide range of noise behavior found in device technologies.

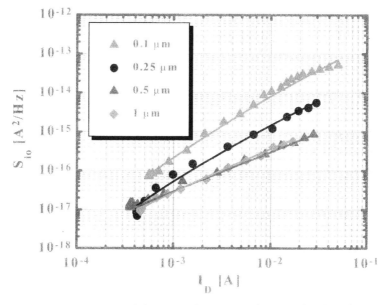

FIGURE 9.6 Output current noise spectral density vs. drain current for a set of n-channel MOSFETs with equal gate widths and different lengths, as given in the legend. Note that all devices exhibit power-law bias dependence, but the exponent increases with decreasing gate length.

parameters are a set of four quantities that characterize the noise of any 2-port (e.g., active device, amplifier, etc.) completely. They are derived from the input referred voltage and current noise sources of the 2-port in the following manner. For a noise voltage v_n and a noise current i_n at the input of a 2-port, the corresponding noise correlation matrix is given by:

$$\hat{C} = \frac{1}{2B} \cdot \begin{bmatrix} \langle v_n^2 \rangle & \langle v_n i_n^* \rangle \\ \langle v_n^* i_n \rangle & \langle i_n^2 \rangle \end{bmatrix} \tag{9.36}$$

Here, B denotes the frequency bandwidth. The voltage and current noise sources used in the expression above are related to the noise spectral densities derived in the previous sections as follows:

$$S_v^{IN} = \frac{\langle v_n^2 \rangle}{B} \tag{9.37a}$$

$$S_i^{IN} = \frac{\langle i_n^2 \rangle}{B} \tag{9.37b}$$

The noise parameters are the minimum noise figure F_{min}, the complex optimum source impedance Z_{opt}, and the noise conductance g_n. They are expressed in terms of the input referred noise voltage and current as follows:

$$F_{min} = 1 + \frac{[v_n + Z_{opt} \cdot i_n]^2}{4kTBZ_{opt}} \tag{9.38a}$$

$$Z_{opt} = \sqrt{\frac{\langle v_n^2 \rangle}{\langle i_n^2 \rangle}} \tag{9.38b}$$

$$g_n = \frac{\langle i_n^2 \rangle}{4kTB} \tag{9.38c}$$

The correlation matrix expressed in terms of the four noise parameters is given by:

$$\hat{C} = 2kT \cdot \begin{bmatrix} g_n |Z_{opt}|^2 & \left[\frac{(F_{min} - 1)}{2} - g_n Z_{opt} \right] \\ \left[\frac{(F_{min} - 1)}{2} - g_n Z_{opt}^* \right] & g_n \end{bmatrix} \tag{9.39}$$

The noise parameters are directly measurable quantities for an active device or circuit and, thus, the noise correlation matrix can be evaluated quantitatively. The overall noise figure of a linear 2-port depends not only on its four noise parameters, but also on the source impedance as given by:

$$F = F_{min} + \frac{g_n}{R_s} \cdot |Z_s - Z_{opt}|^2 \tag{9.40}$$

Here, R_s is the real part of the source impedance Z_s. For a cascade of amplifiers with individual noise figures F_i and gain G_i ($i = 1, 2, \ldots$), the total noise figure is given by:

$$F = F_1 + \frac{(F_2 - 1)}{G_1} + \frac{(F_3 - 1)}{G_1 G_2} + \ldots \tag{9.41}$$

The above expression shows that the primary contribution to the total noise figure of any amplifier comprised of several stages is from the initial stages. A quantity that takes into account both noise figure and gain of an amplifier is the noise measure given by:

$$M = \frac{F - 1}{1 - (1/G)} \qquad (9.42)$$

The noise measure can be used as a figure-of-merit for the overall noise–gain performance of a linear amplifier for analog applications. As described by Eq. 9.41, the primary contribution to amplifier noise can be attributed to the noise of the input active device. Hence, an experimental study of noise in devices provides valuable information for optimum design of low noise amplifiers.

For the purpose of illustrating the main features of amplifier noise, we show measured RF noise behavior of an active device. Figure 9.7 shows the frequency dependence of minimum noise figure, input referred voltage noise, input referred current noise, and noise measure for a BJT. Note that in the frequency range shown in the figure, all quantities increase with frequency corresponding to the impedance coupling regime. The voltage noise is a measure of the thermal noise of the input resistance, and the current noise is a measure of the input referred shot noise. In order to optimize the device technology for low noise applications, it is important to study the relative contributions from the various regions of an active device to the total noise power. This is done most conveniently by extracting a small-signal noise model of the device from measured data. After successively removing the noise contributions from various regions, the resulting noise parameters can be plotted and compared. Figure 9.8 shows the results of such noise modeling of a BJT. The four noise parameters are plotted vs. frequency, with the individual contributions from the various regions of the device illustrated in the figure. Here, it is seen that the

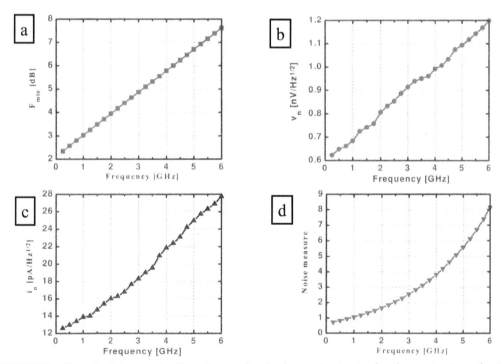

FIGURE 9.7   Plots of minimum noise figure, input referred voltage noise, input referred current noise, and noise measure vs. frequency for an npn BJT with an emitter width of 0.5 μm technology. The plots show the increase of noise with frequency in the impedance coupling regime.

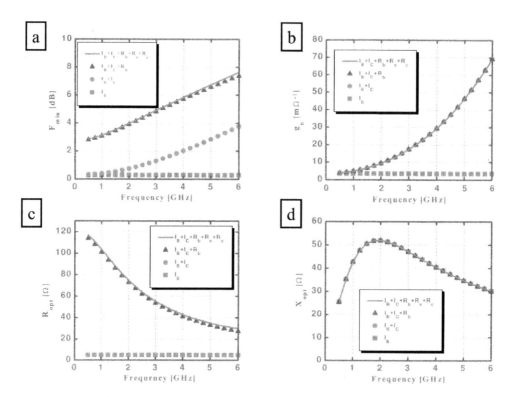

FIGURE 9.8   Plots of the four RF noise parameters vs. frequency obtained from fitting measured data on an npn BJT to a small-signal model of the device. The different curves in each plot indicate the individual contributions from the various regions of the device, as discussed in the text.

major contribution to the minimum noise figure is the base resistance, followed by the base–collector shot noise at higher frequencies. The noise conductance is primarily determined by the base–collector shot noise. The optimum source resistance is given by the thermal noise of the base resistance. Finally, the optimum source reactance is not explicitly determined by any of the noise sources, but by the input impedance of the device.

Oscillator Noise

The major noise source in oscillators is phase noise caused by mixing of device noise (flicker, shot, thermal) with the carrier frequency due to the non-linear operation of the circuits. In general, noise in oscillators is a large-signal phenomenon and, hence, linearization techniques have limited applicability. Several approaches have been published for calculating the phase noise. The simplest is that of Leeson, which estimates the phase noise of oscillators due to non-linear mixing of device flicker noise, and finite bandwidth of the oscillator. More advanced approaches include both analytical and numerical studies, as given in the references. Here, we briefly outline the fundamental features of noise in a negative conductance LCR oscillator driven by an external current source (e.g., an active device). In the presence of small-signal noise, the output voltage of the oscillator with an intrinsic amplitude v_0 and frequency ω_0 is given by:

$$v(t) = v_0 \cdot [1 + a(t)] \cdot \cos\{\omega_0 t - \varphi(t)\} \qquad (9.43)$$

The amplitude fluctuations a(t) and phase fluctuations φ(t) can be evaluated to first order by just taking into account the mixing of the noise with the free oscillations. The spectral density of the amplitude fluctuations (AM noise) is given by:

$$S_a(\Delta\omega) = \frac{S_I(\omega)}{2v_0^2(G_L - G_0)^2 \cdot \left[1 + Q_0^2\left(\frac{\Delta\omega}{\omega_0}\right)^2\right]} \qquad (9.44)$$

Here, G_L is the loss conductance, G_0 the lowest-order term of an expansion of the nonlinear negative conductance, and Q_0 the Q-factor of the oscillator. The frequency offset from the fundamental oscillator is denoted by $\Delta\omega$. The spectral density of the phase fluctuations (PM noise) is given by:

$$S_\varphi(\Delta\omega) = \frac{2S_I(\omega)}{v_0^2(G_L - G_0)^2 \cdot Q_0^2\left(\frac{\Delta\omega}{\omega_0}\right)^2} \qquad (9.45)$$

Both amplitude and phase noise exhibit a $\propto 1/\Delta\omega^2$ dependence as a function of frequency offset from the oscillation frequency. Note that the phase noise is always larger than the amplitude noise and is therefore of most relevance to chip design. Moreover, a frequency-dependent current noise $S_I(\omega) \propto 1/\omega$ (flicker noise of device) gives rise to a $\propto 1/\Delta\omega^3$ dependence of phase noise. Figure 9.9 shows schematically a simplified case of oscillator noise with two typically encountered types of phase noise frequency dependencies. For small offset frequencies, the phase noise is caused by up-conversion of the device flicker noise. At larger offset frequencies, the phase noise shows a behavior depending on the value of the flicker noise corner frequency with respect to that of the oscillator bandwidth.

Timing Jitter

In digital circuits, the timing of pulses exhibits a random fluctuation as a function of time, which is referred to as *jitter*. Jitter is fundamentally caused by the noise of individual components of the circuit

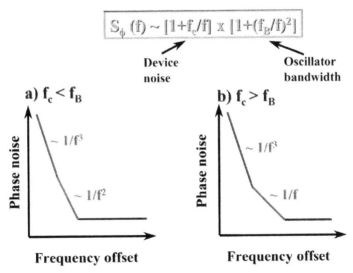

FIGURE 9.9 Schematic of frequency dependence of phase noise vs. offset from the fundamental frequency. In a simplified model, the phase noise can be treated as a product of two terms resulting from device noise and oscillator bandwidth. The two cases shown relate the flicker noise corner frequency f_c to the oscillator bandwidth f_B.

(active devices, resistors, interconnect delays, etc.) and is of primary concern to the design of low-noise oscillators. As signal levels decrease and clock frequencies increase, jitter becomes a serious noise phenomenon in VLSI circuits. Jitter can be expressed in terms of fluctuations of the fundamental period of oscillation. The distribution of these fluctuations has a standard deviation that is used as a measure of jitter:

$$\sigma_T = \frac{\sqrt{\langle T^2 \rangle - \langle T \rangle^2}}{\langle T \rangle} \tag{9.46}$$

For example, the thermal noise of collector load resistors in a differential pair stage of a ring oscillator is shown to result in jitter given by:

$$\sigma_{R_C}^t = \sqrt{2kTC_C} \cdot f(I_E, R_C) \tag{9.47}$$

Note that the above expression contains the voltage noise fluctuations from the collector resistance R_C and shunt capacitance C_C. The thermal noise of transistor input resistances (e.g., base or gate resistance) is shown to give rise to a jitter of the form:

$$\sigma_{R_X}^t = \sqrt{4kTR_X} \cdot f(I_E, C_C) \tag{9.48}$$

Note that here the thermal noise of the input resistance R_x directly affects the jitter of the circuit. Similarly, other noise sources at the oscillator input will modulate the fundamental frequency and contribute proportionally to jitter.

Interconnect Noise

Interconnects in VLSI chips are a source of noise referred to as crosstalk, which originates from capacitive and inductive coupling of signals between transmission lines. A quantitative treatment of crosstalk requires an appropriate modeling of the impedance of the transmission lines. At sufficiently high frequencies, the wavelength of the propagating electromagnetic wave becomes comparable to the length scales of the transmission lines. In this regime, the transmission needs to be treated as a distributed entity. The electrical properties of a distributed transmission line can be characterized by the characteristic impedance Z_0 and the propagation coefficient γ. For a "quasi-TEM" mode of propagation along the line, these quantities are given by:

$$Z_0 = \sqrt{\frac{R + j\omega L}{G + j\omega C}} \tag{9.49a}$$

$$\gamma = \sqrt{(R + j\omega L)(G + j\omega C)} \tag{9.49b}$$

Here, the transmission line is characterized by a lateral resistance per unit length R, a lateral inductance per unit length L, a perpendicular conductance per unit length G, and a perpendicular capacitance per unit length C. The magnitude of the crosstalk between two lines is determined by their mutual capacitance C_m and mutual inductance L_m. Since the capacitive noise is proportional to $C_m \cdot dV/dt$, and the inductive noise is proportional to $L_m \cdot dI/dt$, the crosstalk can be treated as being proportional to V_{DD}/τ and proportional to I_{sat}/τ, where τ is a pulse rise or fall time.

   Another more fundamental source of noise in VLSI interconnects is the thermal noise of transmission lines. It can be shown that by treating the interconnect as a distributed transmission line, the thermal noise can be expressed as:

$$S_v^t = 4kTR \cdot f(\gamma, l) \qquad (9.50)$$

In the above expression, the thermal noise due to the resistance R of the interconnect is modified by a function of only the propagation coefficient γ and the total length l of the interconnect.

9.5   Future Trends

Scaling

The scaling of silicon technology with decreasing minimum feature size is associated with changes in active device parameters as well as changes in passive components and interconnect dimensions. The effect of technology scaling on noise in VLSI chips can be estimated by considering the dependence of the noise of the individual components of the chip on their scaled properties. It turns out that general expressions for technology scaling of noise cannot be formulated, since device dimensions are chosen specific to the chip design. However, one can observe certain trends for active devices. It is found that in MOSFETs, the flicker noise scaling is primarily determined by changes in the gate-oxide quality (trap density in the oxide) and roughness of the oxide–channel interface (carrier mobility in the channel). At RF frequencies, the noise is dependent on device dimensions (base resistance of BJT, gate resistance of MOSFET, etc.) and on small-signal properties (base-collector capacitance, gate-channel capacitance, etc.). Although the specific choice of device dimensions primarily determines its noise, the trend of increasing cutoff frequencies results in improvement of RF noise at a given frequency due to decreased device parasitics. However, the thermal noise of interconnects increases as the technology is scaled to smaller dimensions, due to the associated decrease of cross-sectional area.

Processing

Process development in VLSI technologies is intimately related to device performance. As described above, the RF noise in BJTs is significantly determined by thermal noise in the base resistance and by shot noise from the base–collector junction. These contributions can be minimized by utilizing super-self-aligned device layouts, as shown schematically in Fig. 9.10. In this case, the extrinsic base region is significantly reduced to decrease the base thermal noise. The effect of the shot noise from the base–collector junction is minimized by reducing the base–collector capacitance that acts as a feedback capacitance and couples the collector current shot noise into the base. In MOSFETs, a reduction of the gate resistance is achieved through the use of higher conducting gate stacks (Fig. 9.11)or a multi-finger layout. Moreover, the coupling of channel thermal noise into the gate region can be minimized by reducing the gate–channel capacitance. In general, processing has a strong influence on the flicker noise of semiconductor devices. In BJTs, the passivation of the base–emitter junction region improves flicker noise substantially. In MOSFETs, the quality of the gate-oxide and its interface with the channel

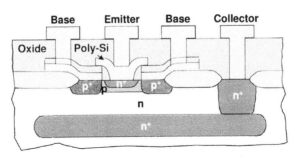

FIGURE 9.10   Schematic of a super-self-aligned BJT layout used to reduce base thermal noise and the effect of base–collector shot noise.

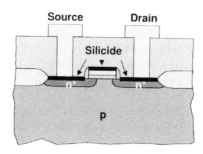

FIGURE 9.11 Schematic of a silicide-gate MOSFET used to reduce the thermal noise from the gate resistance.

affects flicker noise and can be improved through annealing treatments. One of the approaches being studied presently for reducing RF noise in MOSFETs is the use of silicon-on-insulator substrates instead of conducting substrate material, in order to reduce parasitic elements of the device.

Non-Equilibrium Transport

In deep-submicron devices at high electric fields, the assumption of thermal equilibrium between carriers and lattice is not valid. Here, we briefly indicate the type of corrections required for estimating the noise in non-equilibrium transport.

The diffusion coefficient in non-equilibrium transport can be modified by approximating it to be a simple frequency-dependent quantity of the form:

$$D(\omega) = \frac{D(0)}{1 + (\omega\tau)^2} \tag{9.51}$$

The resulting thermal noise is then given by:

$$S_I(\omega) = \frac{S_I(0)}{1 + (\omega\tau)^2} \tag{9.52}$$

In the above expression, the quantity τ is the mean scattering time of the carriers. A further effect of non-equilibrium transport is velocity-velocity correlations, which modify the macroscopic values of the noise spectral densities and need to be considered in detail.

The general expression for shot noise in semiconductors needs to include finite carrier transit times and is given by:

$$S_I(\omega) = 2qI \cdot \left[\frac{\langle N^2 \rangle - \langle N \rangle^2}{\langle N \rangle}\right] \cdot f(\omega, \tau, \tau_T) \tag{9.53}$$

Here, N is the number of carriers, τ the mean scattering time, and τ_T the carrier transit time. We can consider several limiting cases of the above expression. For sufficiently low frequencies and small transit times, such that $\omega \ll 1/\tau \ll 1/\tau_T$, the shot noise is given by:

$$S_I(\omega) = 2qI \cdot \left[\frac{\langle N^2 \rangle - \langle N \rangle^2}{\langle N \rangle}\right] \tag{9.54}$$

The above expression is the full shot noise in a semiconductor sample where all the carriers are traversing the sample along the field direction. For a Poisson distribution function of carrier number, the variance of the fluctuations is equal to the mean of the distribution, and then the simple expression 2qI is recovered for the shot noise. For sufficiently small scattering times, such that $\tau \ll \tau_T$, the shot noise is given by:

$$S_I(\omega) = 2qI \cdot \left[\frac{\langle N^2 \rangle - \langle N \rangle^2}{\langle N \rangle} \right] \cdot \left\{ \frac{2\tau}{\tau_T} \cdot \frac{1}{1 + (\omega\tau)^2} \right\} \qquad (9.55)$$

The above expression indicates that the shot effect is reduced by the randomizing due to scattering of carriers. At high frequencies, the shot noise also shows a decrease due to effective acceleration of carriers along the field direction.

9.6 Conclusions

In this chapter, the issue of noise in VLSI technologies was treated by introducing the microscopic noise mechanisms from fundamental carrier transport in semiconductors. The noise properties of active semiconductor devices and passive components were given in a general form and certain approximations relevant for most applications were outlined. The noise at the chip level was presented in terms of specific circuits where noise-critical applications are realized. Finally, trends in future VLSI technologies and their implications for noise were briefly mentioned.

Acknowledgments

Collaborations with the following organizations within Lucent Technologies are gratefully acknowledged: VLSI Technology Integration Department in Orlando, FL; Analog Design Methodology Group in Reading, PA; Compact Modeling and Measurements Group in Cedar Crest, PA. In addition, we gratefully acknowledge the many helpful discussions with colleagues from the Physical Research, Silicon Research, and Wireless Research Laboratories at Bell Laboratories.

References

Section 9.1

1. Meindl, J. D., "Ultra-Large Scale Integration," *IEEE Trans. Elec. Dev.*, ED-31, 1555, 1984.
2. van der Ziel, A. and Amberiadis, K., "Noise in VLSI," *VLSI Electronics,* Vol. 7, Academic Press, 1984, 261ff.
3. van der Ziel, A., "Noise in VLSI," *VLSI Handbook,* Academic Press, 1985, 603ff.
4. Tsividis, Y., *Mixed Analog-Digital VLSI Devices and Technology,* McGraw-Hill, 1995.
5. Martin, S., Archer, V., Boulin, D., Frei, M., Ng, K., and Yan, R.-H., "Device Noise in Silicon RF Technologies," *Bell Labs Technical Journal,* 2(3), 30, 1997.

Section 9.2

6. van der Ziel, A., *Noise in Measurements*, Wiley, 1976.
7. Gupta, M. S. (ed.), *Electrical Noise: Fundamentals & Sources,* IEEE Press, 1977.
8. Buckingham, M. J., "Noise in Electronic Devices and Systems," Wiley, 1983.
9. Ferry, D. K. and Grondin, R. O., *Physics of Submicron Devices,* Plenum Press, 1991.

Section 9.3

10. Motchenbacher, C. D. and Connelly, J. A., *Low Noise Electronic System Design,* Wiley, 1993.
11. Ng, K., *Complete Guide to Semiconductor Devices,* McGraw-Hill, 1995.
12. Martin, S., Booth, R., Chyan, Y.-F., Frei, M., Goldthorp, D., Lee, K.-H., Moinian, S., Ng, K., and Subramaniam, P., "Modeling of Correlated Noise in RF Bipolar Devices," *IEEE MTT-S Digest,* MTT-Symposium, 1998, 941.
13. See also Refs. 6, 7, and 8.

Section 9.4

14. Hillbrand and Russer, "An Efficient Method for Computer Aided Noise Analysis of Linear Amplifier Networks," *IEEE Trans. Circ. Syst.*, CAS-23, 235, 1976.

15. Abidi, A. and Meyer, R.G., "Noise in Relaxation Oscillators," *IEEE J. Solid-State Circuits*, SC-18, 794, 1983.

16. Brews, J. R., "Electrical Modeling of Interconnections," *Submicron Integrated Circuits* (Ed. R. K. Watts), Wiley, 1989.

17. Vendelin, G. D., Pavio, A. M., and Rohde, U. L., *Microwave Circuit Design*, Wiley, 1990.

18. Gray, P. R. and Meyer, R. G., *Analog Integrated Circuits*, Wiley, 1993.

19. Engberg, J. and Larsen, T., *Noise Theory of Linear and Nonlinear Circuits*, Wiley, 1995.

20. Verghese, N. K., Schmerbeck, T. J., and Allstot, D. J., *Simulation Techniques and Solutions for Mixed-Signal Coupling in Integrated Circuits*, Kluwer, 1995.

21. Schaeffer, D. K. and Lee, T. H., "A 1.5 V, 1.5 GHz CMOS Low Noise Amplifier," *IEEE J. Solid-State Circuits*, 32, 745, 1997.

22. McNeill, J. A., "Jitter in Ring Oscillators," *IEEE J. Solid-State Circuits*, 32, 370, 1997.

23. Restle, P. J., Jenkins, K. A., Deutsch, A., and Cook, P. W., "Measurement and Modeling of On-Chip Transmission Line Effects in a 400 MHz Microprocessor," *IEEE J. Solid-State Circuits*, 33, 662, 1998.

24. Demir, A., Mehrotra, A., and Roychowdhury, J., "Phase Noise in Oscillators: A Unifying Theory and Numerical Methods for Characterization," *IEEE Trans. Circ. Syst.-1: Fundamental Theory and Application*, to be published, 1999.

25. See also Refs. 6 and 8.

Section 9.5

26. Simeon, E. and Claeys, C., "The Low-Frequency Noise Behavior of Silicon-on-Insulator Technologies," *Solid-State Electronics*, 39, 949, 1996.

27. Abou-Allam, E. and Manku, T., "A Small-Signal MOSFET Model for Radio Frequency IC Applications," *IEEE Trans. CAD of IC and Systems*, 437, 1997.

28. Tin, S. F., Osman, A. A., and Mayaram, K., "Comments on "A Small-Signal MOSFET Model for Radio Frequency IC Applications," *IEEE Trans. CAD of IC and Systems*, 17, 372, 1998.

29. See also Refs. 7 and 9.

10

Micromachining

Peter J. Hesketh
The Georgia Institute of Technology

10.1 Introduction

There has been a tremendous growth in activity over the past seven years in exploring the use of micromachining for the fabrication of novel microstructures, microsensors, and microdevices and also their integration with electronic circuits. Specific application areas have developed to such an extent that there are specialist meetings on the following topics: sensors and actuators,[1] microelectromechanical system (MEMS),[2] microchemical analysis systems (μTAS),[3] optical-MEMS,[4] MEMS-electronics,[5] chemical sensors,[7] and microstructures and microfabricated systems.[7] Early examples of MEMS devices and innovations are reviewed in Peterson's[8] classic paper. Up until recently, the MEMS micromachining processes were developed outside the realm of a CMOS line although the advantages of IC fabrication processes and the economies of batch fabrication have been used for production of microdevices at low cost and high volume. The tremendous successes of microfabricated silicon pressure sensors used for blood pressure monitoring and automotive air intake manifold pressure sensing, ink-jet printer heads, and the air bag accelerometer sensors, and most recently projection overhead display systems, demonstrate the tremendous success of this technology. There are several important differences between MEMS processing and IC processing that make this an exciting and rapidly evolving field:

- Wider range of materials
- Wider range of fabrication processes utilized
- Use of three-dimensional structures
- Material properties are not fully characterized
- Interdisciplinary expertise necessary for successful technology implementation
- CAD tools are not yet fully developed for integrated thermal/mechanical, magnetic, optical, and electronic design

In keeping with the general philosophy of this volume, I will emphasize the fundamental principles and discuss CMOS-compatible processing methods. Selected examples of MEMS devices will be given to illustrate some of the exciting applications of this technology. The reader is referred to the comprehensive survey by Göpel et al.[9] and the vast diversity of micromachining and micromanufacturing methods described in recent books by Kovacs,[10] Madou,[11] and Sze.[12] In addition, reference materials include a collection of classic papers by Trimmer,[13] texts on sensors by Middlehoek and Audet,[14] Ristic,[15] Gardener,[16] and bio and chemical sensors texts by Janata,[17] Madou and Morrison,[18] Moseley et al.,[19] and Wilson et al.[20]

10.2   Micromachining Processes

There are a wide range of MEMS processing methods for 2-D and 3-D structures, many of which are incompatible with CMOS fabrication. Micromachining is the process of forming such structures by processes of chemical etching, physical machining, and layering of materials by a diverse range of processes, as listed in Table 10.1. A range of processes for material removal has been developed, including chemical etching,[21] laser ablation,[22] precision machining,[23] focused ion beam etching,[24] ultrasonic drilling and electrodischarge machining,[25] and others. Historically, surface and bulk micromachining have developed independently to produce novel structures and microdevices in two different ranges of dimensional scales. A great deal of work has addressed hybrid approaches that combine bulk and surface processes in novel ways, which will be discussed in the next section. A major obstacle to the further development of MEMS devices is the compatibility of micromachining processes with CMOS circuit processing if integration of electronics is desired. A recent review of micromachining methods for a range of materials of interest is given by Williams and Muller.[26]

10.3   Bulk Micromachining of Silicon

Bulk machining involves etching the bulk single-crystal silicon wafer with respect to a patterned insoluble masking layer that defines the structure on the wafer surface. These methods are either isotropic-independent of direction, or anisotropic-crystal plane selective. Silicon has a diamond lattice as shown in Fig. 10.1(a). The dimensions of the structure in the bulk of the wafer are defined by the crystal plane locations. In particular, the etch rate of the (111) planes are slower than other crystalline directions by typically 35 to 50 times.[27] An SiO_2 or Si_3N_4 mask is used in an etching solution, as shown in Fig. 10.1(b).

Anisotropic etching, which is a function of the crystal planes of the Si substrate, occurs in strong bases. The (111) close-packed silicon lattice plane etches slower than other directions of the crystal.

TABLE 10.1 MEMS Processing Technologies

Process	Physical Dimension Range/Aspect Ratio	Materials	Etch Stop Techniques	Through-put	Cost	Ref.
		Subtractive processes				
Bulk micromachining	μm-cm/1:400	Single-crystal silicon, GaAs glass etching	Dopant-selective electrochemical Buried layer	High	Low	9–11, 27–49
Reactive ion etch	μm-mm/1:100	Wide range of materials	Buried layer	Low	High	62–64
Laser ablation	1-100 μm/1:50	Various	Timed	Low	High	22
Electrodischarge machining	2 μm-mm/*	Si, metals	Timed	Low	Med	25
Precision mechanical cutting	nm-cm/*	PMMA	Tool position	Low	High	23
Focussed ion beam machining	nm-μm	Various	Timed	Low	High	24
Chemical etching	μm/1:10	Metals, semiconductors, insulators	Timed	High	Low	21,26,50
Ultrasonic machining	25 μm-mm/*	Glass, ceramic, semiconductor, metals	Tool position	Moderate	Moderate	—
		Additive processes				
Physical vapor deposition	Wide range of materials	Electron beam or thermal evaporation/sputtering	—	Moderate	High	26
Chemical vapor deposition	Surface micromachining	LPCVD of polysilicon/PSG or sputtered aluminum/photoresist	Selectivity of sacrificial etch to sacrificial layer to structural layer	High	Moderate	67–71,
Laser-assisted CVD	nm-μm	Various	—	Low	High	—
Molecular beam epitaxy	nm	Semiconductors	—	Low	Very High	—
LIGA	μm-cm	PMMA	—	Low	High	51,101
Electroplating into a mold:	μm-mm/1:10	Cu, Ag, Au, Fe, permalloy	—	High	Low	96,98–100
	μm-mm	Polyimide	—	High	Moderate	97
	μm-mm	SU-8	—	High	Low	—
		Thick photoresist	—	High	Low	—

* function of total geometry.

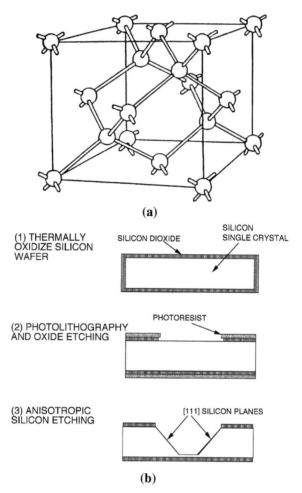

(a)

(1) THERMALLY
OXIDIZE SILICON
WAFER

SILICON DIOXIDE

SILICON
SINGLE CRYSTAL

(2) PHOTOLITHOGRAPHY
AND OXIDE ETCHING

PHOTORESIST

(3) ANISOTROPIC
SILICON ETCHING

[111] SILICON PLANES

(b)

FIGURE 10.1(a-b) (a) Diamond lattice of silicon with principal axis indicated; (b) schematic cross-section of bulk micromachining processing steps to form a cavity in a (100) silicon wafer.

Examples of commonly used formulations are given in Table 10.2. This process has been extremely successful for the fabrication of diaphragms for pressure sensors and other devices, aligning the [110] flat of a (100) wafer to a mask opening a rectangular opening which are produced. Etching into the bulk of the silicon crystal the dimensions of the resulting structure are defined by the slow etching {111} planes (see Fig. 10.1c), which are parallel and perpendicular to the [011] flat on a (100) silicon wafer. Alternatively, on a (110) silicon wafer, a slot is produced, as shown in Fig. 10.1(d). Note here that the shape is bounded by the {111} crystal planes in the vertical direction and the {100} planes or {311} planes at the base of the groove. There is insufficient space here to describe the mechanism of etch chemistry, and the reader is referred to the work of Seidel,[28] Kendall,[29] Palik et al.,[30] Hesketh et al.,[31] and Allongue et al.,[32] the recent review given by Kovacs et al.,[33] and a recent workshop of Wet Chemical Etching of Silicon.[34]

KOH is perhaps the most widely used bulk micromachining wet chemical etchant, although it is a strong ionic contaminant to the CMOS process. Despite these difficulties, it has been demonstrated in post- and pre-processing formalisms with stringent chemical cleaning. The etch has been characterized extensively.[35–37] Table 10.2 lists the plane selectivity of useful concentrations that in addition to the surface roughness are a function of the solution composition. However, roughness appears to be related to hydrogen bubble release from the surface, and work by Bressers et al.[38] has demonstrated

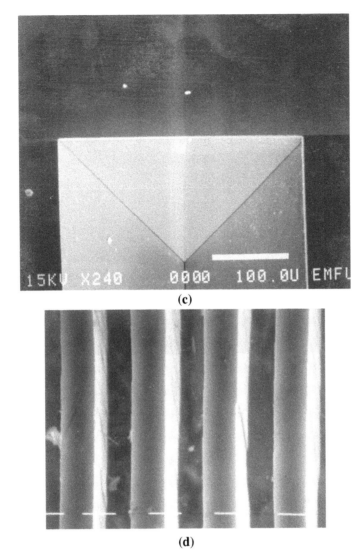

FIGURE 10.1(c-d) (c) Electron micrograph of a cavity etched in (100) silicon wafer, dimensional marker is 100 μm; (d) slots etched in (110) silicon wafer, dimensional marker is 10 μm.

that the addition of ferricyanide ions reduces hillock formation. Note that selectivity to oxide masking layers is not as high as CsOH and TMAH. Alcohol can be added to the etch to improve the surface finish and uniformity.[39]

CsOH etchant has been characterized,[40-42] and the results are summarized in Table 10.2. It has high selectivity to silicon dioxide and is dopant selective, producing smooth membranes at high concentrations. Surface roughness is often a key parameter in device design and is of key importance for technologically useful etches to obtain controlled surface conditions.

EDP has been demonstrated as compatible with CMOS processing. It has high selectivity to dielectric and metallization layers and contains no ionic contaminants. Etch has been characterized by Reisman et al.[43] and Finne and Klein.[44] The popular formulations are as follows: Type-S is 1000 ml ethylenediamine, 160 g pyrocatechol, 133 ml water, and 6 g pyrazine; Type-F is 1000 ml ethylenediamine, 320 g pyrocatechol, 320 ml water, and 6 g pyrazine. These solutions etch thermal oxide at about 55 Å/h at 115°C, there is no detectable attack of LPCVD silicon nitride, and the following metals — Au, Cr, Ag, and Cu — are resistant to EDP. The plane selectivity is listed in Table 10.2.

TABLE 10.2   Bulk Etching Solutions for Silicon

Etching Solution	Concentration (wt%)	Temp. (°C)	Etch rate of (100) plane (μm/hr)[a,b]	Anisotropy[b] $\frac{Rate(110)}{Rate(100)}$	Anisotropy[b] $\frac{Rate(100)}{Rate(111)}$	Selectivity $\frac{Rate\ Si(100)}{Rate\ SiO_2}$ (Thermal)	Selectivity $\frac{Rate\ Si(100)}{Rate\ Si_3N_4}$ (LPCVD)	Metals	Etch stop on p+ silicon	Com.	Ref.
					Isotropic etches						
HNA	250 ml HF/ 500 ml HNO$_3$/ 800ml CH$_3$COOH	20	4–20 [function of stirring]	—	—			—	—		50
					Anisotropic etches						
KOH	45	85	55	~1.5	200	300	40,000	Au	Yes	Not CMOS compatible, inexpensive, safe, widely used	27–34, 36–38
KOH/isopropyl alcohol	26/4	80	66	~0.6	~200	High	Very high	Au	>10×10^{20}/cm^3 decreases rate by 20	Not CMOS compatible, inexpensive, safe, widely used	39
CsOH	50	70	19	0.2–2.5b	50	2000	Very high	Au	Yes	Expensive material, good selectivity with SiO$_2$	40–42
Ethylenediamine/ pyrocatechol/water	255cc/ 45grm/ 120cc	100	66	<1	35	3500	Very high	Au, Cu, Cr, Ag, Ta, Ni	>7 × 10^{19}/cm^3 decreases rate by 50	Carcinogenic, widely used, low anisotropy	43–44
TMAH	4 20	80	54	—	25	5000	Very high	Al [Si doping of solution]	Yes	Flammable, low anisotropy	45–47, 49
TMAH/alcohol	15–25	70–90	11–40	—	15–37	5000	>21,000		Yes		48

[a] Function of temperature.
[b] Function of solution concentration.

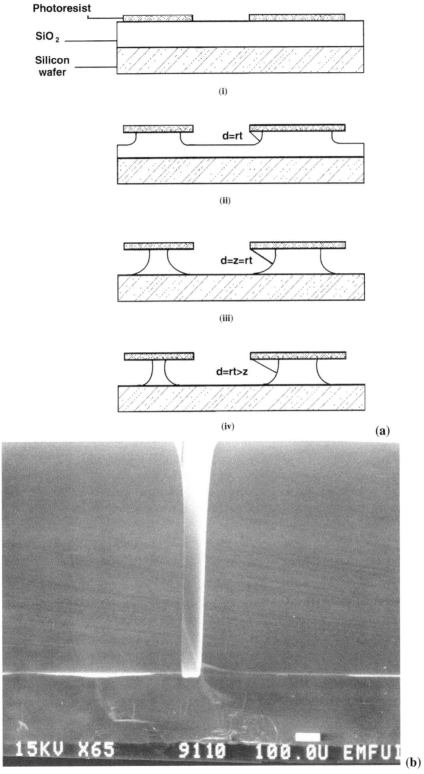

FIGURE 10.2 (a) Schematic diagram of isotropic etching, indicating undercut of mask and definition of isotropy; (b) effect of isotropic etching on grooves sawed into silicon wafer. The etching has rounded the tops of the grooves. (Kasapbasioglu, H. et al., *J. Electrochem. Soc.*, 140, 2319, 1993. With permission.)

TMAH has been studied extensively due to its low ionic contamination with CMOS. The early work on etch characterization was carried out by Tabata.[45] Characteristics are listed in Table 10.2 and the plane selectivity is markedly lower than KOH; however, selectivity to SiO_2 is quite high. The solution also passivates aluminum surfaces when the pH is adjusted into a suitable range.[46] This makes it a strong candidate for CMOS post-processing; however, the etch rate is reduced with silicon doping and there are unresolved issues regarding the solution stability. Ammonium persulphonate was added to the TMAH bath to reduce surface roughness by limiting the formation of hydrogen bubbles, specifically for a 5 wt % solution, 40 g/l silicic acid, and 5 to 10 g/l ammonium persulphate.[47] Changes in the surface morphology and high index plane selectivity are observed with the addition of alcohol[48] and hillock formation has been studied.[49]

Isotropic etching in a mixture of hydrofluoric acid (HF) and nitric acid (HNO_3) produces typically rounded profiles.[50] Figure 10.2 shows a schematic diagram of the profile produced during isotropic etching. The etching takes place in a two-step process, the first being oxidation of the silicon by the HNO_3, and the second being dissolution of the oxidized layer into a soluble H_2SiF_6 silicate. These two processes have different reaction rates and, hence, polishing occurs at low concentrations of HF where it defines the rate-limiting step in the reaction. The etchant can be stabilized by the addition of acetic acid, which helps prevent the dissociation of the nitric acid into NO_3^- and NO_2^-. The etch is dopant selective, having a lower etch rate for a lightly doped region ($<10^{17}/cm^3$) of silicon relative to heavier doping regions.[21] Figure 10.2(b) shows the application of this etching mixture to the rounding of pins that have been produced by mechanical sawing of the silicon.

Etch Stop Methods

There are various methods that have been developed for membrane fabrication at a defined doped layer as shown in Fig. 10.3(a). The heavily doped etch stop occurs when the doping level of p-type silicon is greater than $2–5 \times 10^{19}/cm^3$ for all of the principle etching solutions listed in Table 10.2. It is believed that the etch mechanism is related to the availability of electrons at the surface, which is a participant in the silicon dissolution process. Thus, high p^+ doping results in electron-hole recombination. Doping and etch back has been very successful in producing capacitive pressure sensors and neural recording probes.[52] The etch rate of the heavily doped material has an n^4 dependence on the concentration of the dopant; however, there are some differences in the abruptness of this dependence as indicated in Fig. 10.3(b). This is in agreement with the four electrons per silicon atom dissolution process. The mechanism for dopant-selective behavior is discussed by Collins.[53]

Other methods are listed in Table 10.3. The electrochemical etch stop relies on the formation of an anodic oxide on the silicon, which stops the etching process. The process was first demonstrated by Waggener and has been developed for silicon microstructures by Jackson et al.[54] and Kloeck et al.[55] Figure 10.4 shows a typical experimental set-up in which an n/p junction is reverse-biased in the etching tank and protected in a Teflon holder so that only the p-type surface is exposed to the solution. Because the junction is reverse-biased, the potential on the silicon is close to the open-circuit potential and so etching continues uninhibited. However, once the n-type material is reached, the potential becomes that applied by the external circuit. Care must be taken to ensure that the reverse-bias leakage current is not high enough to allow the potential on the exposed silicon to reaching a point more positive than the passivation potential, as indicated in the figure by the operating potential for the silicon surface. The passivation potential is the potential at which oxide forms rather than dissolution takes place. The current is used as a diagnostic to indicate when the layer is removed in practical etching procedures because a current peak occurs when the oxidation reaction occurs due to the additional electrons required by the reaction. Once a passivating layer of oxide has formed, the current drops to a low level of leakage current through the oxide layer. The oxide dissolution process is slow in these solutions, and hence, if the bias is removed, after a finite time the oxide layer is removed and etching of the silicon continues.

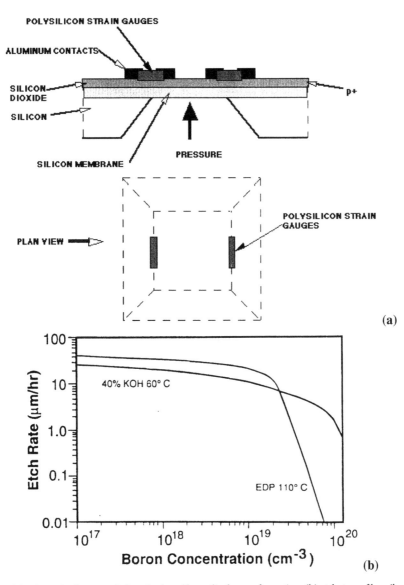

FIGURE 10.3 (a) Schematic diagram of a heavily doped layer diaphragm formation; (b) etch stop of heavily doped silicon for KOH and EDP anisotropic etching solutions. (Collins, S. D., *J. Electrochem. Soc.*, 144, 2242, 1997. With permission.)

TABLE 10.3 Silicon Membrane Fabrication Methods

Method	Thickness Range	Process	Comments	Ref.
Timed etch	5–100s μm	Observation of diaphragm thickness as a function of time	Inaccurate	53
p⁺ heavy doping	0.1–5 μm	Boron doping of ion implant to define location of membrane	Membrane is highly compressive state	52,53
Electrochemical etch stop	0.1–100 μm	p/n junction location of membrane	Electrical contact is necessary to wafer so special fixturing is required	53–55
Buried oxide layer	0.1–1 μm	SIMOX implant of O_2 dose above 10^{18}/cm³	Expensive; annealing damage from surface region after implant critical	53
Si-Si bonding			Alignment of two wafers is required	117–119

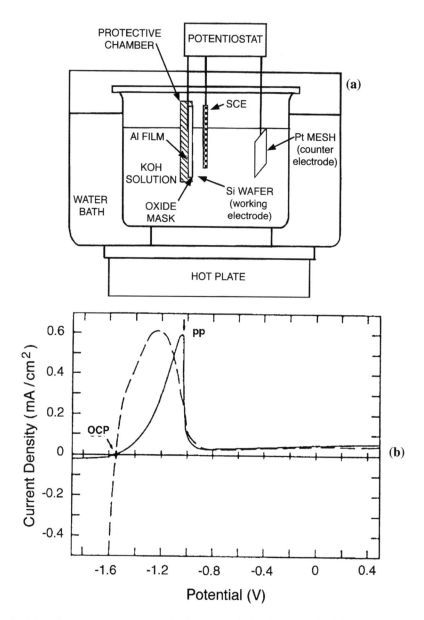

FIGURE 10.4 (a) Typical experimental set-up for electrochemical etch stop method for formation of thin silicon diaphragms (SCE is the Standard Calomel Electrode); (b) current–voltage characteristics of the p-type (solid line) and n-type (dashed line) silicon in 40 wt. % KOH solution at 60°C, where OCP is the open circuit potential and PP passivation potential. (Kloeck, B. et al., *IEEE Trans. Electron. Dev.*, 36, 663, 1989. With permission.)

Electrochemical Machining and Porous Silicon

Tuller and Mlcak[56] have demonstrated that anodic dissolution of silicon offers a controllable method for the fabrication of cantilever beams, membranes, and other structures. Dopant-selective etch stop can be achieved because p-type Si etches in HF solutions under anodic bias, whereas the n-type is stable. However, under strong illumination, porous silicon is formed on the n-type material. Hence, the etch rate and surface morphology are a function of the doping level, current density, and illumination level. The anodic current is a direct measure of the dissolution rate and, by defining an optically opaque insoluble mask on the silicon surface, high aspect ratio structures have been defined in the surface. Porous silicon is under investigation as a MEMS material as discussed by Schöning et al.[57] for its extremely high surface area.

Xenon Difluoride Etching

This dry gas-phase isotropic etch process relies on sublimation of solid XeF_2 at room temperature into a vapor that selectively attacks Si over SiO_2, Si_3N_4, and several metals.[58] Hoffman et al.[59] demonstrated its use for sensors and for CMOS-compatible processing. Chu et al.[60] later confirmed the minimal etch rate of Al, Cr, TiN, and SiC. Alternatively, gas-phase isotropic etching in xenon difluoride, or reactive ion etching methods may be selected. Tea et al.[61] has studied several of these methods and concluded that EDP can attach the Al bonding pads, whereas XeF_2 shows excellent compatibility. Microheaters for chemical sensors and microwave transmission lines were successfully fabricated; however, there was a lack of compatibility with catalytic metals such as Pt, Ir, and Au.

RIE Etching

Dry etching offers the advantages of controlled dimensions independent of the crystal planes in the substrate, in addition to higher dimensional control to sub-micron over wet chemical methods. The process utilized either a dc, RF, microwave, or inductively coupled energy to excite a plasma of reactive ions and accelerate them to the substrate. Depending on the ion acceleration potential utilized and the gas pressure, there are very different process characteristics. Figure 10.5 shows the different processes of (a) sputtering by physical bombardment of the surface, (b) chemical etching which is isotropic, and (c) reactive ion etching in which the ion bombardment enhances the chemical etching rate. This ion-assisted mechanism provides anisotropy, which is useful technologically for the fabrication of high aspect ratio structures. Ion milling represents the processes at high potentials where sputtering is dominant and at the other extreme at low substrate bias plasma etching where chemistry plays the dominant role. The reader is referred to excellent texts on dry etching equipment and processing methods.[62,63] Key factors in the design of dry etching processes for films are the selectivity with respect to the masking layer and with respect to the substrate material. The etch rate, uniformity, and anisotropy are other key process parameters.

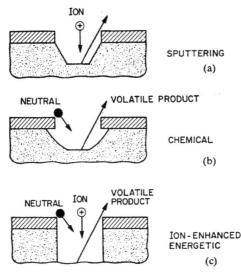

FIGURE 10.5 (a) Sputtering, (b) plasma etching, and (c) reactive ion etching.

Deep-RIE Etching

Deep-RIE etching was developed by the application of a novel etch chemistry and high-density plasmas created by either microwave or inductively coupled sources. Mixed gas chemistries can be utilized to provide sidewall passivation and further increase the anisotropy of the process. Selectivities with respect to a photoresist mask of 200:1 can be reached, allowing one to etch completely through a silicon wafer. Bhardwaj et al.[64] have demonstrated deep reactive ion etching by alternating between deposition of a passivating film and etch chemistry. The surface is coated with a Teflon-like material and the material remains on the sidewalls; however, ion bombardment removes it from the horizontal surfaces. Typically, a low bias voltage is utilized so that the mask erosion rate is minimized. An example of grooves etched with this process is given in Fig. 10.6.

Corner Compensation

Corner compensation of the bulk etched structure is important to maintain the shape of a micromachined structure. For example, microchambers for DNA sequencing by hybridization have been fabricated with corner compensation structures oriented in the [110] direction.[65] A great deal of work has examined the compensation of exposed corners for etching in KOH solutions under different conditions.[66] The shape of the compensation structure is shown in Fig. 10.7. There are several commercially available CAD tools for silicon bulk etching. Figure 10.7(c) is an example of a etched cantilever in CsOH solution. The CAD simulation shows the faceting high-index planes and the general features of the structure.

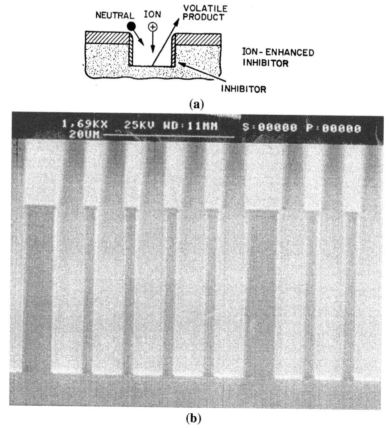

(a)

(b)

FIGURE 10.6 (a) Method of side-wall passivation during RIE etching; (b) slot produced by deep RIE etching process. (Bhardwaj, J. et al. in *Microstructures and Microfabricated Systems-III, Proceedings of the Electrochemical Society*, 97-5, 118, 1997. With permission.)

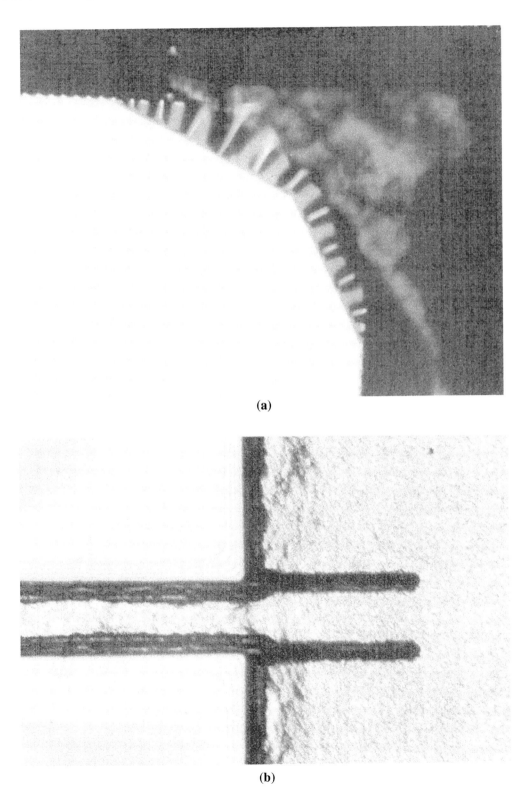

(a)

(b)

FIGURE 10.7(a-b) (a) Exposed corner etched in 50 wt% CsOH for 4 hours; (b) corner compensation structure 20-μm wide after etching for 3 hours in 49 wt% KOH.

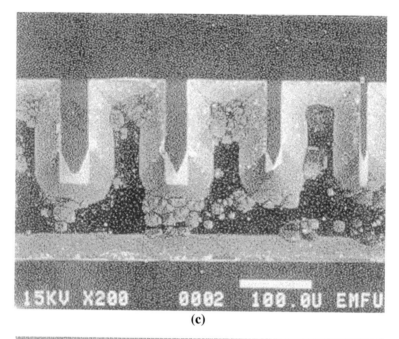

(c)

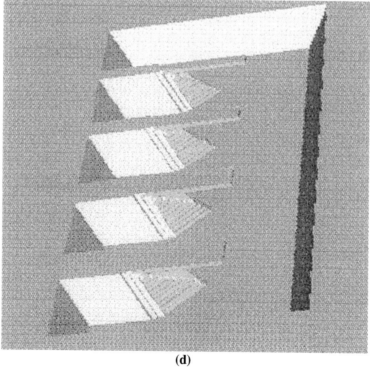

(d)

FIGURE 10.7(c-d) (c) Etched cantilever beams in silicon dioxide by etching in CsOH solution; (d) example of CAD simulation of etched cantilever structures in silicon. (Shih, B. et al., submitted to *J. Electrochem Soc.*, 1998. With permission.)

10.4 Surface Micromachining

Surface micromachining involves depositing thin layers of polysilicon as a structural layer and phosphosilicate glass as a sacrificial layer (see Fig. 10.8), as first demonstrated by Howe and Muller.[67] These layers are typically less than 2 μm in thickness, and etching away the PSG layer in a HF-based etching process. Undercut etch rates have been characterized, and the maximum dimensions of the thin polysilicon beams are limited due to internal stresses and the mechanical strength of the polysilicon.[68] Fine-grained polysilicon deposited at 580°C is subsequently annealed for surface micromachining to produce low stress polysilicon. Guckel et al.[69] have demonstrated that post-deposition anneal conditions are key to defining the low stress characteristics of polysilicon, as shown in Fig. 10.9. Conductive regions are defined by ion implantation. An alternative process is to carry out *in situ* doping with phosphorus and a relatively rapid deposition rate, followed by an RTA at 950°C to produce a low

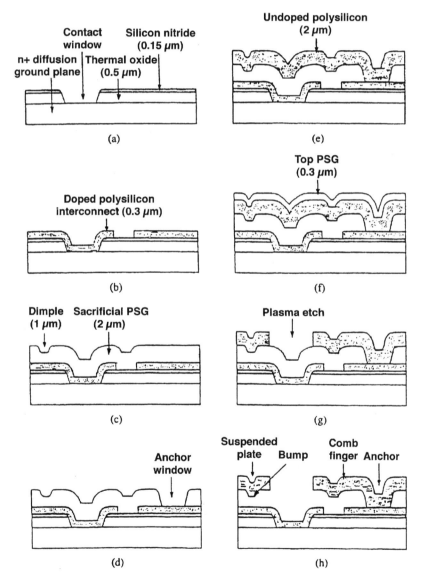

FIGURE 10.8 Cross-sectional diagram of processing steps in the surface micromachining process. (Courtesy of Tong, W. C. et al., *Sensors and Actuators*, 20, 25, 1989. With permission.)

tensile stress film with a small stress gradient through the film.[70] Polysilicon is an extremely stable and good-quality material for producing micromechanical elements, as demonstrated by the success of these devices (examples are given in the next section). Sealing of surface micromachined structures has been developed for absolute pressure sensors by reactive sealing of small openings in the surface or by dry deposition of another layer of material like silicon nitride.[71] Surface micromachining is not limited to poly-Si and PSG. A wide range of material combinations have been studied.[15] Machining with aluminum as the structural layer and a photoresist as the sacrificial layer is discussed by Westerberg et al.[72]

Stiction

Stiction in surface micromachined structures is an issue that must be addressed due to the very small gaps present between surfaces. Tas et al.[73] investigated the origins of stiction in surface micromachined devices and found that roughness plays an important role. The critical dimensions for the mechanical force to snap back a cantilever beam against the stiction force was investigated. The energy of adhesion by liquid bridging found that if the tip touches the substrate, the surface energy plus the deformation energy has a minimum for a detachment length smaller than the beam length. Stiction can occur during processing or during device operation. Various methods have been developed to avoid stiction during processing. Carbon dioxide drying is used because, below the critical temperature of the supercritical carbon dioxide, no liquid vapor interface exists that could cause contact. During device operation, the fail-safe requires that these surfaces do not make contact or an adequate release force can be applied. Surface energy is only partially successful in modifying these interactions and it is important to limit the contact area. Specifically, dimples, bumps, sidewall spacers, or increased surface roughness can define smaller contact areas.

Materials Properties of Thin Films

The properties of thin films differ from bulk material, and measurement techniques have been developed for *in situ* determination of these properties on a silicon wafer. Properties of interest are Young's modulus, Poisson ratio, residual stress, tensile strength, fracture toughness, thermal conductivity, density, and optical absorbance and optical reflectivity. Finding mechanical properties of thin films as a function of deposition conditions has been cataloged for many materials in a properties database available from Intellisense Inc.[74] Table 10.4 summarizes properties of thin-film CMOS and MEMS materials. However, this information should be used with caution given that there is considerable variation in materials properties as a function of the specific growth conditions due to variations in the microstructure and the film thickness. Low-stress polysilicon has been deposited in an epitaxy reactor in a process compatible with bipolar electronics by Gennissen et al.[75] In addition, more recent work by Wenk[76] has demonstrated thick low-stress polysilicon films grown in an epitaxial growth reactor from DCS at 1000°C for a microfabricated accelerometer and inertial gyroscope. Figure 10.9(b) shows a typical biaxial stress plot for the 3-μm polysilicon tested from the MUMPS process by Sharpe et al.[77] Other studies include those of Kahn et al.,[78] Maier-Schneider et al.,[79] and Biebl et al.[80]

Silicon Nitride

Stoichiometric silicon nitride has high tensile stress and this limits the maximum film thickness that can be fabricated on a wafer.[81] Stress relaxation occurs in silicon-rich nitride as a function of the film stoichiometry and has been characterized by Gardemoers et al.,[82] Chu et al.,[83] French et al.,[84] and Habermehl.[85] Figure 10.10(a) shows the intrinsic stress as a function of deposition conditions for silicon-rich nitride growth at several different temperatures and pressures and Fig. 10.10(b) for PECVD material.[86] Other materials have been studied,   including SiC[87] and diamond-like carbon.[88]

TABLE 10.4 Materials Properties of LPCVD Deposited MEMS Materials

Material	Growth Conditions	Film Thickness	Property	Value	Comments	Ref.
			Polysilicon			
MUMPS process		3 μm	Young's modulus Tensile strength	169±6.15 GPa 1.20±0.15 GPa	—	77
Thick polysilicon	1100°C, SiH$_4$/B$_2$H$_6$ or 610°C, Sitty	2.5-10 μm	Young's modulus Fracture toughness	150±3- GPa 2.3±0.1 MPa √m 280 MPa	Undoped film	78
Thin polysilicon	565°C, SiH$_4$, 620°C, SiH$_4$, 100 mTorr	1 μm	As-deposited residual stress			
CMOS		0.33 μm	Young's modulus Tensile strength	168±7 GPa 2.11±0.10 GPa	—	80
			Young's modulus Intrinsic stress Intrinsicicties	162.8±6 GPa −350±12 GPa 162.8±6 GPa	— As deposited After 1000°C anneal	79
			Silicon Nitride			
Standard process	800°C, SiCl$_2$H$_2$]NH$_3$	—	Intrinsic stress	~1.2 GPa	—	81
Si-rich, variable stoichiometry	800, 850°C 200, 410 mTorr SiCl$_2$H$_2$/ NH$_3$		Intrinsic stress	(See Figure 10.10)		82–84
Silicon-rich, variable stoichiometry	850°C 200 mTorr SiCl$_2$H$_2$/NH$_3$	0.25–0.45 μm	Young's modulus			85
PECVD				(190 GPa)		

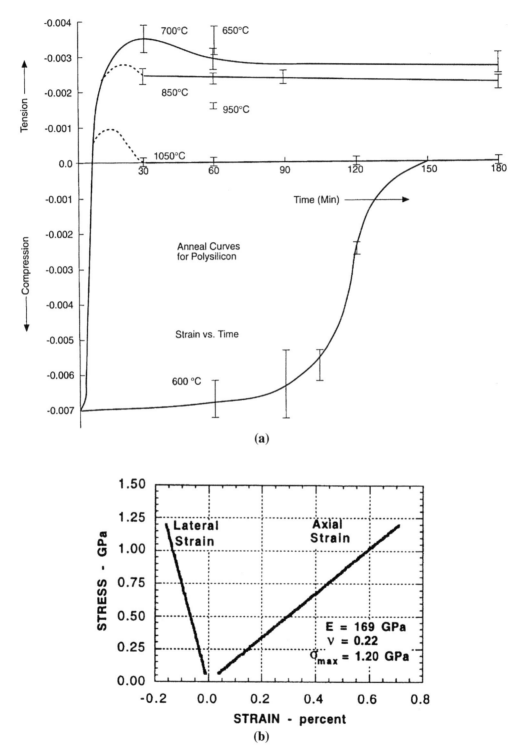

FIGURE 10.9 (a) Stress in fine-grained polysilicon as a function of post-deposition annealing temperature and time (Guckel, H. et al., *Sensors and Actuators A*, 21, 346, 1990. With permission); (b) biaxial stress for polysilicon fabricated in the MUMPS process. (Sharpe, W. N. et al., in *Proceedings of the Tenth Annual Internations Workshop on Micro Electro Mechanical Systems,* Nagoya, Japan, January, 1997, IEEE, New Jersey, Catalog Number 97CH36021, 424. With permission.)

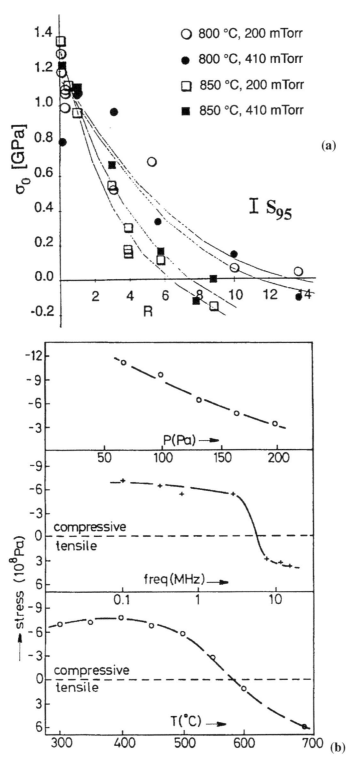

FIGURE 10.10 (a) The effect of deposition parameters on the residual stress of LPCVD deposited silicon rich silicon nitride where R is the ratio of dichlorosilone to ammonia gas flow (Gardeniers, J. G. E. et al., *J. Vac. Sci. Tech. A.*, 14, 2879, 1996. With permission.); (b) intrinsic stress of PECVD Si_3N_4 as a function of processing conditions. (Classen, W. A. P. et al., *J. Electrochem. Soc.*, 132, 893, 1985. With permission.)

10.5   Advanced Processing

We have discussed both surface and bulk machining processes and these offer certain advantages and limitations, as contrasted by French and Sarro.[89] Due to the space limitation, there is only an opportunity to mention a few examples of mixed-technology processes that have combined aspects of surface and bulk machining with other processes like chemical mechanical polishing for planarization.

Chemical Mechanical Polishing

Chemical mechanical polishing is well known in the lapping and polishing of wafers prior to fabrication. It has been recently demonstrated as a key process for achieving multilayers of polysilicon and metallization by Rogers and Sniegowski.[90] They have achieved up to five levels of polysilicon fabrication for micromechanical actuators, linkages, gears, and mechanisms. Examples of interconnected gears, locking pins, and movable elements are shown in Fig. 10.11. The key to this process is planarization of the surface between polysilicon layer deposition processes by chemical mechanical polishing.[91] In this process, planarization is achieved by depositing a thick layer of silicon dioxide over the polysilicon layer and polishing back to achieve planarization.[92] The process flow is shown schematically in Fig. 10.12. The key advantages of this process are: (1) the removal of stringers without the necessity for an over-etch; (2) improvement in line width control given that topographic features are removed; (3) removal of mechanical protrusions between layers that can cause interference; and (4) the possibility to carry out integration by placing the mechanical elements in a etched well in the silicon substrate (as discussed in the next section). Finally, these processing improvements lead to higher process yield.

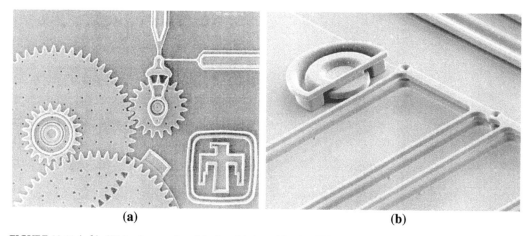

(a)   **(b)**

FIGURE 10.11 (a-b) SEM micrographs of devices fabricated by the SUMMIT process: (a) gears; (b) linear spring.

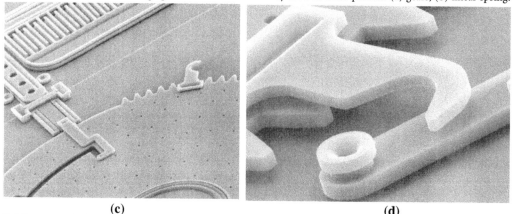

(c)   **(d)**

FIGURE 10.11 (c-d) (c) Mechanism; (d) hook. (Courtesy of Sandia National Laboratories' Intelligent Micromachine Initiative; www.mdl.sandia.gov/Micromachine. With permission.)

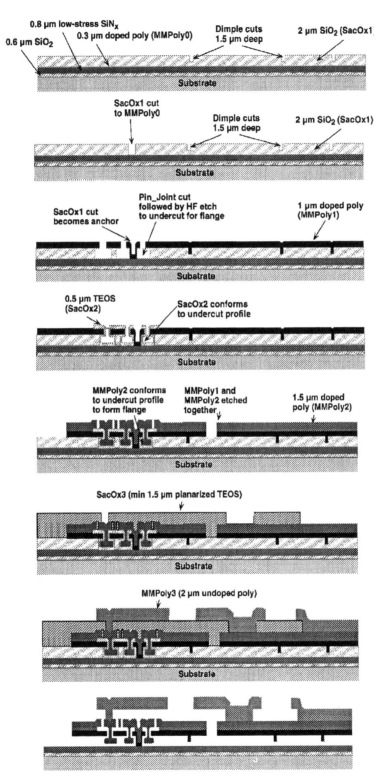

FIGURE 10.12   Schematic diagram of processing steps in the SUMMIT process for polysilicon MEMS. (Courtesy of Sandia National Laboratories' Intelligent Micromachine Initiative; www.mdl.sandia.gov/Micromachine. With permission.)

SCREAM processing utilizes the advantages of the properties of single-crystal silicon for producing a wide range of elegant devices.[93] This method was derived from a process to make isolated islands of submicron silicon by local oxidation. Suspended high aspect ratio structures are fabricated by SF_6 RIE step, followed by sidewall oxidation and isotropic silicon etch to undercut the structure (see Fig. 10.13). Both suspended structures and structures with metallized sides for electrostatic drives can be realized by varying the width of the structure and processes, each with high out-of-plane stiffness, as defined by the ratio of beam height to length, $[h^3/l^3]$. Examples of devices realized by this elegant process include tip-on-tip sensors, microaccelerometers, electrostatic drives, torsional actuators, microloading machines, and STM tips with integrated drive. In addition, beams filled with spin on glass isolation have been demonstrated with high thermal isolation. The planarity of large aspect ratio structures has been investigated by Saif and MacDonald.[94]

HEXSIL is an RIE-etched channel and fill process that utilizes polishing for planarization.[95] The trench controls the material that fills the grooves defined by RIE process. After the groove is coated with an oxide layer, the body of the device is made from undoped polysilicon, followed by a doped layer, and finally, Ni is electrolytically deposited into grooves to define highly conducting regions (see Fig. 10.14 for a process overview). The different groove width provide discrimination between different materials; hence, three types of material combinations are defined. Once the sacrificial oxide is removed, the HEXIL structure is released, micro-tweezers, and other structures have been defined by this method, as shown in Fig. 10.14(b).

Electroplating into Molds

A variety of materials have been utilized for a mold into which metals can be electroplated. These include the LIGA process, thick photoresists, polyimide,[96] and SU-8.[97] Photoresists often have a limited sidewall profile; however, polyimide structures with vertical wall profiles can be processed up to 100 μm in thickness.[98] The process flow is shown in Fig. 10.15 that utilizes a standard optical mask and standard UV exposure system. A spiral inductor structure produced by Ahn et al.[99] is shown in Fig. 10.15(b). A 30-μm thick nickel-iron permalloy magnetic core is wrapped with 40-μm thick multilevel copper conductor lines. [The magnetic circuit consists of an inductor $4 \times 1 \times 0.13$ mm^3 having 33 turns, with an inductance of 0.4 μH at 1 kHz to 1 MHz.] Applications of inductive components are for microsensors, microactuators, and magnetic power devices such as dc-to-dc converters. Layers of copper conductor and high-permeability NiFe are electroplated to form a high-value inductance with parallel and spiral coils.[100] The process SU-8 is a negative-acting epoxy resin that has been fabricated up to 100 μm thick layers in a standard alignment system (see example in Fig. 10.16). Multiple levels of metallization have been developed of ULSI copper plating by the metal is formed by the Damascene process with a photoresist mold. Planarization is achieved by chemical mechanical polishing, which relies on the materials (metals and dielectrics) having similar abrasion rates.

LIGA Process

LIGA is an acronym from the German Lithographie, Galvanoformung, und Abformung, denoting the use of X-ray lithography from a synchrotron source, in thick PMMA layers to define structures with very high aspect ratios, followed by nickel electroplating, and subsequent replication by injection molding.[101] Because LIGA is based on X-ray radiation, it is not compatible with CMOS processing. The intensity of the X-rays produces damage in the dielectric components of the CMOS circuit. An excellent review of recent development in LIGA processing is given by Friedrich et al.[51] Optical gratings have been made by the LIGA process by Cox et al.;[102] specifically, infrared tunable filters driven by magnetic actuators, features 30 μm in height, and a filter period of 8 to 17 μm. (See Fig. 10.17.)

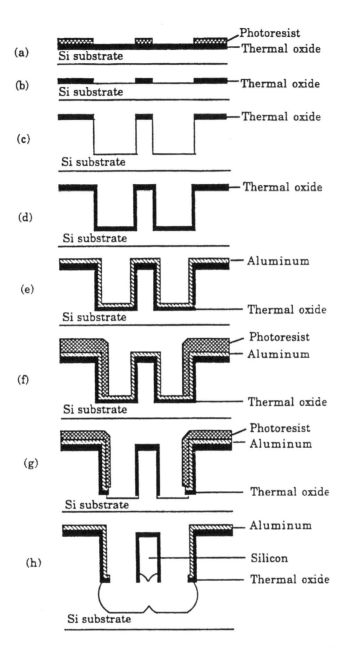

FIGURE 10.13 Cross-sectional diagram of the SCREAM process for the formation of silicon cantilevers with aluminum electrodes adjacent to each side of the beam. The undercut of the silicon sidewalls isolate the aluminum from the substrate. (McDonald, N. C., *Microelectronic Engineering*, 32, 49, 1996. With permission.)

GaAs Micromachining

The properties of GaAs as a MEMS material have been investigated by Hjort et al.[103] The key advantages of GaAs are that it is a direct-bandgap semiconductor also having high electron mobility, so that opto-electronic devices can be realized, and the piezoelectric properties are comparable to that of quartz. Finally, electronic devices can be operated at temperatures up to 350°C. Although the mechanical strength of the material is lower than silicon, the material can be bulk micromachined with HNO_3/HF solutions. Surface-type micromachining processes have also been developed, based on the advanced state-of-the-

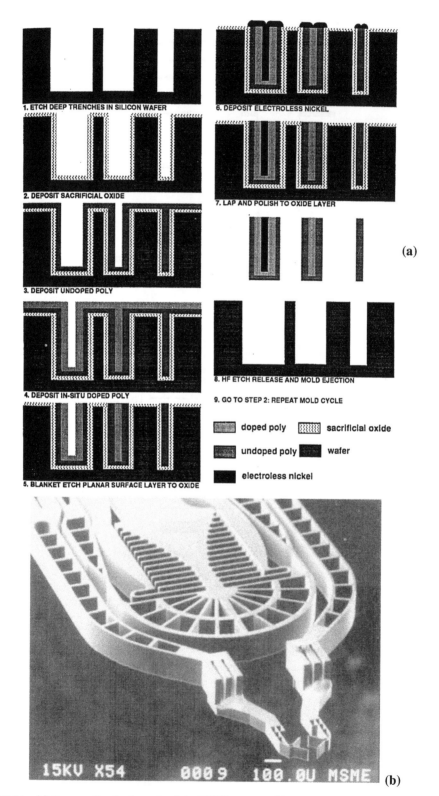

1. ETCH DEEP TRENCHES IN SILICON WAFER

2. DEPOSIT SACRIFICIAL OXIDE

3. DEPOSIT UNDOPED POLY

4. DEPOSIT IN-SITU DOPED POLY

5. BLANKET ETCH PLANAR SURFACE LAYER TO OXIDE

6. DEPOSIT ELECTROLESS NICKEL

7. LAP AND POLISH TO OXIDE LAYER

8. HF ETCH RELEASE AND MOLD EJECTION

9. GO TO STEP 2: REPEAT MOLD CYCLE

doped poly   sacrificial oxide

undoped poly   wafer

electroless nickel

(a)

(b)

FIGURE 10.14 (a) Cross-sectional schematic of the HEXIL process; (b) micro-tweezers produced by the HEXIL process. (Keller, C., *Microfabricated High Aspect Ratio Silicon Flexures*, ISBN 0-9666376-0-7. With permission.)

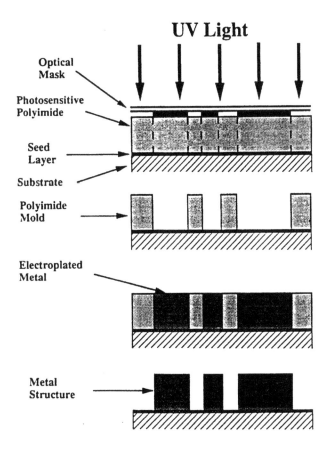

FIGURE 10.15 Schematic representation of the process sequence for fabricating metallic electroplated microstructures using photosensitive polyimide. (Frazier, A. B. et al., *Sensors and Actuators A.*, 45, 47, 1994. With permission.)

art in modulated material doping and selective etching of AlGaAs. Interdigitated electrostatic actuators have been realized in addition to surface acoustic wave sensors (SAW) optical sensors, and piezoelectric sensors (see later section on optical devices).

10.6 CMOS and MEMS Fabrication Process Integration

One of the key challenges in micromachining processes is combining the electronic devices with the mechanical, optical, or chemical function of the MEMS device. Although most work has utilized a hybrid approach in which the MEMS device is fabricated independently of the interface electronics, there are several examples of integrated sensors and other devices. The early work of Parameswaran et al.[104] demonstrated a post-CMOS processing anisotropic etching in KOH solutions that was feasible under limited conditions. The circuit elements and the MEMS devices were defined by a standard CMOS process and then the structures were released from the substrate or thermally isolated from the silicon substrate by a post-processing step. Requirements for such an etch are that it is compatible with exposed silicon dioxide, and silicon nitride, on the chip. Other etching chemistry may be more suitable such as XeF$_2$ or TMAH. Another issue is ionic contamination which leads to failure of the CMOS devices through mobile ions present in the gate dielectric,[105] producing instability in threshold voltages.

FIGURE 10.16 SEM photomicrograph of a high aspect ratio 100-micron thick SU-8 photoresist structure formed by a single spin coating and contact lithography exposure. (Photo courtesy of Electronic Visions, Phoenix, Arizona. With permission.)

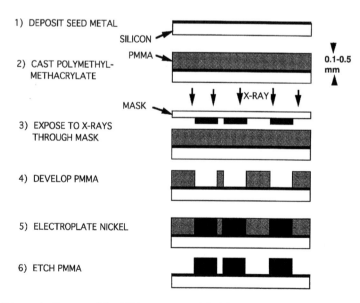

FIGURE 10.17 Schematic diagram of the LIGA process modified for MEMS.

There have been several approaches for combining CMOS circuits with MEMS structures, summarized as follows:

- Post-processing: protecting the CMOS circuit with a chemically resistant film(s) and carrying out the micromachining after the circuits are complete and avoiding any high-temperature steps.
- Combined processing: integration in a custom MEMS/CMOS process or utilizing the CMOS layers themselves for MEMS devices.
- Pre-processing: etching wells in the wafer of a depth equal to the total height for the formation of the MEMS device. Fabrication of MEMS devices and protection with an encapsulation layer that is planar with the silicon surface. CMOS circuit fabrication then follows, and removal of the encapsulating film releases the MEMS structure.

Post-Processing

Micromachined thermally isolated regions have been developed by Parameswaran et al.[106] in a CMOS-compatible process for infrared emitter arrays. Post-processing was carried out following a commercial Northern Telecom Canada Ltd. COMS3 DLM (3-μm 13 mask) process. Openings are defined in layers so that the silicon surface is exposed for the EDP etching. The active devices, both p-MOS and n-MOS, were tested after etching and there was no change in device performance.

Fedder et al.[107] have demonstrated the fabrication of intricate structures with a standard CMOS process. Structures are designed in the metal and polysilicon layers and aligned so that a post-processing dry etch step defines the dimensions and a post-processing etch undercuts the structure. Releasing it from the substrate is carried out by RIE etching (see Fig. 10.18). Electrostatic comb drives actuate microstructures that are 107 μm wide and 109 μm long show a resonance amplitude of 1 μm with an ac drive voltage of 11 V. The effective Young's modulus of the structure was found to be 63 GPa. The design rules are as follows. The scaling factor for the 0.8-μm process is $\lambda = 0.4$ μm. The minimum beam width is 3λ (1.2 μm), and the minimum gap is 3λ (1.2 μm); however, for holes, a minimum dimension of 4 μm is required for release. The CMOS circuit is protected by metal-3 to prevent etching. The metal-1 and -2 collar is inserted underneath the break in metal-3 to prevent the etch from reaching the surface and facilitating electrical interconnects to the MEMS structure.

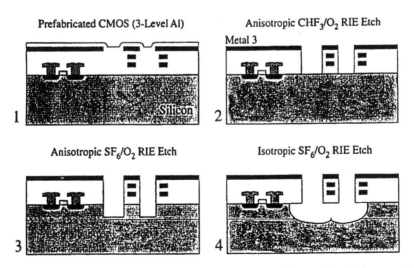

FIGURE 10.18 Example of the use of variable anisotropy dry etch on prefabricated CMOS integrated circuit using the upper level of metallization as the mask: (1) CMOS cross-section, (2) anisotropic CHF_3/O_2 RIE process, (3) anisotropic SF_6/O_2 RIE process, and (4) isotropic SF_6/O_2 RIE process. (Fedder, G. K. et al., in *Proceedings of the International Meeting on MicroElectroMechanical Systems*, IEEE, p. 13, 1996. With permission.)

Finally, planarization techniques have been developed by Lee et al.[108] to prepare a foundry-fabricated chip for post-processing by other low-temperature methods, including electroforming, LIGA, and reactive ion etching.

Mixed Processing

The surface micromachined accelerometer manufactured at analog devices utilizes an integrated process. The BiCMOS process is interleaved with the micromachining so that the higher-temperature steps are completed first; then the lower temperature steps and metallization complete the device fabrication.

A different approach has been taken by the group at the University of California at Berkeley, using high-temperature metallization of tungsten and titanium silicide and TiN barrier layers to replace the aluminum.[109] A double polysilicon single metal, n-well CMOS technology is fabricated first, encapsulated with PSG and low-stress nitride, as shown schematically in Fig. 10.19.

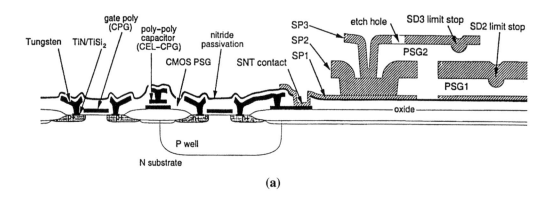

(a)

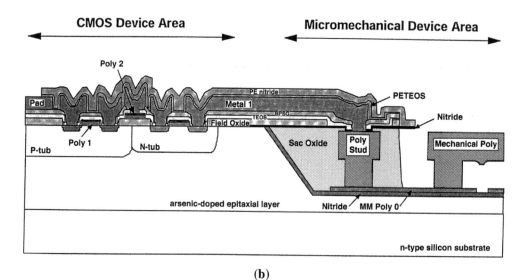

(b)

FIGURE 10.19 (a) Schematic cross-section of the modular integration of CMOS and microstructures technology using post-IC MEMS fabrication with tungsten interconnect technology; (b) a cross-sectional schematic of the subsurface, embedded MEMS integrated technology. (Sniegowski, J. J., in *Microstructures and Microfabricated Systems-IV*, Vol. 98-14, Ed: P.J. Hesketh, H. Hughes, and W. E. Bailey, The Electrochemical Society, Pennington, New Jersey, 1998. With permission.)

Pre-Processing

Smith et al.[110] pioneered the pre-processing of fabricating the MEMS device in a buried well and then encapsulation with a protective passivation film, which is later be removed once the CMOS process is complete. The key to this process is the use of CMP to planarize the surface after MEMS device fabrication and before the CMOS fabrication is begun. The release etch must be highly selective to materials in the MEMS structure and not damage the CMOS outer layers of material (see Fig. 10.19). This process has been very successful in fabricating structures such as pressure sensors, electronic oscillators, microaccelerometers, and gyroscopes. Also, Gianchandani et al.[111] have demonstrated pre-processing integration of thick polysilicon microstructures with a CMOS process. Silicon-to-silicon bonding has been utilized for sensor integration with a pre-etched sealed cavity process,[112] shown in Fig. 10.19(c). The thin membrane formed undergoes plastic deformation, and as a result, the proper design of the cavity geometry is critical to control the gas pressures during bonding. Pressure sensors, accelerometers, and gas flow shear stress sensors have been demonstrated with and without integrated electronics. Lowering the bonding temperature is of key interest to allow more widespread use of this bonding method — because of its incompatibility with many materials and processes.

10.7 Wafer Bonding

Anodic bonding was first demonstrated by Wallis and Pemerantz[113] and can be carried out between a range of glass/metal sealing combinations. The key to obtaining a reliable bond is to minimize the stress; thus, the glass must be selected very carefully to match the expansion coefficient with the silicon. For example, the expansion coefficient of silicon is a function of temperature and that of Pyrex glass is much less so (Fig. 10.20). Sodium ions that become mobile at elevated temperatures (>300°C) produce a depletion layer at the silicon/glass interface.[114] The resulting electrostatic attraction between these two charge layers brings the surfaces into intimate contact, and a chemical bonding takes place that is irreversible and strong. This bond is hermetic and is widely used in microsensor technology for pressure sensors and other devices, such as fluid pumps and valves.[115] The process compatibility with CMOS is

TABLE 10.5 Wafer Bonding Techniques

Bonding Technique	Materials	Surface Treatment	Process	Time	Bond Strength/Comments	Ref.
Anodic bonding	Silicon/7740 Pyrex glass	Clean	350–450°C ~500-1000V	~1-10 min	1-3 MPa[a]/uniform reliable hermetic bond formed	113,116
Silicon-silicon	Si-Si SiO_2-Si and SiO_2/SiO_2	Hydrophobic Hydrophilic	500–1100	Hrs	Difficult to avoid voids unless processed at higher temperatures	117,118
Borosilicate glass	Si/SiO_2 and Si_3N_4		450	30 min	—	120,122
Eutectic	$Si-Au-SiO_2$	Clean and oxide-free	~350	—	148 MPa[b]/Nonuniform bonding area	119
Solder	SiO_2-Pb/Sn/Ag-SiO_2	Needs solder flux	250–400	min	Large difference in thermal expansion coefficient can lead to mechanical fracture	119
Glass frit	SiO_2-glass Ag mixture-SiO_2	Clean	~350	<hr	Difficult to form thin layers	121

under investigation.[116] The circuit is protected from the large electrostatic fields by shorting the gate regions together with a polysilicon strip. After bonding, regions were opened up in this area to facilitate etching. In addition, cavities were drilled ultrasonically in the Pyrex wafer to reduce the electric field over the active circuits.

The key advantage of silicon-silicon direct bonding is that the same material is used so there are minimal thermal stresses after bonding. The wafers must be flat, scrupulously clean, and in prime condition to achieve a reliable bond. First, the wafers are chemically cleaned and surface-activated in a nitric acid solution. Then the wafers are bonded at room temperature in a special jig that has been demonstrated to improve the bonding yield, as shown in Fig. 10.21.[117] A subsequent anneal step increases the bonding strength through a chemical reaction that grows a very thin silicon dioxide layer at the interface (Fig. 10.21(b)). The wafers can be inspected for voids utilizing an infrared microscope or an ultrasonic microscope. High yield has been achieved and the community that developed silicon-on-insulator technology have published conference proceedings on these methods.[118]

Other bonding methods are listed in Table 10.5; these include eutectic,[119] low-temperature glass,[120] glass frit,[121] and borosilicate glass.[122] Materials are selected to minimize the stresses in the bond by selecting a match in the thermal expansion coefficients, or a compliant layer is utilized at the interface.

10.8 Optical MEMS

There is great interest in taking advantage of MEMS devices for the manipulation of optical beams for which they are naturally suited, because of the ease of batch fabrication into large arrays, the small forces required, and the high speed of operation. Examples of devices that have been demonstrated include chopper beam steering, diffraction gratings, optical scanners, Fabre-Perot interferometer, sensors, and spectrometers on a chip. This demonstrates the high speed of operation that is achievable with MEMS technology. A recent excellent review of MEMS devices is given by Wu[123] and an earlier paper by Motamedi.[124]

Components

In the first generation of microfabricated optical components, hybrid assembly into optical devices was carried out rather than integrated functionality. Various optical components were reviewed by Motamedi et al.,[125] including diffractive optical components integrated with infrared detectors, refractive micro-optics, and microlenses. However, the integration of optical components into a micro-optical bench offers several advantages over guided wave approaches; in particular, high spatial bandwidth, independent optical routing 3-D interconnects, and optical signal processing. Moving the individual optical elements out of the plane has greatly expanded the utility of this method.[126] Micro-electrostatic deformable mirrors have also been demonstrated as an effective method for reducing aberrations.[127]

Tunable semiconductor laser diodes have been demonstrated by Uenishi et al.[128] with anisotropically etched (110) silicon substrates and hybrid assembly. A cantilever 1.7 mm long and 8 μm wide defines a (111) surface reflecting mirror normal to the substrate provided modulation from 856 to 853 nm, with a drive of 12.5 V. The etching conditions in KOH solutions were optimized for minimum surface roughness of the (111) surface.

External mirrors for edge emitting lasers can also be produced in GaAs by micromachining. Larson et al.[129] have demonstrated a Fabry-Perot mirror interferometer by combining a GaAs-AlAs vertical cavity laser VCSEL with a suspended movable top mirror membrane. The bottom mirror is a 12.5 period GaAs/AlAs distributed Bragg reflector of 640 Å/775 Å thickness with a center wavelength of 920 nm. The GaAs laser had a center wavelength of 950 nm and a cavity of 2580 Å thick GaAs layer. The top electrode was 2230 Å SiN$_x$H$_y$ with a nominal 200-Å Au reflector. The air gap thickness is modulated around $3\lambda/4$ with electrostatic means to provide a 40-nm tuning range with an 18-V drive. Figure 10.22(b) shows the reflectance spectra from the device. Fabry-Perot tuning was also applied to a photodiode detector by Wu et al.[123] A DBR mirror is defined on top of a movable cantilever. A 30-nm tuning range was achieved

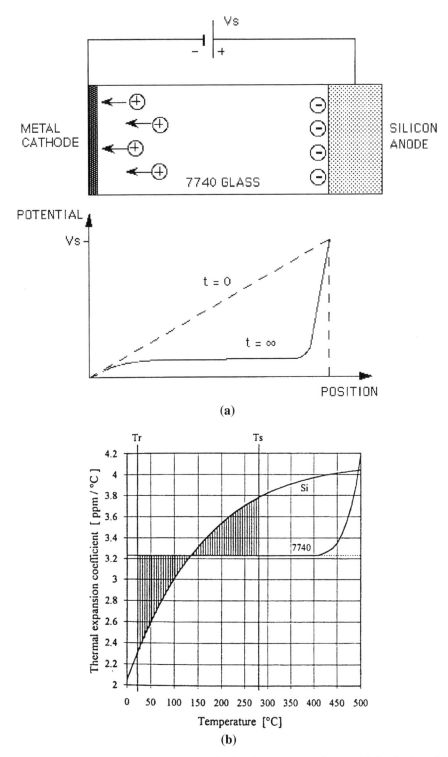

FIGURE 10.20 (a) Schematic diagram showing anodic bonding process and potential distribution; (b) thermal expansion coefficient of Si and Pyrex glass as a function of temperature. (Peeters, E., Process Development for 3D Silicon Microstructures with Application to Mechanical Sensors Design, Ph.D. thesis, Catholic University of Louvain, Belgium, 1994. With permission.)

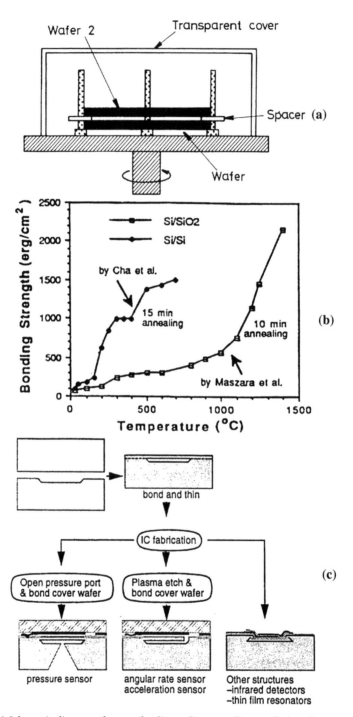

FIGURE 10.21   (a) Schematic diagram of set-up for direct silicon-to-silicon wafer bonding process (Cha, G. et al., in *Proc. First Int. Symp. Semicond. Wafer Bonding Sci. Tech. Appl.*, Eds., Gösele, U. et al., *The Electromechanical Society*, Pennington, NJ, p. 249, 1992 and Masgara, W. P. et al., *J. Appl. Phys.*, 64, 4943, 1989. With permission.); (b) bond strength versus anneal temperature for silicon-silicon direct bond (Mitani, K. and Gösele, U.M., *J. Electron. Mat.*, 21, 669, 1992. With permission.); (c) method for formation of silicon diaphragm by silicon to silicon bonding. (Parameswaran, L. et al., in *Meeting Abstracts of the 194th Meeting of the Electrochemical Society*, Boston, Nov. 1-6th, Abst #1144, 1998. With permission.)

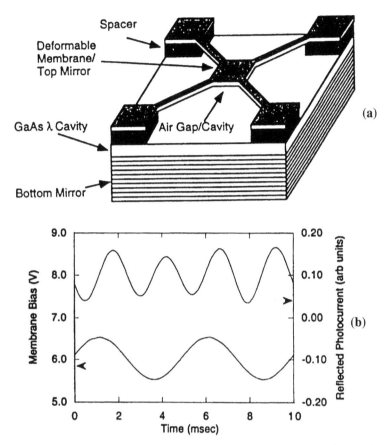

FIGURE 10.22 (a) Schematic diagram of coupled-cavity microinterferometer; (b) membrane voltage and reflected photocurrent traces for the device acting as an intensity modulator for an active wavelength of 933 nm. (Larson, M. C. et al., *IEEE Phot. Tech. Lett.*, 7, 382, 1995. With permission.)

with a 7-V drive and 17-dB extinction ratio. The DBR comprised a top reflector of 18 pairs n-doped $Al_{0.6}Ga_{0.4}As$-$Al_{0.1}Ga_{0.9}$As and a fixed portion was two pairs p^+-doped $Al_{0.6}Ga_{0.4}As$-$Al_{0.1}Ga_{0.9}As$. The bottom DBR was 13 pairs of n-doped AlAs-GaAs mirror grown onto an n^+-doped GaAs substrate.

Modulators and Active Gratings

Micromechanical optical modulators offer certain advantages over traditional means, specifically, maintenance of wavelength coherence, reliability, and temperature insensitivity. Goossen et al.[130] have developed a silicon mechanical antireflectance switch based on a silicon nitride membrane. The thickness of the nitride membrane can be defined precisely so that an antireflection condition occurs when $\lambda/4$ film is brought into contact with no gap. The air gap defines a second coupled cavity at $m\lambda/4$. High transmission is achieved for m even and reflection for m odd. Contrast ratios of 24 dB were obtained with maximum response times of 250 ns.

Sene et al.[131] developed a grating light modulator based on a surface micromachined polysilicon layer. Each polysilicon beam in the array is individually addressed and deflected electrostatically. When the grating is moved, it diffracts the optical beam from the zero order to the +/- first order (see Fig. 10.23). The design incorporates drive plates at the edges of the grating so that the grating lines are not part of the electrostatic actuation. Two anti-sag support lines keep the grating lines parallel during actuation.

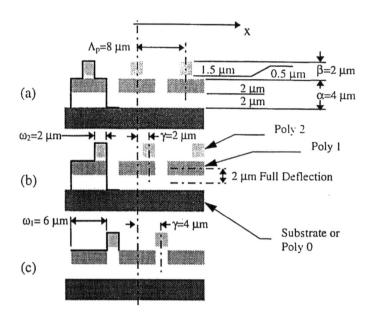

FIGURE 10.23   Schematic representation of variable diffraction grating. The value of γ will depend on the position of the upper, poly-2 grating and the three positions span a full period as indicated by the solid lines, for the following values of γ: (a) γ = 0, (b) γ = 2 μm, and (c) γ = 4 μm. (Sene, D. E. et al., in *Proceedings of the Ninth Annual International Workshop on Micro Electro Mechanical Systems*, San Diego, CA, February 1996, 222. With permission.)

There are also 0.75-μm deep dimples in the upper electrode to prevent stiction. Thermally actuated beams are used to assemble these structures (see Section 10.9).

Burns and Bright[132] have developed microelectromechanical variable blaze gratings operated by adjusting the blaze angle of each slat so that specular reflection of the incident light matches a particular grating diffraction order. Figure 10.24 shows a grating element that was fabricated with polysilicon using the MUMPS process surface micromachining available at MCNC.[45] Both electrostatic and thermal actuators were studied. Light beams of diameters greater than 1 mm and power levels of 1 W have been directed. Measurements with 20 mW HeNe (632.8 nm) produced diffraction efficiency in the far field of 55%, which agreed with model results. Devices with gold metallization demonstrate improved reflectivity; however, they are not fully compatible with CMOS processing.

Scanning Mirrors

Miller et al.[133] fabricated a magnetically actuated scanner with a 30-turn coil on an 11-μm thick permalloy layer. The external magnetic field provided deflection while the coil provided fine control and/or fast motion. Asada et al.[134] fabricated optical scanners with bulk micromachining and a magnetic drive. The electroplated copper used photoresist mold with a period of 50 μm. Coils formed on the *x*-axis and *y*-axis plate of the Pyrex glass plate. Spring constants were evaluated for the *x*- and *y*-axes at 6.48×10^{-4} Nm and 12.8×10^{-4} Nm, respectively, and resonant frequencies of 380 Hz and 1450 Hz, respectively. Judy and Muller[135] demonstrated a torsional mirror scanner moved with a magnetic field. They electroplated a nickel mirror 450-μm square on a polysilicon flexure over a 10-turn coil. With a current of 500 mA and field of ~5 kA/m, the mirror moved more than 45°. Micromachined electromagnetic scanning mirrors have also been fabricated and demonstrated by Miller and Tai.[136] One advantage of magnetic actuators is that both attractive and repulsive forces can be generated. The mirrors are permalloy coated ($Ni_{90}Fe_{10}$) and formed on a silicon substrate with a 20-μm thick epitaxial layer for etch stop. Copper coils are electroplated into a photoresist mold. The mirror is shown schematically in Fig. 10.25(a), and the deflection as a function of the external magnetic field in Fig. 10.25(b). Utilizing these mirrors, holographic

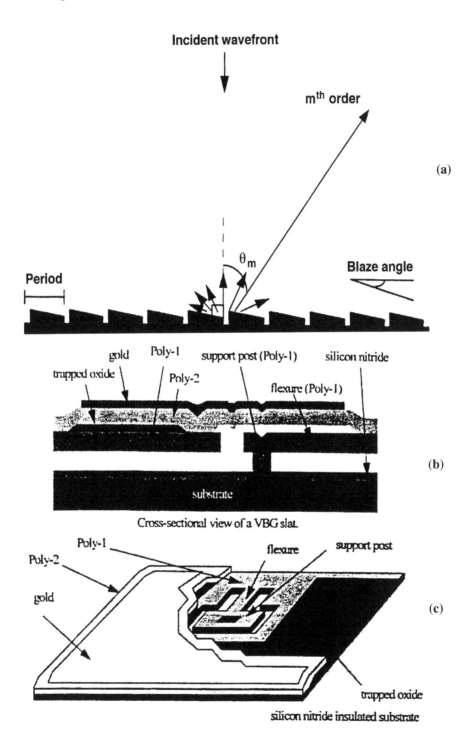

FIGURE 10.24 (a) Schematic diagram of reflective blazed grating illuminated at normal incidence; (b) cross-sectional view of the slat support posts and flexure used in the electrostatically actuated variable blaze grating; (c) the embossing present in the cross-section view of the gold layer. (Burns, D. B. and Bright, V. M., *Sensors and Actuators A*, 64, 7, 1998. With permission.)

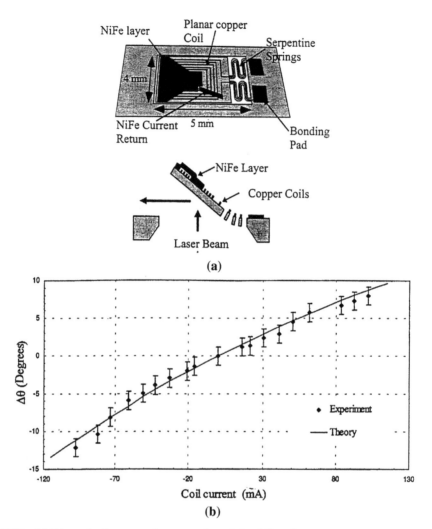

FIGURE 10.25 (a) Schematic diagram and cross-section of the deflected magnetic micromirror; (b) change in deflection angle from bias position for a variable coil current with external field of 994 Oe. (Miller, R. and Tai, Y.-C., *Opt. Eng.*, 36, 1399, 1997. With permission.)

data storage has been demonstrated. Scanners are widely used in printers, display devices, graphic storage systems, and bar code readers.

Kiang et al.[137] have developed polysilicon hinged structures for scanners which utilized an electrostatic drive. The 200×250-μm mirror was rotated 12° with a drive voltage of 20 V, and the device had a resonant frequency of 3 kHz.

Fischer et al.[138] have utilized electrostatic means for mirror deflection with integrated p-well CMOS drive circuits. Two types of torsional mirrors, comprising a polysilicon layer with double-beam suspension and a reflecting area of 75×41 μm^2, have been realized. The mirrors were integrated by post-processing a layer of polysilicon at 630°C, implanting with phosphorus, dose 5×10^{15}/cm^2, and annealing at 900°C, resulting in a resistivity of 100 Ω/sq. The electronics included a demultiplexer circuit for addressing the mirrors and a drive circuit producing 30 V for the electrostatic deflection. Bühler et al.[139] have also developed an electrostatically driven mirror made of aluminum in arrays for low-cost applications. The CMOS-compatible process consisted of modifying the second metal layer deposited process into two successive passes. The first, of 1.1 μm, established a thick metal layer for the mirror plate and the second, of 0.3 μm, a thin metal layer for the hinges. .Deposition was carried out by sputtering at 250°C for

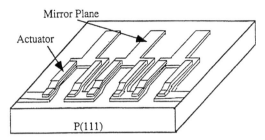

FIGURE 10.26 Perspective view of the ZnO-driven cantilever array. (Huang, Y. et al., in *Digest of Technical Papers, Solid-State Sensor and Actuator Workshop*, Hilton Head, South Carolina, June 1996, 191. With permission.)

improved step coverage; however, a roughness of ~53 nm resulted. Smooth reflecting surface room-temperature depositions were preferred, resulting in roughness ~12.5 nm. The mirrors were released by sacrificial aluminum and oxide etching. They were deflected with a drive voltage of 11 V for a pixel area of $30 \times 40 \ \mu m^2$.

Ikeda et al.[140] fabricated a scanning system that had a two-dimensional array with integrated photo-detectors and piezoresistors. A bulk piezoelectric actuator moves the silicon nitride bridges with a torsional spring scanning angle of 40° and 30° bending and twisting and a resonant frequency of 577 Hz in bending and 637 Hz in torsional motion.

Huang et al.[141] have demonstrated piezoelectrically actuated ZnO cantilevers for application in projection displays, as shown in Fig. 10.26 One of the key advantages of piezoelectrically controlled motion is that the displacement is linearly proportional to the applied voltage. Although sputtered ZnO films have a lower piezoelectric coefficient than PZT, the fabrication process is compatible with CMOS processing. Calculations show that for a beam length of 150 μm, the tip deflection is 0.06°/V or 0.12 μm/V. The ZnO is fabricated with a sacrificial spin-on-glass process, the upper and lower electrodes formed from aluminum. The release step involves a HF vapor etch at low concentration to avoid attack of the Al and ZnO thin film. Tip displacements were measured with a laser interferometer and, in order to distinguish between any thermal contribution to the measured deflections, a drive waveform of unbiased square wave was selected. The frequency response was over 80 Hz with a 1 μm air gap and limited to about 10 Hz with a 0.5-μm air gap, indicating the dominance of squeeze film damping. The thermal deflection was about 2 to 3 orders of magnitude less than the piezoelectric response.

Spectrometer on a Chip

Surface micromachining has been demonstrated by Lin et al.[142] for out-of-plane assembly of optical elements. The hinge mechanism allows the plate to be moved to a vertical position and locked into place with a spring latch (see Fig. 10.27(a)). Micro-Fresnel lenses with a 280-μm diameter and an optical axis 254 μm above the plane of the silicon have been realized. The slide latch precisely defines the angle of the element (see Fig. 10.27(b)). It has a 'V' shape 2 μm wide in the center. In addition, rotating structures can be realized, such as a rotating mirror. Surface micromachined free space optical components have been demonstrated by Lee et al.[143] for the collimation and routing of optical beams. Microgratings with 5-μm pitch are fabricated on flip-up structures metallized with a thin layer of aluminum. A diffraction pattern was imaged with a CCD camera and beams directed to a second grating into the zero-order beam. Recent progress in micro-optical systems is reviewed by Bright et al.,[144] including mirrors, Fresnel lenses, gratings, and larger systems. All of these structures were fabricated by surface micromachining in the MCMC processes[145] using an electrostatically actuated gold surfaced mirror.

Micromechanical Displays

Miniature display systems have been commercialized by Texas Instruments in projection television systems.[146] They offer the advantage of cold operation and high contrast (>100:1) compared to the

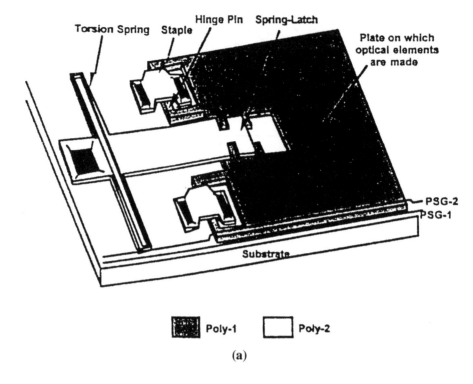

Torsion Spring Staple Hinge Pin Spring-Latch

Plate on which optical elements are made

PSG-2
PSG-1

Substrate

Poly-1   Poly-2

(a)

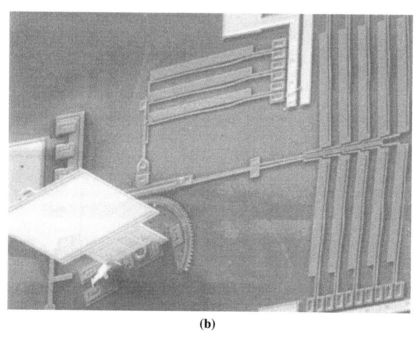

(b)

FIGURE 10.27 (a) Schematic diagram of the three-dimensional micro-optics element. After release etch, the micro-optical plate can be rotated out of the substrate plane and locked by the spring latches (Bright, V. M. et al., *IEICE Trans. Electron.*, E80-C, 206, 1997. With permission.); (b) SEM micrograph of the micro-Fresnel lens in the micro-XYZ stage integrated with eight scratch drive actuators. (Lee, S. S. et al., *Appl. Phys. Lett.*, 67, 2135, 1995. With permission.)

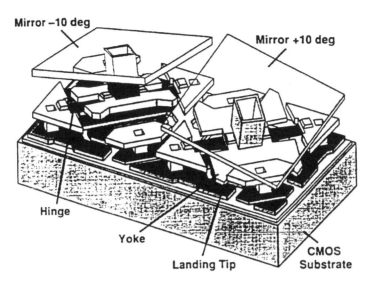

FIGURE 10.28   A two digital micro-mirror device pixels mirrors are shown as transparent for clarity in the diagram. (Hornbeck, L. J., in *Symposium Micromachining and Microfabrication, Proceedings of SPIE,* Vol. 2783, Austin, TX, 1996, 2. With permission.)

cathode-ray tube (CRT). A CMOS-like process over a CMOS memory element defined aluminum mirrors, each $16 \times 16 \ \mu m^2$ in area, that can reflect light in two directions. The hinges and support structure are positioned under the reflector element. The display element can be moved by up to +/-10° and at speeds up to 10 μs, which make them suitable for standard rate NTSC video. Figure 10.28 shows the structure of one element in the 124×124 elements. The underlying memory cell operates on 5 V. Eight-bit pulse width modulation of the mirror state produces a gray scale or color image. The fabrication process is compatible with CMOS processing; however, to date, only the display has been fabricated on chip, and hybrid packaging is used for the drive electronics.

Optical Disk Pick-Up Head

The technology for fabrication of optical pick-up heads is currently limiting the track access speed and maximum data rate. Ukita et al.[147] developed a flying optical head consisting of a laser diode and a photodiode. The recording medium was a phase change material SbTeGe on a 130-mm diameter glass substrate. The read/write head was mounted on a gold electrode on the slider. Light reflected from the medium was fed back into the active region, with head-to-disk spacing of typically <1 μm. The 1.3-μm wavelength InGaAsP laser diode has a spot diameter constrained by the ridged waveguide shape. The reflection of the recording medium was reduced by an antireflection coating of 0.24-μm SiN_xH_y. Recording was achieved by producing a change in the reflectivity of the phase change medium, based on crystalline to amorphous states of the film, typically at a power of 20 mW and data rates of up to 1 MHz. Reading is achieved at lower power and higher rates. A single chip which integrates a photodetector, several Fresnel lenses, and a semiconductor laser has been demonstrated by Lin et al.[148] Surface micromachining allows the integration of a 45° reflector, Fresnel lens, rotary beam splitter, and photodetector on a chip. A 980-nm laser source was attached to the surface with an optical axis 245 μm above the silicon surface. The beam splitter has a 20-nm gold layer and the reflectors and mirrors have a thicker layer of gold. The focusing lens, with a NA of 0.17, results in a spot with FWHM of 6.1 μm in the *x*-direction and 2.6 μm in *y*-direction. The advantages of this system are small size, light weight, and potentially low-cost integration of actuation on chip to achieve track-to-track alignment.

10.9   Actuators for MEMS Optics

A variety of actuation principles have been demonstrated that are compatible with optical elements on chip. A limited number of these are compatible with CMOS processing. Further developments in integration of miniature optical components with functional optical MEMS devices are expected in the near future.[149]

Electrostatic Comb Drive Actuators

The electrostatic force generated between two conductive plates provides a compact, efficient actuation principle. These actuators are often limited to small displacements, and the force depends on the capacitance of the electrode. Interdigitated comb drives provide increased force and extended linear range, compared to parallel plate design. Recently, a large displacement actuator has been designed at Sandia National Laboratories using multiple stages of gearing (see Fig. 10.29). In general, high voltages of ~100 V are required to drive electrostatic actuators. Circuits have been designed to work in conjunction with micromachining processes to integrate the actuators and drive circuit on the same chip.[150]

Linear Microvibromotor

The linear microvibromotor is based on impact momentum to produce small displacements. Each impact from the comb drive produces a step of typically 0.27 μm. Although this is an impulsive drive, the standard deviation between steps is 0.17 μm. A maximum speed to 1 mm/s has been demonstrated and used for a slide-tilt mirror and alignment of beams in a fiber coupler.[151]

Scratch Drive

The scratch drive is based on applying pulses to a plate and allowing the successive bending and release to produce lateral motion of the bushing to move out.[152] During release, the non-symmetric functional forces produce an incremental motion DX. Microactuators and XYZ stages have been developed for a micro-optical bench.[153] A comb drive is used to drive the torsional z actuator with displacements up to 140 μm. Figure 10.29(b) is a schematic diagram of the microactuator stage. The lower 45° mirror is moved to achieve lateral adjustment of the beam. The translation stages are defined in the first (poly-1) layer, and the second (poly-2) layer defines the optical elements. The scratch drive actuator is particularly suited for this application because of the high forces and small step size (~10 nm) at moderate drive voltages of 87 V. Two-dimensional optical beam scanning has been demonstrated of several mm in the far field at a distance of 14 cm utilizing a HeNe laser source. A micro-Fresnel lens has been integrated into the actuator with eight scratch drive actuators

Thermal Actuator

Thermal actuator arrays for positioning surface micromachined elements have been demonstrated, as well as automated assembly of polysilicon mirrors and other elements with thermal actuators.[154–156] The thermal actuator was designed for vertical and horizontal motions. The horizontal actuator is shown schematically in Fig. 10.29(c). These structures were fabricated with MUMPS processes.[145] It comprises a hot and a cold arm of polysilicon. Initially, the components are on the surface of the substrate; however when current is passed, one side becomes hotter than the other. The deflection of up to 16 μm is produced at moderate power levels of 16 mW and forces of 7 μN. In a second mode of operation if the actuator is heated above that for maximum deflection the hot arm becomes shorter than before and a negative deflection results with power off condition (see Fig. 10.29). The vertical actuator consists of two polysilicon beams separated by a 0.75-μm air gap. The lower arm is wider than the top one. When current is applied, the upper arm becomes hotter, providing higher electrical and higher thermal resistance, and thus a higher temperature driving the arm downward toward the substrate. Back-bending of the vertical actuator is particularly useful for clamping components in automatic assembly operations without the continuous application of current. These actuators are suitable for forming arrays of devices; designs of linear motors have been described. A self-engaging locking mechanism is also described which takes advantage of a tether from the upper polysilicon layer interacting with a key hole on a movable plate. When the movable plate is rotated out of the plane of the wafer, the tether slides into the wide section of the opening, which is wider than the tether. For example, the assembly of a polysilicon mirror 104 × 108 μm is achieved with two vertical actuators and a linear motor. Bending of the actuators is achieved with current levels of 4.2 mA at a voltage of ~14 V and the linear motor is operated with 24 mA at 5.5 V.[156]

Magnetically Driven Actuators

Magnetically driven micromirrors have been developed by Shen et al.[157] and one key advantage of magnetic drive is that actuation can be achieved in two directions by reversing the current flow, unlike electrostatic drive. A CMOS-compatible process was utilized to fabricate a suspended plate approximately 200 × 200 μm on a side. The deflection of the plate was ±1.5° with a single suspension; however, with multiple suspension arms, up to ±27° was observed at a drive current of 20 mA. Other work on magnetically driven mirrors was discussed in the previous section.

10.10 Electronics

For electronic applications of MEMS, the compatibility of the micromachining processes with IC processing is key for integration with active electronic components. There has been considerable work on fabrication of passive components by micromachining, specifically capacitors, inductors, and microwave transmission lines, and other components. The key advantages for passive component integration are ease of manufacturability for the higher frequency range of 100 to 1000 GHz where characteristic dimensions are mm to sub-mm range compatible with.micromachining. This offers the opportunity to fabricate components and packaging in an integrated approach. Applications include test instruments, communications systems, radar, and others.

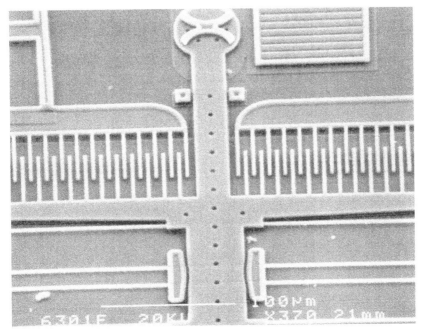

FIGURE 10.29(a) Electron micrograph of an interdigitated electrostatic drive. (Courtesy of Sandia National Laboratories' Intelligent Micromachine Initiative; www.mdl.sandia.gov/Micromachine. With permission.)

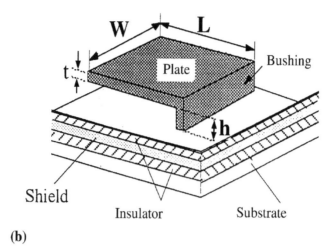

(b)

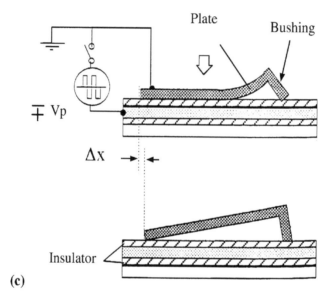

(c)

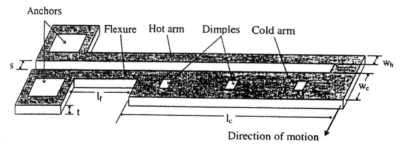

FIGURE 10.29(b-c) (b) Scratch drive actuator. (Fukuta, Y. et al., in *Proceedings of the 10th Annual International Workshop on Micro Electro Mechanical Systems*, Nagoya, Japan, IEEE, New Jersey, 1997, 477. With permission.); (c) schematic diagram of the lateral thermal actuator. Typical dimensions are given in the text. (Comtois, J. H. and Bright, V. M., in *Digest of Technical Papers, Solid-State Sensor and Actuator Workshop*, Hilton Head, South Carolina, June 1996, 152. With permission.)

TABLE 10.6 Microrelays

Application	Fabrication Process	Drive	Contact On-Resistance	Maximum Current	Off- Resistance/ Breakdown voltage	Switching Time	Insertion Loss	Ref.
Electrostatic								
Automated test equipment	Bulk micromachining and anodic bonding	<100 V	<3 Ω	—	—	<20 μs	—	172
Switching	CMOS compatible	1-10 V with DC bias of 30-54 V	—	—	—	<1 ms	—	176
RF to microwave	Surface micromachining	28 V at >50 nA	~0.22 Ω	200 mA	—	—	0.1 db at 4 GHz	173
RF to microwave	Surface micromachining on GaAs	~30 V	—	—	—	—	0.3 dB at 20 GHz	174
Switching	Electroplated metal films	24 V	0.05 Ω (initial)	5 mA (single contact); 150 mA (multiple contacts)	>100 V	—	—	170–171
Thermal								
Small-signal RF	Surface micromachining	20–100 V [10 μW]	10–80 Ω	1 mA	—	2.6– 20 μs	-	175
Switching	MUMPS	7-12 V	2.4 Ω	80 mA	—	—	—	178
Magnetic								
RF impedance matching	Surface micromachining in polysilicon	12 mW	2.1–35.6 Ω	>1mA	>10^{12} Ω 400 V	<0.5 ms	—	177
Electrical control circuits	Polyimide mold and electroplated metals	180 mA (33 mW)	0.022 Ω	1.2 A	—	0.5-2.5 ms	—	179

RF and Microwave Passive Components

Large suspended inductors were fabricated by Chang et al.[158] to demonstrate the integration of such components with electronic circuits. A 100-nH spiral with a self-resonance at 3 GHz integrated with a balanced cascade tuned amplifier with gain of 14 dB centered at 770 MHz, implemented with standard digital 2-μm CMOS process. The amplifier had a noise figure of 6 dB and a power dissipation of 7 mW, operated from a 3-V supply.

A 1-GHz CMOS RF front end has been demonstrated by Rofougaran et al.[159] The application is for direct conversion wireless receiver, or zero RF, and frequency shift-keying receivers. The building blocks consist of low-noise RF amplifier, down-conversion mixer, and the contact pad is modified to reduce electric noise by connecting metal layer-1 to RF ground and metal-2 to RF input. The MEMS aspects are in the tuned amplifier's 50-nH spiral inductor, which would normally self-resonate at 700 MHz due to capacitive coupling to the substrate through the 1-μm field oxide, so that the use of standard CMOS inductors is generally limited to 5 to 10 nH. In this work, a gas-phase isotropic etch provides removal of the substrate from under the coil, allowing 50 nH to be realized.

López-Villegas et al.[160] have studied integrated RF passive components fully integrated with CMOS processing. Spiral inductors of 10 to 20 turns with 10-μm line width were characterized over the frequency range 50 MHz to 40 GHz. The influence of the material was key to defining the self-resonance of the structure, typically 1 to 6 GHz. Interdigitated capacitors were also fabricated with values of 0.5 to 1.35 pF and had self-resonances of 6 to 7 GHz.

Microwave Waveguides

Rectangular waveguides have been fabricated by McGrath et al.[161] by a bulk micromachining process and characterized over the frequency range 75 to 110 GHz. Slots are formed in a (110) silicon wafer, which were subsequently coated with 250 Å Cr and 5000 Å Au to form a plating base for a further 3 μm of electroplated Au. Losses measured in a 2.5-cm guide were comparable to commercial waveguides at about 0.024 dB/m. Active and passive components could be integrated into these structures.

Circuit components have also been fabricated by bulk micromachining with the added advantage of an integrated package by Franklin-Drayton et al.[162,163] and Katehi et al.[164] A series open-end tuning stub and a stepped impedance low-pass filter were realized for the frequency range 10 to 40 GHz. The method of design is based upon a quasi-static model utilizing TEM or quasi-TEM approximation, following this with a finite difference time-domain technique to evaluate the performance. The mesh was carefully selected to reduce truncation errors and grid dispersion errors, typically less than 1/20 of the shortest wavelength. Electrical conductors are assumed perfect conductors and, at dielectric interfaces, the average of the two permitivities was taken. Metallization comprised Ti/Au/Ti with subsequent electroplating to a final thickness of 3 μm Au. Alignment between cavities and waveguide structures was achieved via windows etched through the wafer thickness. Figure 10.30(a) shows a five-section, stepped impedance low-pass filter in which the 100- and 20-ohm impedance steps are formed by conductor widths of 20 μm and 380 μm, with slot widths of 210 μm and 30 μm, respectively. Figure 10.30(b) shows the integrated packaging topology and micrographs of the fabricated structures.

Microwave transmission lines have also been fabricated by Milanović et al.[165] with a commercial CMOS process with post-processing micromachining. The transmission lines were designed to operate in TEM mode with 50 and 120 Ω nominal characteristic impedance with standard layout tools. The post-processing etch was used to remove the silicon from underneath the conductive aluminum transmission lines to lower the losses. Figure 10.30(c) shows the simplified layout of the co-planar waveguides where the open areas are shaded in black. The open areas are first etched with a xenon difluoride, followed by anisotropic chemical etching with EDP. The cavities connect beneath the aluminum conductors, but enough material remains for mechanical stability. A fully formed trench also lowers electromagnetic coupling to the substrate. Measurements for test chip with three different lengths, 0.8 to 3.7 mm, with open and short stubs were made between 1 and 40 GHz. Insertion loss was calculated based on transmission line measurements, as shown in Fig. 10.30(d).

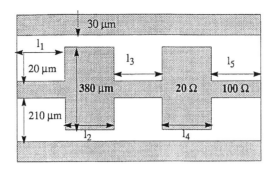

FIGURE 10.30(a) Integrated packaging. (Drayton, R. F. et al., *IEEE Trans. Microwave Theory and Tech.*, 46, 900, 1998. With permission.)

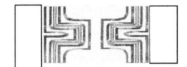

Upper Cavity Package   **Circuit Layout**

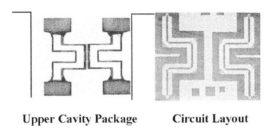

Lower Cavity Package

FIGURE 10.30(b) Microfabricated two-stage coupler. Chip measures 7.6 × 10 cm. Upper cavity shows probe windows as dark region. Conductor is dark region in circuit layout. (Franklin-Drayton, R. et al., *The International Journal of Microcircuits and Electronic Packaging*, 18, 19, 1995. With permission.)

Tuning Fork Oscillator

A tuning fork-based oscillator has been fabricated by surface micromachining in polysilicon with integrated electronics by Roessig et al.[166] The device has an output frequency of 1.0022 MHz and exhibits a noise floor of –88 dBc/Hz at a distance of 500 Hz from the carrier. Previous surface micromachined oscillators used resistors to detect motion — which limits the noise floor. In this device, the capacitive detection is employed and, hence, the large impedance at the sensing node introduces a smaller input current noise than the resistive method. Figure 10.31 shows the integrated device with a double-ended tuning fork with tines 2 μm wide, 60 μm long, and 2 μm thick. These are fabricated in an etched well prior to the CMOS circuit, as discussed in Section 10.6.

Thermal Devices

A great deal of progress has been made in the integration of thermal sensors, infrared sensors, and gas flow sensors with on-chip CMOS electronics. Baltes et al.[167] describe the fabrication and operation of a thermoelectric air-flow sensor and an infrared sensor, in addition to measurements of the thermophysical properties of material components of CMOS electronics.

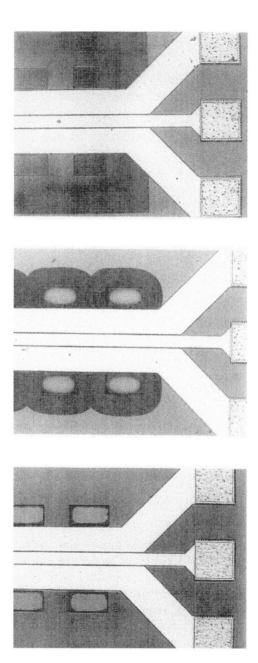

FIGURE 10.30(c) SEM micrograph of the 50 Ω transmission lines, of width 130 μm. (a) after CMOS fabrication; (b) after isotropic etch; and (c) after combined etch. (Milanovic, V. et al., *IEEE Trans. Microwave Theory and Techniq.*, 45, 630 1997. With permission.)

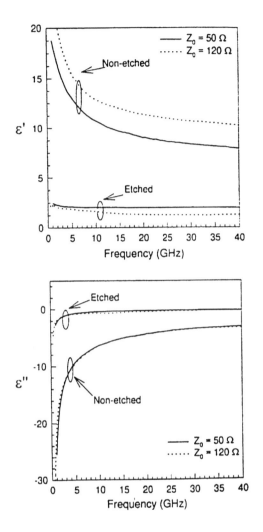

FIGURE 10.30(d) Measured effective dielectric constant of transmission lines before and after etching. (Milanovic, V. et al., *IEEE Trans. Microwave Theory and Techniq.*, 45, 630 1997. With permission.)

Bandgap Voltage Reference

Reay et al.[168] have demonstrated thermally isolated bandgap voltage reference temperature sensors. The reference has a 5300°C/W thermal resistance isolation from the substrate silicon, a 2.5-ms thermal time constant, and uses 1.5 mW at 25°C ambient temperature. This regulation achieved a reduction of the temperature coefficient from 400 to 9 ppm/°C for an ambient temperature range of 0 to 80°C. A schematic of the circuit is shown in Fig. 10.32(a). The servo-amplifier adjusts the reference voltage until the currents in the two branches are equal and thus generate the bandgap voltage.

RMS Converter

Measurement of true RMS voltage is complicated by the fact that the waveform shape is important and the peak value is only utilized the waveform shape must be known. Commercial true RMS meters utilize thermal heat to evaluate the power in the signal with specialized components. Klassen et al.[169] have developed a CMOS fabrication process in which a suspended, thermally isolated platform is utilized for this purpose. On the platform, a resistive heater and diode-connected vertical BJT are formed, temperature sensitivity of 2 mV/K. The beams, 85 to 225 μm in length, are defined in an oxide-nitride diaphragm and undercut with a bulk-silicon etch in TMAH, as shown in Fig. 10.32(b). The beams had a maximum

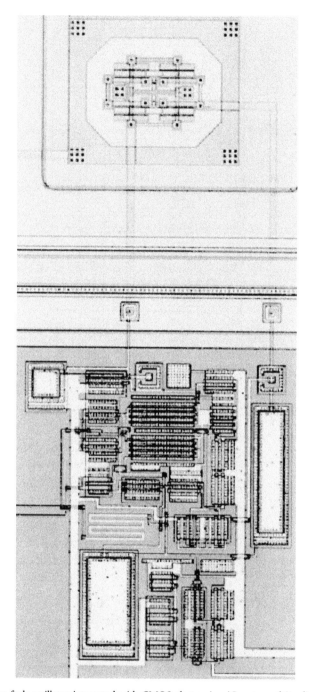

FIGURE 10.31   Tuning fork oscillator integrated with CMOS electronics. (Courtesy of Sandia National Laboratories' Intelligent Micromachine Initiative; www.mdl.sandia.gov/Micromachine and T. Roessig U.C. Berkeley. With permission.)

thermal resistance of 37,000 K/W in air. The cascade CMOS operational amplifier, followed by a source follower to provide up to 50 mA of current for the heating element, operated from a 5-V supply. The quiescent power consumption of the amplifier was 950 μW and the –3-dB frequency was 415 MHz. With a sinusoidal input signal at 1 kHz, the measured dynamic range of the system was from 2.4 mV to 1.1

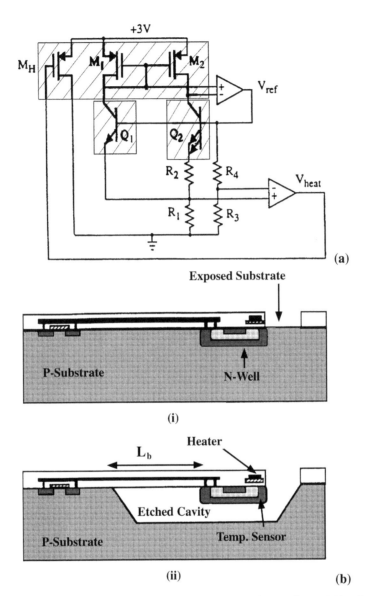

FIGURE 10.32 (a) Schematic diagram of a complete bandgap reference showing the PMOS heating transistors and thermal control loop. The shaded regions are thermally separated isolated n-wells; (b) cross-sectional view of a thermoclement for an RMS converter, at different stages in the fabrication process: (i) upon completion of the CMOS process, and (ii) after the post processing step of etching in TMAH. (Reay, R. J. et al., *IEEE J. Solid-State Circuits*, 30,1374, 1995. With permission.)

V r.m.s. (53 dB). Measurements of nonlinearity were 4% compared to a Hewlett-Packard HP3478A, which had a specified nonlinearity of 0.2% for low-frequency inputs.

Microrelays

An important illustrative example of MEMS process integration in which electronic and mechanical function are combined is the microrelay. There has been considerable interest in relays and switches for high-impedance isolation of circuit components, and for RF and microwave switching. There is insufficient space in this chapter to give a comprehensive overview of these activities; however, Table 10.6 summarizes some of the work that has been directed toward the success of these microdevices — grouped by actuation method. These devices typically have lifetimes of greater than 10^6 cycles. Zavracky et al.[170] and McGruer et al.[171] have built electrostatic relays with multiple contacts to increase the maximum current-carrying capacity. Micrographs of the electroplated thick film of the relay are shown in Fig. 10.33. Other electrostatic designs demonstrate low power consumption,[172–175] and CMOS-compatible microrelays have been demonstrated by Grétillat et al.[176] Novel latching surface micromachined devices have been demonstrated and an example[177] is shown in Fig. 10.34(a).[178] Alternative actuation schemes are thermal and magnetic. Finite element modeling of the actuator and the magnetic circuit has been carried out by Taylor et al.[179] to provide improved design methods for these devices (see Fig. 10.34(b)). The thickness of the permalloy layer must be large enough to avoid saturation of the magnetic field. Minimum switching current and optimum coil spacing for operation at under 100 mA were achieved in these devices. The hold force is high — to provide low contact resistance. Contact resistance is a critical issue and has been studied in macroscopic relays by Holm[180] and Schimkat[181] with forces in the µN range.

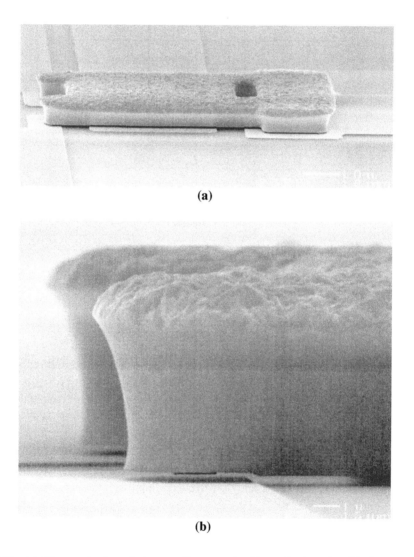

(a)

(b)

FIGURE 10.33   (a) Micrograph of an electrostatically actuated gold metal microrelay; (b) close up view of the contact area. (McGruer, N. E. et al., in *Digest of Technical Papers, Solid-State Sensor and Actuator Workshop*, Hilton Head, South Carolina, June 1998, 132. With permission.)

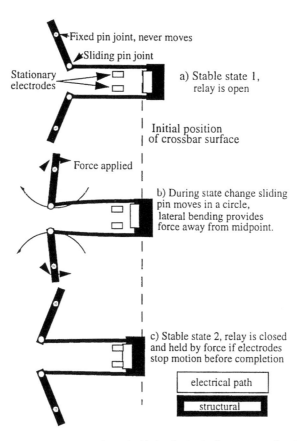

FIGURE 10.34(a) Bistable action in the relay frame hold the device in the open or closed state without actuation. (Kruglick, E. J. J. and Pister, K. S. J., in *Digest of Technical Papers, IEEE Solid-State Sensor and Actuator Workshop,* Hilton Head, SC, 1998, 333. With permission.)

Integrated Ink-Jet Printers

Integration of fluidic elements with a drive circuit on a single chip has been demonstrated by Krause et al.[182] Figure 10.35 shows the integrated structure. The fluid cavity is bulk micromachined from the back of the wafer, while the (2 μm-process) CMOS electronics are fabricated on the front side. The devices were fabricated on 4 in. dia. p-type 30-50 Ω-cm, 400 μm thick (110) silicon to achieve high aspect ratio slots. The heating elements for bubble formation are integrated into the structure with 30 μJ required to eject each ink droplet. The heaters have a possible output of 1 GW/m², but, in this application, generate 7.5 W in a 50-nozzle array firing at 5 kHz. The chip is covered with an electroplated layer, shielding the device from mechanical, chemical, and electric damage comprising 4-μm nickel and 1 μm gold on a Ti/Cu adhesion layer.

10.11 Chemical Sensors

There has been great success in developing chemical sensors; however, one of the key stumbling blocks is the compatibility of chemically sensitive layers with IC processes, thus limiting the possibilities for process integration. When one considers it the processing complexity required to integrate the electronics and chemically sensitive layers on chip is justified, there are a number critical issues:

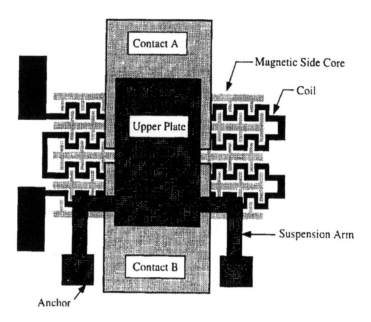

FIGURE 10.34(b) Schematic top view of the magnetic microrelay, illustrating the relative positions of the upper moveable plate, contact, side cores, and coil. The conductor width in the coil is 80 μm. (Taylor, W. P. et al., *J. Microelectro. Mech. Syst.*, 7, 181, 1998. With permission.)

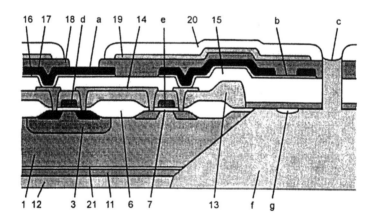

FIGURE 10.35 Structure of the backshooter microsystem ink-jet print head (not to scale) illustration: 1 - substrate, 6 - field oxide, 7 - gate oxide, 11 - etch stop layer, 12 - PECVD SiO_2, 13 - BSG, 14 - first Al layer, 15 - undoped silica glass, 16 - heater layer, 17 - second Al layer, 18 - thermal throttle layer PECVD Si_3N_4, 19 - galvanic adhesive layer Ti/Cu, 20 - carrier layer Ni/ASu, 21 - thermal SiO_2. Elements a - bond pad (Al), b - heating element, c - nozzle, d - p-MOS transistor, e - NMOS transistor, f - ink-jet chamber, g - vapor bubble formed. (Krause, P. et al., *Proceedings of Transducers 95*, Stockholm, Sweden, 1995. With permission.)

- Is the cost per packaged functional sensor lowered by integration?
- Does the application require integration (i.e., is small size essential?)?
- Is the sensing function improved by integration?

An excellent review of recent work on chemical sensors has been published by Janata et al.[183] Most work has focused on hybrid designs in which the electronics and chemical sensor arrays are fabricated separately and then interconnected. The work of Madou et al.[184] in a blood gas sensor for the measurement of pH, CO$_2$, and O$_2$ *in vivo* is an example of this approach. Here, the device was fabricated by bulk micromachining on a thin silicon piece approximately 350 μm wide and bonded to an associated interface circuit chip that was made at an IC foundry.

ISFET

The chemically selective FET developed by Janata[185] and Bergveld[186] demonstrates specific analyte selectivity based on an FET structure with the gate coated with a chemically sensitive layer exposed to the solution. The sensing mechanism is based on a variety of surface-specific interactions.[187] These devices may be configured as gas-sensitive devices, for hydrogen detection,[188] ion-selective devices,[189] enzyme FETs, and most recently, suspended gate structures. Figure 10.36 shows a schematic diagram of an ISFET. The important characteristic of these sensors is that the gate potential and, hence, the channel threshold voltage are defined by the potential applied at the reference electrode and the interfacial potentials. This potential is related to the activities of participating ions rather than their concentrations. Hydrogen ion sensitivity is intrinsic to the dielectric material coating the gate. Bousse[190] has developed an OH site-binding theory to account for the pH dependence of the FET for oxide and nitride gates. The most stable gate dielectric choices are TaO$_2$ and Al$_2$O$_3$. Advanced concepts for back-side contact FETs are reviewed by Cane et al.[191] One of the advantages of ISFET technology is that it can be made compatible with CMOS processing. An integrated CHEMFET, demonstrated by Domanský et al.,[192] is capable of measurement of work function and bulk resistance changes. Here, a carbon black/organic polymer composite film for the detection of solvents covers the gate, as shown in Fig. 10.36(b).

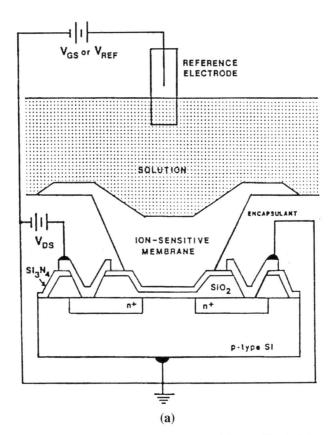

(a)

FIGURE 10.36(a) Schematic diagram of an ISFET. (Janata, J., in *Solid State Chemical Sensors*, J. Janata and R. J. Huber, Eds., Academic Press, New York, 1985. With permission.)

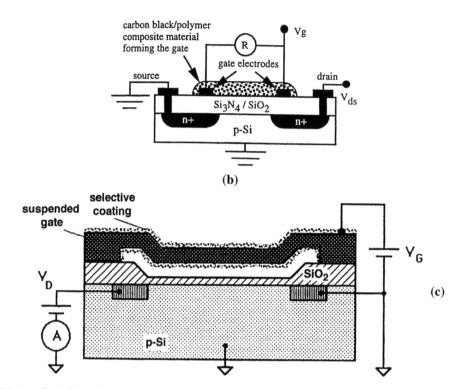

FIGURE 10.36(b-c) (b) Carbon black impregnated gate FET. (Domanský, K. et al., in *Digest of Technical Paper, IEEE Solid-State Sensors and Actuators Workshop,* Hilton Head, SC, 1998, 187. With permission.); and (c) suspended gate FET. (Mosowicz, M., and Janata, J., in *Chemical Sensor Technology,* T. Seiyama, Ed., Elsevier, New York, 1988. With permission.)

Chemically sensitive layers are, in general, not process compatible with CMOS circuit fabrication. Approaches that have been made in this area include fabrication of the electronics first, followed by deposition of the chemically sensitive membranes while the CMOS circuit is covered with a passivation coating.

Hydrogen Sensor

The Pd gate MOS hydrogen sensitive FET was invented by Lundström et al.[188] The device is shown schematically in Fig. 10.37(a). Upon exposure to hydrogen, dipoles are created at the SiO_2/Pd interface producing a shift in the threshold voltage. A hydrogen chemical sensor has been successfully integrated with electronics components at Sandia National Laboratories (Rodriguez et al.[193]). This sensor utilizes two Pd/Ni layers, one as a chemiresistor and the second as the gate of an FET. Figure 10.37(b) shows a picture of the sensor with integrated heaters and temperature sensors. The FET shown in Fig. 10.37(b) is more sensitive at low concentration ranges and has a logarithmic response; however, the conductimetric sensor has a square-root dependence on the hydrogen concentration. The sensors are operated at an elevated temperature of approximately 100°C to increase the reaction kinetics and improve reversibility. Typical response data are shown in Fig. 10.37(c) with a 1% hydrogen concentration, resulting in a response time of a few seconds. The sensor combination has an exceptionally wide dynamic range of six orders of magnitude, and response time was within 5 seconds. Heating is achieved through two power transistors and temperature monitoring with an array of nine p/n junction diodes. Typical die size is 270 × 120 mils. Devices have been demonstrated with stabilities of over 60 days and show reversible behavior. Sensors are being commercialized for detection of hydrogen in aerospace applications. Advanced versions of this sensor have also been produced with fully integrated op-amps and control electronics, including analog capacitors, high-value polysilicon resistors, current mirrors, and operational amplifiers.

Gas Sensors

Microhotplates have been developed by Suehle et al.[194] for tin oxide chemical sensors. These devices are conductimetric sensors for reducing gases and operate at elevated temperatures, typically ~350°C. A suspended sandwich structure of CVD oxides, encapsulating a polysilicon heater, and integrated with an aluminum layer to provide thermal diffusion, is shown in Fig. 10.38. Post-processing was carried out after a standard CMOS process by EDP etching with added aluminum hydroxide to ensure passivation of any exposed Al conductors. Heating current (mA) was provided to the polysilicon layer and temperature sensing from van der Pauw aluminum layer with a temperature coefficient of resistance typically 0.003667/°C. The hotplates were effectively thermally isolated, showing efficiency of 8°C/mW in air, thermal time constant of 0.6 ms, and maximum operating temperature of 500°C.

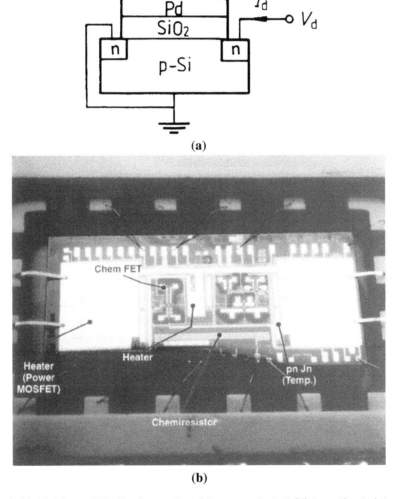

FIGURE 10.37(a-b) (a) Pd-gate FET. (Lundström, I. and Svensson, C., in *Solid State Chemical Sensors*, J. Janata and R. J. Huber, Eds., Academic Press, New York, 1985. With permission.); (b) Photograph of the robust hydrogen sensor with integrated temperature sensors, Pd gate FET, chemiresistor, and heater elements. (Rodriguez, J. L. et al., *IEDM Tech. Digest*, IEEE, San Francisco, CA, Dec. 1992, 521–524. With permission.)

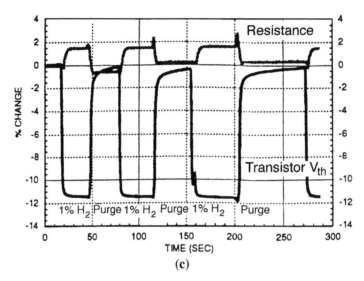

(c)

FIGURE 10.37(c) Response of sensor to hydrogen. (Rodriguez, J. L. et al., *IEDM Tech. Digest*, IEEE, San Francisco, CA, Dec. 1992, 521–524. With permission.)

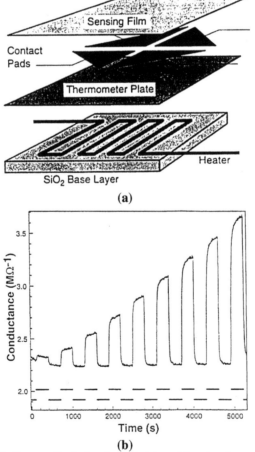

(b)

FIGURE 10.38 (a) Schematic diagram of a single microhot plate and functional cross-section of component parts; (b) static response at 130°C of a Pt/SnO$_2$ microsensor to on/off CO exposures, into dry air, of increasing concentrations from 5 to 45 ppm. (Suehle, J. S. et al., *IEEE Elec. Dev. Lett.*, 14, 118 1993. With permission.)

SnO$_2$ was deposited onto the hotplate by reactive sputter deposition in ultrahigh vacuum; and by heating the platform during deposition, selective control of the local material properties was achieved (such as grain size and the conductivity). After deposition, annealing was also carried out selectively *in situ* on the hotplates. The selectivity of these devices can be further modified by addition of catalytic metals such as Pt, Pd, or Ir. Semancik and Cavicchi[195] have demonstrated kinetic sensing on micro-hotplates by modulation of the sensor temperature to enhance analyte discrimination. Microhotplates were also fabricated with tungsten metallization so they could operate up to 800°C. The response of a Pt-doped SnO$_2$ sensor operating at 130°C to CO gas is shown in Fig. 10.38(b). The stability of high-temperature micromachined TiO$_x$ gas sensors has been investigated by Patel et al.[196] for measurements of hydrogen and propylene in the presence of oxygen. The temperature played a key role in defining the sensor response to hydrogen at 100°C and propylene when above 350°C.

Artificial Nose

Microfabrication technology lends itself to the construction of arrays of sensors with differing chemical selectivities. Capacitive-based gas sensors having selectivity to different classes of chemical species[197] along

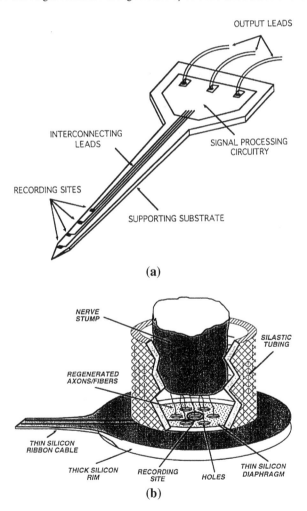

FIGURE 10.39 (a) Schematic diagram of boron doped etch stopped neural probe. (Najafi, K., *Handbook of Microlithography, Micromachining, and Microfabrication, Vol. 2: Micromachining and Microfabrication*, Ed. P. Rai-Choudhury, SPIE, Washington, 1997, 517. With permission.); (b) schematic diagram of the neural interface structure. (Akin, T. and Najafi, K., *IEEE Trans. Biomed. Eng.*, 41, 305, 1994. With permission.)

with pattern recognition[198] have been demonstrated as a viable scheme for the realization of the electronic nose. The polymer coatings produce characteristic dielectric constant, mass, or conductivity changes when the analyte is adsorbed. This work on chemiresistor arrays has also been integrated with CMOS interface circuits for applications in food quality and odor identification.

Neural Probes

Najafi[199] has reviewed his extensive work on neural probes with integrated electronics. The early design involved four masks and had a high yield. A boron diffusion defined the thickness of the structure. Three-dimensional multielectrode systems were later developed to improve electrode positioning. Each array of neural probes is inserted into a silicon machined substrate and electrical connections are made between the probe and support chip by electroplating nickel. On the chip, preamplifiers are followed by analog multiplexers and a broad-band output buffer to drive the external data line. Later developments included a NMOS and CMOS integrated circuit with ten recording sites of gold electrodes on 100-μm centers. The circuit operated with a 5-V supply and consumes 5 mW. A 32-electrode version also has an integrated multiplexing for 32-to-8 switching array. Preamp specifications were 150 to 300 V/V, –3 dB bandwidth 100 Hz to 10 kHz. Akin and Najafi[200] have developed novel sieve structures for stimulation electrodes. They include a silicon ribbon cable that allows connections with minimal mechanical hindrance of the implant while maintaining electrical connections. Neural probes for recording brain activity have also been fabricated by Kewley et al.,[201] with integration of the buffer electronics with the probe tip electrode. The advantage of a dry-etch process is a small, well-defined tip radius of 0.25 μm in this case. He integrated 18-channel preamplifiers in a MOSIS 2-μm, low-noise analog process, each having a total gain of 150 V/V. Probe tips of iridium were fabricated with PECVD layers of silicon nitride low-stress material over the electrodes, achieving a parasitic capacitance of 20 pF and an electrode capacitance of 40 pF; stable, low-leakage current of less than 0.25 pA at a 5-V bias in buffered saline solutions, in addition to maintaining a well-adhered film necessary to maintain the tip electrode integrity.

References

1. Biannual meetings in June: (1) International Meeting on Transducers; (2) Sensors and Actuators Workshop held at Hilton Head, South Carolina.
2a. *Institute of Electrical and Electronic Engineers (IEEE) MEMS Conference* every year in Jan/Feb.
2b. Annual Meeting of the American Society of Mechanical Engineers (ASME).
3. *Proceedings of the Micro Total Analysis Systems Workshop*; most recent meeting held in Banff, Canada, Oct. 13-16th, 1998, published by Kluwer, 1998.
4. The Society for Photo and Instrumentation Engineers (SPIE), Bellingham, WA.
5. IEEE, IEDM Meeting, New York.
6. International Meeting on Chemical Sensors.
7. Fall and Spring Symposia at Meetings of The Electrochemical Society, Pennington, New Jersey.
8. Peterson, K. E., Silicon as a mechanical material, *Proc. IEEE*, 70, 420, 1982.
9. Göepel, W. et al., *Sensors a Comprehensive Survey, Fundamentals and General Aspects*, John Wiley & Sons, New York, 1989.
10. Kovacs, G. T. A., *Micromachined Transducers Sourcebook*, McGraw-Hill, New York, 1998.
11. Madou, M. J., *Fundamentals of Microfabrication*, CRC Press, Boca Raton, FL, 1997.
12. Sze, S. M., *Semiconductor Sensors*, John Wiley & Sons, Somerset, NJ, 1994.
13. Trimmer, W. S., *Micromechanics and MEMS: Classic and Seminal Papers to 1990*, IEEE Press, New York, 1997.
14. Middelhoek, S. and Audet, S. A., *Silicon Sensors*, Academic Press, Boston, MA, 1989.
15. Ristic, L. J., *Sensor Technology and Devices*, Artech House, London, 1994.
16. Gardner, J. W., *Microsensors: Principles and Applications*, John Wiley & Sons, Chichester, West Sussex, U.K., 1994.

17. Janata, J., *Principles of Chemical Sensors*, Plenum Press, New York, 1989.

18. Madou, J. M. and Morrison, J. R., *Solid-State Chemical Sensors*, Plenum Press, New York, 1989.

19. Moseley, P. T., Norris, J., and Williams, D. E., *Techniques and Mechanisms in Gas Sensing*, Adam Higler, New York, 1991.

20. Turner, A. P. F., Karube, I., and Wilson, G., *Biosensors Fundamentals and Applications*, Oxford University Press, New York, 1987.

21. Vossen, J. L. and Kern, W., *Thin Film Processes*, Academic Press, 1978.

22. Brannon, J., *Eximer Laser Ablation and Etching, AVS Monograph Series*, M-10, American Vacuum Society, New York, 1993.

23. Friedrich, C. R., Warrington, R., Bacher, W., Bauer, W., Coane, P. J., Göttert, J., Hanemann, T., Haußelt, J., Heckele, M., Knitter, R., Mohr, J., Piotter, V., Ritzhaupt-Kleissl, H.-J., and Ruprecht, R., High Aspect Ratio Processing, in *Handbook of Microlithography, Micromachining, and Microfabrication*, Vol. 2: Micromachining and Microfabrication, Ed. P. Rai-Choudhury, SPIE, Washington, 1997, 299.

24. Stewart, D. K. and Casey, J. D., Focused ion beams for micromachining and microchemistry, in *Handbook of Microlithography, Micromachining and Microfabrication*, Vol. 2: Micromachining and Microfabrication, Ed. P. Rai-Choudhury, SPIE, Washington, 1997, 153.

25. Allen, D. M., *The Principles and Practice of Photochemical Machining and Photoetching*, Adam Hilger, Bristol, 1986.

26. Williams, K. R. and Muller, R. S., Etch rates for micromachining processing, *J. Microelectromech. Syst.*, 5, 256, 1996.

27. Kendall, D. L. and Shoultz, R. A., Wet chemical etching of Silicon and SiO_2 and ten challenges for micromachiners, in *Handbook of Microlithography, Micromachining, and Microfabrication*, vol. 2: *Micromachining and Microfabrication*, Ed. P. Rai-Choudhury, SPIE, Washington, 1997, 41.

28. Seidel, H., Cspregi, L., Heuberger, A., and Baumgartel, H., *J. Electrochem. Soc.*, 137, 3612, 1990; and Seidel, H., Cspregi, L., Heuberger, A., and Baumgartel, H., *J. Electrochem. Soc.*, 137, 3626, 1990.

29. Kendall, D. L., Vertical etching of silicon at a very high spect ratios, *Ann. Rev. Mater. Sci.*, 9, 373, 1979.

30. Palik, E. D., Glembocki, O. J., and Heard, J. I., Study of bias-dependent etching of Si in aqueous KOH, *J. Electrochem. Soc.*, 134, 404, 1987.

31. Hesketh, P. J., Ju, C., Gowda, S., Zanoria, E., and Danyluk, S., A surface free energy model of silicon anisotropic etching, *J. Electrochem. Soc.*, 140, 1080, 1993.

32. Allongue, P. V., Costa-Kiedling, and Gerishcher, H., Etching of silicon in NaOH solutions, *J. Electrochem. Soc.*, 134, 404, 1987.

33. Kovacs, G. T. A., Mauluf, N. I., and Petersen, K. E., Bulk micromachining of silicon, *Proc. IEEE*, 86, 1536, 1998.

34. *Proceedings of Workshop of Physical Chemistry of Wet Chemical Etching of Silicon*, Holten, The Netherlands, May, 1998.

35. Clark, L. D. and Edell, D. L., *Proceedings IEEE Micro Robots and Teleoperators Workshop*, Hyannis, MA, 1987.

36. Bean, K., Anisotropic etching of silicon, *IEEE Trans. Electron. Dev.*, 25, 1185, 1978.

37. Glembocki, O. J., Palik, E. D., de Guel, G. R., and Kendall, D. L., Hydration model for the molarity dependence of the etch rate of si in aqueous alkali hydroxides, *J. Electrochem. Soc.*, 138, 1055, 1991.

38. Bressers, P. M. M. C., Kelly, J. J., Gardeniers, J. G. E., and Elwenspoek, M., Surface morphology of p-type (100) silicon etched in aqueous alkaline solutions, *J. Electrochem. Soc.*, 143, 1744, 1996.

39. Price, J. B., Anisotropic etching of silicon with $KOH-H_2O$-isopropyl alcohol, in *Semiconductor Silicon*, Eds. H. R. Huff and R. R. Burgess, Softbound Proceedings of the ECS, 1973, 339.

40. Clark, J. D., Lund, J. L., and Edell, D. J., Cesium hydroxide [CsOH]: a useful etchant for micromachining silicon, in *Technical Digest of Papers, Solid-State Sensor and Actuator Workshop*, Hilton Head, South Carolina, 1988, 5.

41. Ip Yam, J. D., Santiago-Aviles, J. J., and Zemel, J. N., An investigation of the anisotropic etching of (100) silicon using cesium hydroxide, *Sensors and Actuators A*, 29, 121, 1991.

42. Ju, C. and Hesketh, P. J., Measurements of the anisotropic etching of silicon in aqueous cesium hydroxide, *Sensors and Actuators A*, 33, 191, 1992.

43. Reisman, A., Berkenbilt, M., Chan, A. A., Kaufman, F. B., and Green, D. C., The controlled etching of silicon in catalyzed ethylene-diamine-pyrocatechol-water solutions, *J. Electrochem. Soc.*, 126, 1406, 1979.

44. Finne, R. M. and Klein, D. L., A water-amine complexing agent system for etching in silicon, *J. Electrochem. Soc.*, 114, 965, 1967.

45. Tabata, O., Asahi, R., Funabashi, H., Shimaoka, K., and Sugiyama, S., Anisotropic etching of silicon in TMAH solutions, *Sensors and Actuators A*, 34, 51, 1992.

46. Pandy, A., Landsberger, L., Nikpour, B., Paranjape, M., and Kahrizi, M., Experimental investigation of high Si/Al selectivity during anisotropic etching in tetra-methyl ammonium hydroxide, *J. Vacuum Sci. Tech. A*, 16, 868, 1998.

47. Klaassen, E. H., Reay, R. J., Storment, C., Audy, J., Henry, P., Brokaw, A. P., and Kovacs, G. T. A., Micromachined thermally isolated circuits, in *Digest of Technical Papers, Solid-State Sensors and Actuators Workshop*, Hilton Head, South Carolina, June, 1996, pg. 127.

48. Merlos, A., Acco, M., Bao, M. H., Bausells, J., and Esteve, J., TMAH/IPA anisotropic etching characteristics, *Sensors and Actuators A*, 37-38, 737, 1993.

49. Landsberger, L. M., Naseh, S., Kahrizi, M., and Paranjape, M., On hillocks generated during anisotropic etching of Si in TMAH, *J. Microelectromechanical Syst.*, 5, 106, 1996.

50. Schwartz, B. and Robbins, H. R., Chemical etching of silicon. III. A temperature study I the acid system, *J. Electrochem. Soc.*, 108, 365 1961.

51. Friedrich, C. R. et al., High aspect ratio processing, in *Handbook of Microlithography, Micromachining and Microfabrication*, vol. 2: *Micromachining and Microfabrication*, Rai-Chowdhury, P., Ed., SPIE Optical Engineering Press, Bellingham, WA, 1997, Chap. 6.

52. Najafi, K., Wise, K. D., and Mochizuki, T., A high-yield IC-compatible multichannel recording array, *IEEE Trans. Elec. Dev.*, 32, 1206, 1985.

53. Collins, S. D., Etch stop techniques for micromachining, *J. Electrochem. Soc.*, 144, 2242, 1997.

54. Jackson, T., N., Tischler, M. A., and Wise, K. D., An electrochemical p-n junction etch stop for the formation of silicon microstructures, *IEEE Electron. Dev. Lett.*, 2, 44, 1981.

55. Kloeck, B., Collins, S. D., de Rooij, N. F., and Smith, R. L., Study of electrochemical etch-stop for high-precision thickness control of silicon membranes, *IEEE Trans. Electron. Dev.*, 36, 663, 1989.

56. Tuller, H. L. and Mlcak, R., Photo-assisted silicon micromachining: opportunities for chemical sensing, *Sensors and Actuators B*, 35, 255, 1996.

57. Schöning, M. J., Ronkel, F., Crott, M., Thust, M., Schultze, J. W., Kordos, P., and Lüth, H., Miniaturization of potentiometric sensors using porous silicon microtechnology, *Electrochimica Acta*, 42, 3185, 1997.

58. Winters, H. F. and Coburn, J. W., The etching of silicon with XeF_2 vapor, *Appl. Phys. Lett.*, 34, 70, 1979.

59. Hoffman, E., Warneke, B., Kruglick, E., Weingold, J., and Pister, K. S. J., 3D structures with piezoresistive sensors in standard CMOS, in *Proceedings of the IEEE International Meeting on Micro Electro Mechanical Systems*, Amsterdam, The Netherlands, Jan. 29-Feb. 2, 1995, 288.

60. Chu, P. B., Chen, J. T., Yeh, R., Lin, G., Huang, J. C. P., Warneke, B. A., and Pister, K. S. J., Controlled pulse-etching with xenon difluoride, in *Proceedings of Transducers '97, Int. Conf. Solid-State Sensors and Actuators*, Chicago, IL, June 16-19, 1997, 665.

61. Tea, N. H., Milanovic, V., Zincke, C. A., Suehle, J. S., Gaitan, M., Zaghloul, M. E., and Geist, J., Hybrid postprocessing etching for CMOS-compatible MEMS., *J. Micromechanical Systems*, 6, 363, 1997.

62. Manos, D. M. and Flamm, D. L., *Plasma Etching: An Introduction*, Academic Press, New York, 1989.

63. Sugawara, M., *Plasma Etching*, Oxford Science Publications, New York, 1998.

64. Bhardwaj, J., Ashraf, H., and McQuarrie, A., Dry silicon etching for MEMS, in *Microstructures and Microfabricated Systems-III, Proceedings of the Electrochemical Society*, vol. 97-5, 118, 1999.

65. Shih, B., St.Clair, L., Hesketh, P. J., Naylor, D. L., and Yershov, G. M., Corner compensation for CsOH micromachining of a silicon fluidic chamber, submitted to *J. Electrochem Soc.*, 1999.

66. Offereins, H. L., Sandmaier, H., Marusczyk, K., Kühl, K., and Plettner, A., Compensating corner under-cutting of (100) silicon in KOH, *Sensors and Materials*, 3, 127, 1992.

67. Howe, R. T. and Muller, R. S., Polycrystalline and amorphous silicon micromechanical beams: annealing and mechanical properties, *Sensors and Actuators*, 4, 447, 1983.

68. Maier-Schneider, D., Maibach, J., Obermeier, E., and Schneider, D., Variations in Young's modulus and intrinsic stress of LPCVD-polysilicon due to high-temperature annealing, *J. Micromech. Microeng.*, 5, 121, 1995.

69. Guckel, H., Sniegowski, J. J., Christenson, T. R., and Raissi, F., The application of fine-grained, tensile polysilicon to mechanically resonant transducers, *Sensors and Actuators A*, 21, 346, 1990.

70. Krulevitch, P. A., Micromechanical Investigations of Silicon and Ni-Ti-Cu Thin Films, Ph.D. thesis, University of California Berkeley, 1994.

71. Guckel, H. and Burns, D. W., Planar processed polysilicon sealed cavities for pressure transducer arrays, in *Proceedings of the IEEE International Electron Devices Meeting*, San Francisco, CA, Dec. 9-12th, 1984, 223.

72. Westberg, D., Paul, O., Anderson, G. I., and Baltes, H., Surface micromachining by sacrificial aluminum etching, *J. Micromech. Microeng.*, 6, 376, 1996.

73. Tas, N., Sonnenberg, T., Jansen, H., Legtenberg, R., and Elwenspoek, M., Stiction in surface micromachining, *J. Micromech. Microeng.*, 6, 385, 1996.

74. Intellisense Inc., Cambridge, MA.

75. Gennissen, P. T. J., Bartek, M., French, P. J., and Sarro, P. M., Bipolar-compatible epitaxial poly for smart sensors: stress minimization and applications, *Sensors and Actuators*, 62, 636, 1997.

76. Wenk, B., Thick polysilicon based surface micromachining, in *Microstructures and Microfabricated Systems -IV, Proceedings of the Electrochemical Society*, vol. 98-14, 12, 1998.

77. Sharpe, W. N., Yuan, B., Vaidyanathan, R., and Edwars, R. L., Measurements of Young's modulus, Poisson's ratio, and tensile strength of polysilicon, *Proceedings of the Tenth Annual International Workshop on Micro Electro Mechanical Systems*, Nagoya, Japan, January 1997, IEEE, NJ, Catalog Number 97CH36021, 424.

78. Kahn, H., Stemmer, S., Nandakumar, K., Heuer, A. H., Mullen, P. L., Ballarini, R., and Huff, M. A., Mechanical properties of thick, surface micromachined polysilicon films, in *Proceedings of the Ninth Annual International Workshop of Micro Electo Mechanical Systems*, San Diego, CA, February 1996, IEEE, NJ, Cat. Number 96CH35856, 343.

79. Maier-Schneider, D., Maibach, J., Obermeier, E., and Schneider, D., Variations in Young's modulus and intrinsic stress of LPCVD-polysilicon due to high-temperature annealing, *J. Micromech. Microeng.*, 5, 121, 1995.

80. Biebl, M. and Philipsborn, Fracture strength of doped and undoped polysilicon, in Digest of Technical Papers, *The 8th International Conference on Solid-State Sensors and Actuators, and Eurosensors IX*, Stockholm, Sweden, June, 1995, 72.

81. Adams, A. C., *Dielectric and Polysilicon Film Deposition*, Chapter 6, in VLSI Technology, Editor S. M. Sze, McGraw-Hill, New York, 1983, 93.

82. Gardeniers, J. G. E., Tilmans, H. A. C., and Visser, C. C. G., LPCVD silicon-rich silicon nitride films for applications in micromechanics, studied with statistical experimental design, *J. Vac. Sci. Tech. A.*, 14, 3879, 1996.

83. Chou, B. C. S., Shie, J.-S., and Chen, C.-N., Fabrication of low-stress dielectric thin-film for microsensor applications, *IEEE Electron Device Letters*, 18, 599, 1997.

84. French, P. J., Sarro, P. M., Mallée, R., Fakkeldij, E. J. M., and Wolffenbuttel, Optimization of a low-stress silicon nitride process for surface-micromachining applications, *Sensors and Actuators A*, 58, 149, 1997.

85. Habermehl, S., Stress relaxation in Si-rich silicon nitride thin films, *J. Appl. Phys.*, 83, 4672, 1998.

86. Classen, W. A. P. et al., Influence of deposition temperature, gas pressure, gas phase composition, and RF frequency on composition and mechanical stress of plasma silicon nitride layers, *J. Electrochem. Soc.*, 132, 893, 1985.

87. Sarro, P. M., deBoer, C. R., Korkmaz, E., and Laros, J. M. W., Low-stress PECVD SiC thin films for IC-compatible microstructures, *Sensors and Actuators A*, 67, 175, 1998.

88. Pan, L. S. and Kania, D. R., *Diamond: Electronic Properties and Applications*, Kluwer Academic Pub., 1995.

89. French, P. J. and Sarro, P. M., Surface versus bulk micromachining: the contest for suitable applications, *J. Micromech. Microeng.*, 8, 45, 1998.

90. Rogers, M. S. and Sniegowski, J. J., 5-level polysilicon surface micromachine technology: application to complex mechanical systems, in *Digest of Technical Papers Solid-State Sensor and Actuator Workshop*, Hilton Head, SC, June 1998, 144.

91. Sniegowski, J. J., Miller, S. L., LaVigne, G. F., Rogers, M. S., and McWhorter, P. J., Monolithic geared-mechanisms driven by a polysilicon surface-micromachined on-chip electrostatic microengine, in *Digest of Technical Papers, Solid-State Sensors and Actuators Workshop*, Hilton Head, SC, June, 1996, 178,

92. Sniegowski, J. J., Chemical mechanical polishing: an enabling fabrication process for surface micromachining technologies, in *Microstructures and Microfabricated Systems-IV*, Ed., P. J. Hesketh, H. Hughes, and W. E. Bailey, Proceedings of the Electrochemical Society, vol. 98-14, 1, 1998.

93. McDonald, N. C., SCREAM MicroElectroMechanical Systems, *Microelectronic Engineering*, 32, 49, 1996.

94. Saif, M. T. A. and MacDonald, N. C., Planarity of large MEMS, *J. Microelectromech. Syst.*, 5, 79, 1996.

95. Keller, C. G. and Howe, R. T., Nickel-filled hexsil thermally actuated tweezers, in *Proceedings of Transducers*, 95.

96. Soane, D. S. and Martynenko, Z., *Polymers in Microelectronics Fundamentals and Applications*, Elsevier, New York, 1989.

97. Lorenz, H., Despont, M., Fahrni, N., LaBianca, N., Renaud, P., and Vettiger, P., SU-8: a low-cost negative resist for MEMS, *J. Micromech. Microeng.*, 7, 121, 1997.

98. Frazier, A. B., Ahn, C. H., and Allen, M. G., Development of micromachined devices using polyimide-based processes, *Sensors and Actuators A*, 45, 47, 1994.

99. Ahn, C. H., Kim, Y. J., and Allen, M. G., A fully integrated planar toroidal inductor with a micromachined nickel-iron magnetic bar, *IEEE Trans. Comp. Packag. & Manuf. Tech. A*, 17, 3, 1994.

100. Ahn, C. H. and Allen, M. G., A planar micromachined spiral inductor for integrated magnetic microactuator applications, *J. Micromech. Microeng.*, 3, 37, 1993.

101. Becker, E. W. et al., Fabrication of microstructures with high aspect ratios and great structural heights by synchrotron radiation lithography, galvanoforming, and plastic molding (LIGA Process), *Microelectronic Eng.*, 4, 35, 1986.

102. Cox, J. A., Zook, J. D., Ohnstein, T., and Dobson, D. C., Optical performance of high-aspect LIGA gratings, *Opt. Eng.*, 36, 1367, 1997.

103. Hjort, K., Söderkvist, J., and Schweits, J.-A., Gallium arsenide as a mechanical material, *J. Micromech. Microeng.*, 4, 1, 1994.

104. Parameswaran, M., Baltes, H. P., Ristic, L.J., Dhaded, A. C., and Robinson, A. M., A new approach for the fabrication of micromechanical structures, *Sensors and Actuators*, 19, 289, 1989.

105. Nicollian, E. H. and Brews, J. R., *MOS (Metal Oxide Semiconductor) Physics and Technology*, Addison-Wesley, Reading, MA, 1990.

106. Parameswaran, M., Robinson, A. M., Blackburn, D. L., Gaitan, M., and Geist, J., Micromachined thermal radiation emitter from a commercial CMOS process, *IEEE Elec. Dev. Lett.*, 12, 57, 1991.

107. Fedder, G. K., Santhanam, S., Reed, M. L., Eaggle, S. C., Guillou, D. F., Lu, M. S.-C., and Carley, L. R., Laminated high-aspect-ratio microstructures in a conventional CMOS process, in *Proceedings of the Ninth Annual International Workshop of Micro Electro Mechanical Systems*, San Diego, CA, February 1996, IEEE, NJ, Cat. Number 96CH35856, 13.

108. Lee, J.-B., English, J., Ahn, C.-H., and Allen, M. G., Planarization techniques for vertically integrated metallic MEMS on silicon foundry circuits, *J. Micromach. Microeng.*, 7, 44, 1997.

109. Bustillo, J. M., Fedder, G. K., Nguyen, C. T.-C., and Howe, R. T., Process technology for the modular integration of CMOS and polysilicon microstructures, *Microsystem. Technol.*, 1, 30, 1994.

110. Smith, J., Montague, S., Sneigowski, J., Murry, J., and McWhorter, P., Embedded micromechanical devices for the monolithic integration of MEMS with CMOS, *IEDM Tech. Digest '95*, 1995, 609.

111. Gianchandani, Y. B., Kim, H., Shinn, M., Ha, B., Lee, B., Najafi, K., and Song, C., A MEMS-first fabrication process for integrating CMOS circuits with polysilicon microstructures, in *Proceedings IEEE the Eleventh Annual International Workshop on Micro Electro Mechanical Systems*, Heidelberg, Germany, January 1998, 257

112. Parameswaran, L., Hsu, C. H., and Schmidt, M. A., Integrated sensor technology, in *Meeting Abstracts of the 194th Meeting of the Electrochemical Society*, Boston, Nov. 1-6, Abst # 1144, 1998.

113. Wallis, G. and Pomerantz, Field assisted glass-metal sealing, *J. Appl. Phys.*, 40, 3946, 1969.

114. Albaugh, K. B. and Cade, E., Mechanism of anodic bonding of silicon to pyrex glass, *Digest of Technical Papers, IEEE Solid-State Sensors and Actuators Workshop*, Hilton Head Island, SC, 1988, 109.

115. Cunneen, J., Lin, Y.-C., Caraffini, S., Boyd, J. G., Hesketh, P. J., Lunte, S. M., and Wilson, G. S., A positive displacement micropump for microdialysis, *Mechatronics Journal*, 8, 561, 1998.

116. van der Groen, S., Rosmeulen, M., Baert, K., Jansen, P., and Deferm, L., Substrate bonding techniques for CMOS processed wafers, *J. Micromech. Microeng.*, 7, 108, 1997.

117. Shimbo, M. et al., Silicon-to-silicon direct bonding method, *J. Appl. Phys.*, 60, 2987, 1986.

118. Tong, Q. Y. and Gösele, U., *Semiconductor Wafer Bonding*, John Wiley & Sons, 1998.

119. Knecht, T. A., Bonding techniques for solid-state pressure sensors, *Proc. Transducers '87*, Tokyo, 95-98.

120. Legtenberg, R., Bouwstra, S., and Elwenspoek, M., Low-temperature glass bonding for sensor applications using boron oxide thin films, *J. Micromech. Microeng.*, 1, 157-160, 1991.

121. Nguyen, M. N., Low stress silver-glass die attach material, *IEEE Trans. Comp. Hybrid, Manuf. Tech.*, 13, 478, 1990.

122. Hanneborg, A., Nese, M., and Ohlckers, P., Silicon-to-silicon anodic bonding with a borosilicate glass layer, *J. Micromech. Microeng.*, 1, 139, 1991.

123. Wu, M. C., Micromachining for optical and optoelectronic systems, *Proc. IEEE*, 85, 1833, 1997.

124. Motamedi, M. E., Micro-opto-electro-mechanical systems, *Optical Engineering*, 33, 3505, 1994.

125. Motamedi, M. E., Wu, M. C., and Pister, K. S. J., Micro-opto-electro-mechanical devices and on-chip optical processing, *Opt. Eng.*, 36, 1282, 1997.

126. Pister, K. S. J., Judy, M. W., Burgett, S. R., and Fearing, R. S., Microfabricated hinges, *Sensors and Actuators A*, 33, 249, 1992.

127. Roggeman, M. C., Bright, V. M., Welsh, B. M., Hick, S. R., Roberts, P. C., Cowan, W. D., and Comtois, J. H., Use of micro-electro-mechanical deformable mirrors to control aberrations in optical systems: theoretical and experimental results, *Opt. Eng.*, 36, 1326 1997.

128. Usenishi, Y., Tsugari, M., and Mehregany, M., Micro-opto-mechanical devices fabricated by anisotropic etching of (110) silicon, *J. Micromech. Microeng.* 5, 305, 1995.

129. Larson, M. C., Pezeshki, B., and Harris, J. S., Vertical coupled-cavity microinterferometer on GaAs with deformable-membrane top mirror, *IEEE Phot. Tech. Lett.*, 7, 382, 1995.

130. Gossen, K. W., Walker, J. A., and Arney, S. C., Silicon modulator based on mechanically-active anti-reflection layer with 1 Mbit/sec capability for fiber-in-the-loop applications, *IEEE Photon. Tech. Lett.*, 6, 1119, 1994.

131. Sene, D. E., Grantham, J. W., Bright, V. M., and Comtois, J. H., Development and characterization of micro-mechanical gratings for optical modulation, in *Proceedings of the Ninth Annual International Workshop on Micro Electro Mechanical Systems*, San Diego, CA, February 1996, 222.

132. Burns, D. B. and Bright, V. M., Development of microelectromechanical variable blaze gratings, *Sensors and Actuators A*, 64, 7, 1998.

133. Miller, R., A., Burr, G. W., Tai, Y.-C., and Psaltis, D., Electromagnetic MEMS scanning mirrors for holographic data storage, in *Digest of Technical Papers, Solid-State Sensor and Actuator Workshop*, Hilton Head, South Carolina, June 1996, 183.

134. Asada, N., Matsuki, H., Minami, K., and Esashi, M., Silicon micromachined two-dimensional galvano optical scanner, *IEEE Trans. Mag.*, 30, 4647, 1994.

135. Judy, J. W. and Muller, R. S., Batch-fabricated, addressable, magnetically actuated microstructures, in *Digest of Technical Papers, Solid State Sensors and Actuators Workshop*, Hilton Head, SC, June 1996, 187.

136. Miller, R. and Tai, Y.-C., Micromachined electromagnetic scanning mirrors, *Opt. Eng.*, 36, 1399, 1997.

137. Kiang, M. H., Solgaard, O., Muller, R. S., and Lau, K. Y., Surface-micromachined electrostatic-comb drive scanning micromirrors for barcode scanners, in *Proceedings of the Ninth Annual International Workshop on Micro Electro Mechanical Systems*, San Diego, CA, February 1996, 192.

138. Fischer, M., Nägele, M., Eichner, D., Schöllhorn, C., and Strobel, R., Integration of surface-micromachined polysilicon mirrors and a standard CMOS process, *Sensors and Actuators A*, 52,140, 1996.

139. Bühler, J., Steiner, F.-P., Hauert, R., and Baltes, H., Linear array of complementary metal oxide semiconductor double-pass metal micromirrors, *Opt. Eng.*, 35, 1391, 1997.

140. Ikeda, M., Goto, H., Sakata, M., Wakabayashi, S., Imanaka, K., Takeuchi, M., and Yada, T., Two dimensional silicon micromachined optical scanner integrated with photo detector and piezoresistor, in *Digest of Technical Papers, The 8th International Conference on Solid-State Sensors and Actuators, and Eurosensors IX*, Stockholm, Sweden, 1995, 293.

141. Huang, Y., Zhang, H., Kim, E. S., Kim, G. K., and Joen, Y. B., Piezoelectrically actuated microcantilever for actuated mirror array application, in *Digest of Technical Papers, Solid-State Sensor and Actuator Workshop*, Hilton Head, South Carolina, June 1996, 191.

142. Lin, L. Y., Lee, S. S., Pister, K. S. J., and Wu, M. C., Micro-machined three-dimensional micro-optics for integrated free-space optical systems, *IEEE Photon. Tech. Lett.*, 6, 1445, 1994.

143. Lee, S. S., Lin, L. Y., and Wu, M. C., Surface-micromachined free-space micro-optical systems containing three-dimensional microgratings, *Appl. Phys. Lett.*, 67, 2135, 1995.

144. Bright, V. M., Comtois, J. H., Reid, J. R., and Sene, D. E., Surface micromachined micro-opti-electro-mechanical systems, *IEICE Trans. Elec.*, E80, 206, 1997.

145. Markus, K. W. and Koester, D. A., Multi-user MEMS process (MUMPS) introduction and design rules, MCNC Electronics Tech. Div., 3201 Cornwallis Road, Research Triangle Park, NC, Oct. 1994.

146. Hornbeck, L. J., Digital light processing™ and MEMS: reflecting the digital display needs of the networked society, in *Symposium Micromachining and Microfabrication*, Proc. SPIE vol. 2783, Austin, TX, 1996, 2.

147. Ukita, H., Sugiyama, Y., Nakada, H., and Katagiri, Y., Read/write performance and reliability of a flying optical head using a monolithically integrated LD-PD, *Appl. Optic.*, 30, 3770, 1991.

148. Lin, L. Y., Shen, J. L., Lee, S. S., and Wu, M. C., Realization of novel monolithic free-space optical disk pickup heads by surface micromachining, *Optics Letters*, 21, 155, 1996.

149. Ukita, H., Micromechanical photonics, *Optical Review*, 4, 623, 1997.

150. Maluf, N. I., Reay, R. J., and Kovacs, G. T. A., High-voltage devices and circuits fabricated using foundry CMOS for use with electrostatic MEM actuators, *Sensors and Actuators A*, 52, 187, 1996.

151. Daneman, M. J., Tien, N. C., Solgaard, O., Lau, K. Y., and Muller, R. S., Linear vibromotor-actuated micromachined microreflector for integrated optical systems, in *Digest of Technical Papers, Solid-State Sensor and Actuator Workshop*, Hilton Head, SC, June 1996, 109.

152. Fukuta, Y., Collard, D., Akiyama, T., Yang, E. H., and Fujita, H., Microactuated self-assembling of 3D polysilicon structures with reshaping technology, in *Proceedings of the 10th Annual International Workshop on Micro Electro Mechanical Systems*, Nagoya, Japan (IEEE, NJ, 1997), 477.

153. Lin, L. Y., Shen, J. L., Lee, S. S., Su, G. D., and Wu, M. C., Microactuated micro-xyz stages for free-space micro-optical bench, in *Proceedings of the 10th Annual International Workshop on Micro Electro Mechanical Systems*, Nagoya, Japan, 1997, 43.

154. Comtois, J. H. and Bright, V. M., Surface micromachined polysilicon thermal actuator arrays and applications, in *Digest of Technical Papers, Solid-State Sensor and Actuator Workshop*, Hilton Head, SC, June 1996, 152.

155. Reid, J. R., Bright, V. M., and Comtois, J. H., Arrays of thermal micro-actuators coupled to micro-optical components, *Proc. SPIE*, 2865, 74, 1996.

156. Reid, J. R, Bright, V. M., and Butler, J. T., Automated assembly of flip-up micromirrors, *Sensors & Actuators A*, 66, 292, 1998.

157. Shen, B. et al., Cantilever micromachined structures in CMOS technology with magnetic actuation, *Sensors and Materials*, 9, 347, 1997.

158. Chang, J. Y.-C., Abidi, A. A., and Gaitan, M., Large suspended inductors on silicon and their use in a 2-μm MOS RF amplifier, *IEEE Electron Device Letters*, 14, 246, 1983.

159. Rofougaran, A., Chang, J. Y. C., Rofougaran, M., and Abidi, A., A GHz CMOS RF front-end IC for a direct-conversion wireless receiver, *IEEE J. Solid-State Circuits*, 31, 880, 1996.

160. López-Villegas, J. M., Samitier, J., Bausells, J., Merlos, A., Cané, and Knöchel, R., Study of integrated RF passive components performed using CMOS and Si micromachining technologies, *J. Micromach. Microeng.*, 7, 162, 1997.

161. McGrath, W. R., Walker, C., Yap, M., and Tai, Y.-C., Silicon micromachined waveguides for milli-meter-wave and submillimeter-wave frequencies, *IEEE Microwave and Guided Wave Letters*, 3, 61, 1993.

162. Franklin-Drayton, R., Dib, N. I., and Katehi, L. P. B., Design of micromachined high frequency circuit components, *International Journal of Microcircuits and Electronic Packaging*, 18, 19, 1995.

163. Drayton, R. F., Henderson, R. M., and Katehi, L. P. B., Monolithic packaging concepts for high isolation in circuits and antennas, *IEEE Trans. Microwave Theory and Tech.*, 46, 900, 1998.

164. Katehi, L. P. B., Rebeiz, G. M., Weller, T. M., Drayton R. F., Cheng, H.-J., and Whitaker, J. F., Micromachined circuits for millimeter- and sub-millimeter-wave applications, *IEEE Antennas and Propagation Magazine*, 35, 9, 1993.

165. Milanovic, V., Gaitan, M., Bowen, E. D., and Zaghloul, M. E., Micromachined microwave trans-mission lines in CMOS technology, *IEEE Trans. Microwave Theory and Techniques*, 45, 630 1997.

166. Roessig, T. A., Howe, R. T., Pisano, A. P., and Smith, J. H., Surface-micromachined 1Mhz oscillator with low-noise pierce configuration, in *Digest of Technical Papers, IEEE Solid-State Sensors and Actuators Workshop*, Hilton Head Island, SC, 1988, 328.

167. Baltes, H., Paul, O., and Brand, O., Micromachined thermally based CMOS microsensors, *Proceedings of the IEEE*, 86, 1660, 1998.

168. Reay, R. J., Klaassen, E. K., and Kovacs, G. T. A., A micromachined low-power temperature-regulated bandgap voltage reference, *IEEE J. Solid-State Circuits*, 30, 1374, 1995.

169. Klaassen, E. H., Reay, R. J., and Kovacs, G. T. A., Diode-based thermal r.m.s. converter with on-chip circuitry fabricated using CMOS technology, *Sensors and Actuators A*, 52, 33, 1996.

170. Zavracky, P. M., Majumder, S., and McGruer, N. E., Micromechanical switches fabricated using nickel surface micromachining, *J. Microelectromechanical Systems*, 6, 3, 1997.

171. McGruer, N. E., Zavracky, P. M., Majumder, S., Morrison, R., and Adams, G. G., Electrostatically actuated microswitches; scaling properties, in *Digest of Technical Papers, IEEE Solid-State Sensors and Actuators Workshop*, Hilton Head Island, SC, 1988, 132.

172. Drake, J., Jerman, H., Lutze, B., and Stuber, M., An electrostatically actuated micro-relay, in *Proceedings of Transducers '95, Int. Conf. Solid-State Sensors and Actuators*, 1995, 384.

173. Yao, J. J. and Chang, M. Frank, A surface micromachined miniature switch for telecommunications applications with signal frequencies from DC up to 4 Ghz, in *Proceedings of Transducers '95, Int. Conf. Solid-State Sensors and Actuators*, 1995, 384.

174. Randall, J. N., Goldsmith, C., Denniston, D., and Lin, T.-H., Fabrication of micromechanical switches for routing radio frequency signals, *J. Vac. Sci. Tech. B*, 14, 3692, 1996.

175. Schiele, I., Huber, J., Hillerich, B., and Kozlowski, F., Surface-micromachined electrostatic microre-lay, *Sensors and Actuators A*, 66, 345, 1998.

176. Grétillat, M.-A., Thiebaud, P., Linder, C., and de Rooij, N. F., Integrated circuit compatible electrostatic polysilicon microrelays, *J. Micromech. Microeng.*, 5, 156, 1995.

177. Sun, X., Q., Farmer, K. R., and Carr, W. N., A bistable microrelay based on two-segment multimorph cantilever actuators, in *Proceedings IEEE the Eleventh Annual International Workshop on Micro Electro Mechanical Systems*, Heidelberg, Germany, January 1998, 154.

178. Kruglick, E. J. J. and Pister, K. S. J., Bistable MEMS relays and contact characterization, *Digest of Technical Papers, IEEE Solid-State Sensors and Actuators Workshop*, Hilton Head Island, SC, 1988, 333.

179. Taylor, W. P., Brand, O., and Allen, M. G., Fully integrated magnetically actuated micromachined relays, *J. Microelectro. Mech. Syst.*, 7, 181, 1998.

180. Holm, R., *Electric Contacts Handbook*, Springer-Verlag, Berlin, 1958.

181. Schimkat, J., Contact materials for microrelays, in *Proceedings IEEE the Eleventh Annual International Workshop on Micro Electro Mechanical Systems*, Heidelberg, Germany, January 1998, 190.

182. Krause, P., Obermeier, E., and Wehl, W., Backshooter — a new smart micromachined single-chip ink jet printhead, in *Proceedings of Transducers '95, Int. Conf. Solid-State Sensors and Actuators*, 1995.

183. Janata, J., Josowicz, Vanýsek, P., and DeVaney, D. M., Chemical Sensors, *Anal. Chem.*, 70, 179R, 1998.

184. Madou, M., Otagawa, T., Joseph, J., Hesketh, P., and Saaman, A., *Catheter-Based Micromachined Electrochemical Sensors, SPIE Optics, Electro-Optics, and Laser Applications in Science and Engineering*, Los Angeles, CA, January 1989.

185. Janata, J., Chemically sensitive field-effect transistors, in *Solid State Chemical Sensors*, J. Janata and R. J. Huber, Eds., Academic Press, New York, 1985.

186. Bergveld, P., Development of an ion-sensitive solid-state device for neurophysiological measurements, *IEEE Trans Biomed. Eng.*, 17, 70, 1970.

187. Janata, J., Chemical selectivity of field-effect transistors, *Sensors and Actuators*, 12, 121, 1987.

188. Lundström, I. and Svensson, C., Gas-sensitive metal gate semiconductor devices, in *Solid State Chemical Sensors*, J. Jananta and R. J. Huber, Eds., Academic Press, New York, 1985.

189. Sibald, A., Recent advances in field-effect chemical microsensors, *J. Molecular Electronics*, 2, 51, 1986.

190. Bousse, L. and Bergveld, P., The role of buried OH-sites in the response mechanism of inorganic-gate pH-sensitive ISFETs, *Sensors and Actuators*, 6, 65, 1984.

191. Cané, C., Grácia, I., and Merlos, A., Microtechnologies for pH ISFET chemical sensors, *Microelectronics Journal*, 28, 389, 1997.

192. Domanský, K., Zapf, V. S., Grate, J. W., Ricco, A. J., Yelton, W. G., and Janata, J., Integrated chemiresistor and work function microsensor array with carbon black/polymer composite materials, in *Digest of Technical Papers, IEEE Solid-State Sensors and Actuators Workshop*, Hilton Head Island, SC, 1988, 187.

193. Rodriguez, J. L., Hughes, R. C., Corbett, W. T., and McWhorter, P. J., Robust, wide range hydrogen sensor, *IEDM Tech. Digest*, IEEE, Cat # 92CH3211-0, San Francisco, CA, 1992, 521.

194. Suehle, J. S., Cavicchi, R. E., Gaitan, M., and Semancik, S., Tin oxide gas sensor fabricated using CMOS micro-hotplates and in-situ processing, *IEEE Elec. Dev. Lett.*, 14, 118 1993.

195. Semancik, S. and Cavicchi, R., Kinetically controlled chemical sensing using micromachined structures, *Acc. Chem. Res.*, 31, 279, 1998.

196. Patel, S. V., Wise, K. D., Gland, J. L., Zanini-Fisher, M., and Schwank, J. W., Characteristics of silicon-micromachined gas sensors based on Pt/TiO_x thin films, *Sensors and Actuators B*, 42, 205, 1997.

197. Cornila, C., Hierlemann, A., Lenggenhager, R., Malcovati, P., Baltes, H., Noetzel, G., Weimar, U., and Göpel, W., Capacitive sensors in CMOS technology with polymer coating, *Sensors and Actuators B*, 24, 357, 1995.

198. Hierlemann, A., Schwiezer-Berberich, M., Weimar, U., Kraus, G., Pfau, A., and Göpel, W., Pattern recognition and multicomponent analysis, in *Sensors Update*, H. Baltes, W. Göpel, and J. Hesse, Eds., VCH Pub., Weinheim, Germany, 1996.

199. Najafi, K., Micromachined systems for neurophysiological applications, in *Handbook of Microlithography, Micromachining, and Microfabrication*, vol. 2: Micromachining and Microfabrication, Ed., P. Rai-Choudhury, SPIE, Washington, 1997, 517.

200. Akin, T. and Najafi, K., A micromachined silicon sieve electrode for nerve regeneration applications, *IEEE Trans. Biomed. Eng.*, 41, 305, 1994.

201. Kewley, D. T., Hills, M. D., Borkholder, D. A., Opris, I. E., Maluf, N. I., Storment, C. W., Bower, J. M., and Kovacs, G. T. A., Plasma-etched neural probes, *Sensors and Actuators A*, 58, 27, 1985.

11

Microelectronics Packaging

Krishna Shenai
Pankaj Khandelwal
University of Illinois at Chicago

11.1 Introduction

Packaging of electronic circuits is the science and the art of establishing interconnections and a suitable operating environment for predominantly electrical circuits. It supplies the chips with wires to distribute signals and power, removes the heat generated by the circuits, and provides them with physical support and environmental protection. It plays an important role in determining the performance, cost, and reliability of the system. With the decrease in feature size and increase in the scale of integration, the delay in on-chip circuitry is now smaller than that introduced by the package. Thus, the ideal package would be one that is compact, and should supply the chips with a required number of signal and power connections, which have minute capacitance, inductance, and resistance. The package should remove the heat generated by the circuits. Its thermal properties should match well with semiconductor chips to avoid stress-induced cracks and failures. The package should be reliable, and it should cost much less than the chips it carries[1] (see Table 11.1).

TABLE 11.1 Electronic Packaging Requirements

Speed	Size
• Large bandwidth	• Compact size
• Short inter-chip propagation delay	
Thermal & Mechanical	Test & Reliability
• High heat removal rate	• Easy to test
• A good match between the thermal coefficients of the dice and the chip carrier	• Easy to modify
	• Highly reliable
	• Low cost
Pin Count & Wireability	Noise
• Large I/O count per chip	• Low noise coupling among wires
• Large I/O between the first and second level package	• Good-quality transmission line
	• Good power distribution

11.2 Packaging Hierarchy

The semiconductor chip is encapsulated into a package, which constitutes the first level of packaging. A printed circuit board is usually employed because the total circuit and bit count required might exceed that available on a single first-level package. Further, there may be components that cannot be readily integrated on a chip or first-level package, such as capacitors, high power resistors, inductors, etc. Therefore, as a general rule, several levels of packaging will be present. They are often referred to as a packaging hierarchy. The number of levels within a hierarchy may vary, depending on the degree of integration and the totality of packaging needs[2] (see Fig. 11.1).

In the past, the packaging hierarchy contained more levels. Dies were mounted on individual chip carriers, which were placed on a printed circuit board. Cards then plugged into a larger board, and the boards were cabled into a gate. Finally, the gates were connected to assemble the computer. Today, higher levels of integration make many levels of packaging unnecessary, and this improves the performance, cost, and reliability of the computers. Ideally, all circuitry one day may be placed on a single piece of semiconductor. Thus, packaging evolution reflects the integrated circuits progress.[3,4]

11.3 Package Parameters

A successful package design will satisfy all given application requirements at an acceptable design, manufacturing, and operating expense. .As a rule, application requirements prescribe the number of

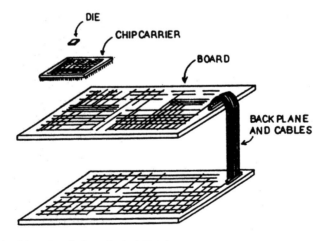

FIGURE 11.1 Packaging hierarchy of a hypothetical digital computer.

logic circuits and/or bits of storage that must be packaged, interconnected, supplied with electric power, kept within a proper temperature range, mechanically supported, and protected against the environment. Thus, IC packages are designed to accomplish the following three basic functions:[5]

- Enclose the chip within a protective envelope to protect it from the external environment
- Provide electrical connection from chip to circuit board
- Dissipate heat generated by the chip by establishing a thermal path from a semiconductor junction to the external environment

To execute these functions, package designers start with a fundamental concept and, using principles of engineering, material science, and processing technology, create a design that encompasses:

1. Low lead capacitance and inductance
2. Safe stress levels
3. Material compatibility
4. Low thermal resistance
5. Seal integrity
6. High reliability
7. Ease of manufacture
8. Low cost

Success in performing the functions outlined depends on the package design configuration, the choice of encapsulating materials, and the operating conditions.[6,7] Package design is driven by performance, cost, reliability, and manufacturing considerations. Conflicts between these multiple criteria are common. The design process involves many tradeoff analyses and the optimization of conflicting requirements.

While designing the package for an application, the following parameters are considered.

Number of Terminals

The total number of terminals at packaging interfaces is a major cost factor. Signal interconnections and terminals constitute the majority of conducting elements. Other conductors supply power and provide ground or other reference voltages.

The number of terminals supporting a group of circuits is strongly dependent on the function of this group. The smallest pinout can be obtained with memory ICs because the stream of data can be limited to a single bit. Exactly the opposite is the case with groups of logic circuits which result from a random partitioning of a computer. The pinout requirement is one of the key driving parameters for all levels of packaging: chips, chip carriers, cards, modules, cables, and cable connectors.

Electrical Design Considerations

Electrical performance at the IC package level is of great importance for microwave designs and has gained considerable attention recently for silicon digital devices due to ever-increasing speed of today's circuits and their potentially reduced noise margins.[8] As a signal propagates through the package, it is degraded due to reflections and line resistance. Controlling the resistance and the inductance associated with the power and ground distribution paths to combat ground bounce and the simultaneous switching noise has now become essential. Controlling the impedance environment of the signal distribution path in the package to mitigate the reflection-related noise is becoming important. Reflections, in addition, cause an increase in the transition time, and may split the signal into two or more pulses with the potential of causing erroneous switching in the subsequent circuit and thus malfunction of the system. Controlling the capacitive coupling between signal traces in the signal distribution path to reduce crosstalk is gaining importance. Increased speed of the devices demands that package bandwidth be increased to reduce undue distortion of the signal. All these criteria are related through geometric variables, such as conductor cross-section and length, dielectric thickness, and the dielectric constant of the packaging body. These problems are usually handled with transmission line theory.[9]

Thermal Design Considerations

The thermal design objective is to keep the operating junction temperature of a silicon chip low enough to prevent triggering the temperature-activated failure mechanisms. Thus, the package should provide a good medium for heat transfer from junction to the ambient/heat sink. It is generally recommended to keep the junction temperature below 150°C to ensure proper electrical performance and to contain the propensity to fail.[10,11]

Thermal expansion caused by heating up the packaging structure is not uniform — it varies in accordance with the temperature gradient at any point in time and with the mismatches in the thermal coefficient of expansion. Mechanical stresses result from these differences and are one of the contributors to the finite lifetime and the failure rate of any packaging structure.[12]

In a simplistic heat transfer model of a packaged chip, the heat is transferred from the chip to the surface of the package by conduction, and from the package surface to the ambient by convection and radiation.[13,14] Typically, the temperature difference between the case and ambient is small, and hence radiation can be neglected. This model also neglects conduction heat transfer out of the package terminals, which can become significant. A multilayer example, which models the heat transfer from a region in the silicon device to the ambient, is shown in Fig. 11.2. The total thermal resistance from the junction to the ambient is given by:

$$R_{\theta ja} = R_{\theta jc} + R_{\theta cs} + R_{\theta sa} \tag{11.1}$$

The resulting junction temperature, assuming a power dissipation of P_d, is

$$T_j = P_d(R_{\theta jc} + R_{\theta cs} + R_{\theta sa}) + T_a \tag{11.2}$$

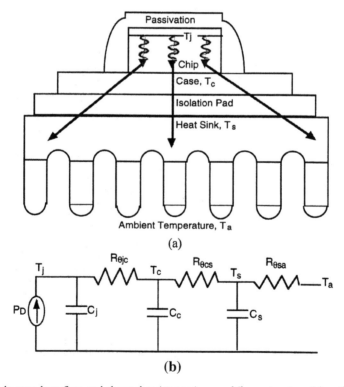

FIGURE 11.2 Steady-state heat flow and thermal resistance in a multilayer structure (a) path of heat flow; (b) equivalent electrical circuit based on thermal resistance.

in analogy with electric circuits. If there are parallel paths for heat flow, the thermal resistances are combined in exactly the same manner as electrical resistors in parallel.

$R_{\theta cs}$, the conductive thermal resistance, is mainly a function of package materials and geometry. With the higher power requirements, one must consider the temperature dependence of materials selected in design. T_j depends on package geometry, package orientation in the application, and the conditions of the ambient in the operating environment. The heat sink is responsible for getting rid of the heat of the environment by convection and radiation. Because of all the many heat transfer modes occurring in a finned heat sink, the accurate way to obtain the exact thermal resistance of the heat sink would be to measure it. However, most heat sink manufacturers today provide information about their extrusions concerning the thermal resistance per unit length.

Reliability

The package should have good thermomechanical performance for better reliability. A variety of materials of widely differing coefficients of thermal expansion (CTEs) are joined to create interfaces. These interfaces are subject to relatively high process temperatures and undergo many temperature cycles in their useful life as the device is powered on and off. As a result, residual stresses are created in the interfaces. These stresses cause reliability problems in the packages.[15,16]

Testability

Implicit in reliability considerations is the assumption of a flawless product function after its initial assembly — a zero defect manufacturing. Although feasible in principle, it is rarely practiced because of the high costs and possible loss of competitive edge due to conservative dimensions, tolerances, materials, and process choices. So, several tests are employed to assess the reliability of the packages.[17,18]

11.4 Packaging Substrates

An IC package falls into two basic categories. In the first, a single-layer type, the package is constructed around the IC chip on a lead frame. In the second, a multilayer type, the IC chip is assembled into a prefabricated package.

In a single-layer technology, the IC chip is first mechanically bonded to a lead frame, and then electrically interconnected with fine wires from the chip bond pads to the corresponding lead-frame fingers. The final package is then constructed around the lead-frame subassembly. Two single-layer technologies are used in the industry: molded plastic and glass-sealed pressed ceramic.

Plastic Packaging

Plastic is a generic term for a host of man-made organic polymers. Polymer materials are relatively porous structures, which may allow absorption or transport of water molecules and ions.[19] The aluminum metallization is susceptible to rapid corrosion in the presence of moisture, contaminants, and electric fields. So, plastic packages are not very reliable. Impurities from the plastic or other materials in the construction of the package can cause threshold shifts or act as catalysts in metal corrosion. Fillers can also affect reliability and thermal performance of the plastic package.

Ceramic Packaging

Pressed ceramic technology packages are used mainly for economically encapsulating ICs and semiconductor devices requiring hermetic seals. Hermeticity means that the package must pass both gross and fine leak tests and also exclude environmental contaminants and moisture for a long period of time. Further, any contaminant present before sealing must be removed to an acceptable level before or during the sealing process.[20] Silicon carbide (SiC), aluminum nitride, beryllia (BeO), and alumina

TABLE 11.2A Thermal and Electrical Properties of Materials Used in Packaging

Metals

Metals	Coefficient of Thermal Expansion (CTE) (10^{-6} K^{-1})	Thermal Conductivity (W/cm-K)	Specific Electrical Resistance $10^{-6}\Omega$-cm
Aluminum	23	2.3	2.8
Silver	19	4.3	1.6
Copper	17	4.0	1.7
Molybdenum	5	1.4	5.3
Tungsten	4.6	1.7	5.3

Substrates

Insulating Substrates	Coefficient of Thermal Expansion (CTE) (10^{-6} K^{-1})	Thermal Conductivity (W/cm-K)	Dielectric Constant
Alumina (Al_2O_3)	6.0	0.3	9.5
Beryllia (BeO)	6.0	2.0	6.7
Silicon carbide (SiC)	3.7	2.2	42
Silicon dioxide (SiO_2)	0.5	0.01	3.9

Semiconductors

Semiconductors	Coefficient of Thermal Expansion (CTE) (10^{-6} K^{-1})	Thermal Conductivity (W/cm-K)	Dielectric Constant
Silicon	2.5	1.5	11.8
Germanium	5.7	0.7	16.0
Gallium arsenide	5.8	0.5	10.9

TABLE 11.2B Some Properties of Ceramic Packaging Materials

Property	BeO	AlN	Al_2O_3 (96%)	Al_2O_3 (99.5%)
Density (g/cm^3)	2.85	3.28	3.75	3.8
CTE (ppm/K)	6.3	4.3	7.1	7.1
TC (W/cm-K)	285	180	21	25.1
Dielectric const.	6.7	10	9.4	10.2
Loss tangent	0.0001	0.0005	0.0001	0.0001

(Al_2O_3) are some of the ceramics used in electronic packaging. In comparison with other ceramics, SiC has a thermal expansion coefficient closer to silicon, and as a result less stress is generated between the *dice* and the substrate during temperature cycling. In addition, it has a very high thermal conductivity. These two properties make SiC a good packaging substrate and a good heat sink that can be bonded directly to silicon *dice* with little stress generation at elevated temperatures. Its high dielectric constant, however, makes it undesirable as a substrate to carry interconnections. Alumina and BeO have properties similar to SiC.[21]

11.5 Package Types

IC packages have been developed over time to meet the requirements of high speed and density. The history of IC package development has been the continuous battle to miniaturize.[22,23] Figure 11.3 illustrates the size and weight reduction of IC packages over time.

Several packages can be classified as follows.

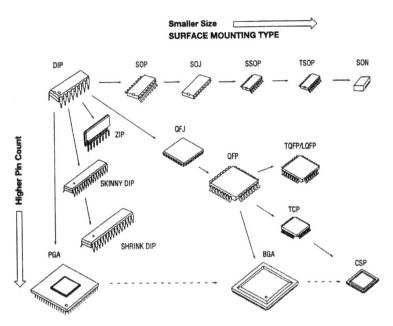

FIGURE 11.3 Packaging trends.

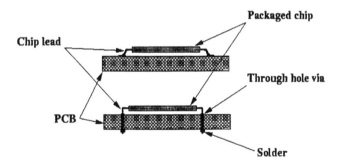

FIGURE 11.4 A generic schematic diagram showing the difference between the surface-mount technology (upper) and through hole mounting (lower).

Through Hole Packages

Through-the-board hole mounting technology uses precision holes drilled through the board and plated with copper. This copper plating forms the connections between separate layers. These layers consist of thin copper sheets stacked together and insulated by epoxy fiberglass. There are no dedicated via structures to make connections between wiring levels; through holes serve that purpose. Through holes form a sturdy support for the chip carrier and resist thermal and mechanical stresses caused by the variations in the expansions of components at raised temperatures. Different types (see Fig. 11.5) of through hole packages can be further classified as:

Dual-in-Line Packages (DIPs)

A dual-in-line package is a rectangular package with two rows of pins in its two sides. Here, first the die is bonded on the lead frame and in the next step, chip I/O and power/ground pads are wire-bonded to the lead frame, and the package is molded in plastic. DIPs are the workhorse of the high-volume and general-purpose logic products.

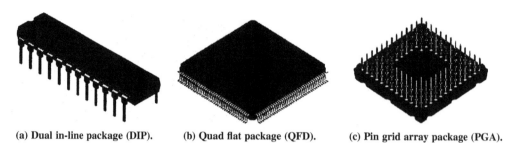

(a) Dual in-line package (DIP).   (b) Quad flat package (QFD).   (c) Pin grid array package (PGA).

FIGURE 11.5 Different through mount packages.

Quad Flat Packages (QFPs)

With the advances in VLSI technology, the lower available pin counts of the rectangular DIP became a limiting factor. With pins spaced 2.4 mm apart on only two sides of the package, the physical size of the DIP has become too great. On the other hand, the physical size of an unpackaged microelectronic circuit (bare die) has been reduced to a few millimeters. As a result, the DIP package has become up to 50 times larger than the bare die size itself, thus defeating the objective of shrinking the size of the integrated circuits. So, one solution is to provide pins all around. In QFPs, pins are provided on all four sides. Thin QFPs are developed to reduce the weight of the package.

Pin Grid Arrays (PGA)

A pin grid array has leads on its entire bottom surface rather than only at its periphery. This way it can offer a much larger pin count. It has cavity-up and cavity-down versions. In a cavity-down version, a die is mounted on the same side as the pins facing toward the PC board, and a heat sink can be mounted on its backside to improve the heat flow. When the cavity and the pins are on the same side, the total number of pins is reduced because the area occupied by the cavity is not available for brazed pins. The mounting and wire bonding of the dice are also more difficult because of the existence of the pins next to the cavity. High pin count and larger power dissipation capability of PGAs make them attractive for different types of packaging.

Surface-Mounted Packages

Surface mounting solves many of the shortcomings of through-the-board mounting. In this technology, a chip carrier is soldered to the pads on the surface of a board without requiring any through holes. The smaller component sizes, lack of through holes, and the possibility of mounting chips on both sides of the PC board improve the board density. This reduces package parasitic capacitances and inductances associated with the package pins and board wiring. Various types of surface-mount packages are available on the market and can be divided into the following categories. (See Fig. 11.6.)

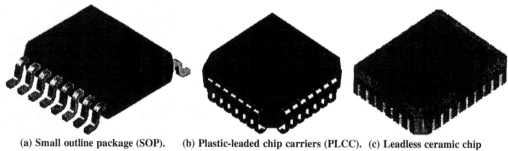

(a) Small outline package (SOP).   (b) Plastic-leaded chip carriers (PLCC). (c) Leadless ceramic chip
                                                                                  carriers (LCCC).

FIGURE 11.6 Different surface-mount packages.

Small-Outline Packages (SOPs)

The small-outline package has gull-wing shaped leads. It requires less pin spacing than through-hole-mounted DIPs and PGAs. SOP packages usually have small lead counts and are used for discrete, analog, and SSI/MSI logic parts.

Plastic-leaded Chip Carriers (PLCCs)

Plastic-leaded chip carriers, such as gull-wing and J-leaded chip carriers, offer higher pin counts than SOP. J-leaded chip carriers pack denser and are more suitable for automation than gull-wing leaded carriers because their leads do not extend beyond the package.

Leadless Ceramic Chip Carriers (LCCCs)

Leadless ceramic chip carriers take advantage of multilayer ceramic technology. The conductors are left exposed around the package periphery to provide contacts for surface mounting. *Dice* in leadless chip carriers are mounted in cavity-down position, and the back side of the chip faces away from the board, providing a good heat removal path. The ceramic substrate also has a high thermal conductivity. LCCCs are hermetically sealed.

Flip-Chip Packages

The length of the electrical connections between the chip and the substrate can be minimized by placing solder bumps on the *dice*, flipping the chips over, aligning them with the contacts pads on the substrate, and reflowing the solder balls in a furnace to establish the bonding between the chips and the package. This method provides electrical connections with minute parasitic inductance and capacitance. In addition, contact pads are distributed over the entire chip surface. This saves silicon area, increases the maximum I/O and power/ground terminals available with a given die size, and provides more efficiently routed signal and power/ground interconnections on the chips.[24] (See Fig. 11.7.)

Chip Size Packages (CSPs)

To combine the advantages of both packaged chip and bare chip in one solution, a variety of CSPs have been developed.[25,26] CSPs can be divided into two categories: the fan-in type and the fan-out type.

Fan-in type CSPs are suitable for memory applications that have relatively low pin counts. This type is further divided into two types, depending on the location of bonding pads on the chip surface; these are the center pad type and the peripheral pad type. This type of CSP keeps all the solder bumps within the chip area by arranging bumps in area array format on the chip surface.

The fan-out CSPs are used mainly for logic applications: because of the die size to pin count ratio, the solder bumps cannot be designed within the chip area.

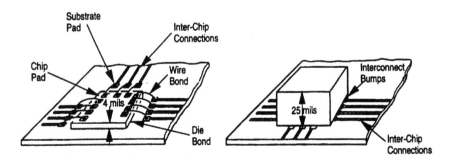

FIGURE 11.7 Flip-chip package and its interconnections.

Multi-Chip Modules (MCMs)

In a multi-chip module, several chips are supported on a single package. Most multi-chip packages are made of ceramic. By eliminating one level of packaging, the inductance and capacitance of the electrical connections among the *dice* are reduced. Usually, the *dice* are mounted on a multilayer ceramic substrate via solder bumps, and the ceramic substrate offers a dense interconnection network.[27,28] (See Fig. 11.8.) There are several advantages of multi-chip modules over single-chip carriers. The multi-chip module minimizes the chip-to-chip spacing and reduces the inductive and capacitive discontinuities between the chips mounted on the substrate by replacing the die-bump-interconnect-bump-die path. In addition, narrower and shorter wires on the ceramic substrate have much less capacitance and inductance than the PC board interconnections.

3-D VLSI Packaging

The driving forces behind the development of three-dimensional packaging technology are similar to the multi-chip module technology, although the requirements for the 3-D technology are more aggressive. These requirements include the need for significant size and weight reductions, higher performance, small delay, higher reliability, and potentially reduced power consumption.

Silicon-on-Silicon Hybrid

A silicon substrate can also be used as an interconnection medium to hold multi-chips as an alternative to ceramic substrates. This is called silicon-on-silicon packaging or, sometimes, hybrid wafer-scale integration. Thin-film interconnections are fabricated on a wafer and separately processed, and test dice are mounted on this silicon substrate via wire bonding, TAB, or solder bumps. Using this technique, chips fabricated in different technologies can be placed on the same hybrid package. The silicon substrate can also potentially contain active devices that serve as chip-to-chip drivers, bus and I/O multiplexers, and built-in test circuitry.[29]

11.6 Hermetic Packages

In hermetic packages, the packaged cavity is sealed in such a way that all external chemical species are permanently prevented from entering into it. In practice, however, a finite rate of leakage occurs through diffusion and permeation. Moisture is the principal cause of device failures. Moisture by

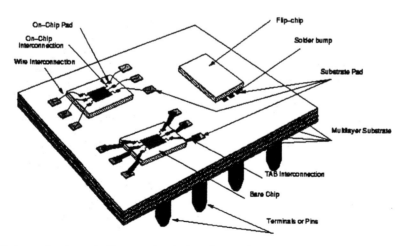

FIGURE 11.8 A generic schematic diagram of an MCM, showing how bare *dice* are interconnected to an MCM substrate using different interconnection technologies.

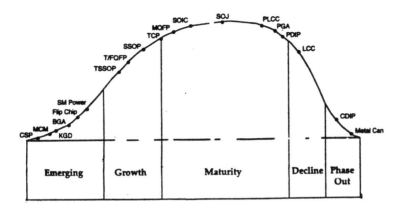

FIGURE 11.9 IC packaging life cycle.

itself does not cause electronic problems when trapped in an electronic package, because it is a poor electrical conductor. However, water can dissolve salts and other polar molecules to form an electrolyte, which, together with the metal conductors and the potential difference between them, can create leakage paths as well as corrosion problems. Moisture is contributed mainly by the sealing ambient, the absorbed and dissolved water from the sealing materials, lid and the substrate, and the leakage of external moisture through the seal. No material is truly hermetic to moisture. The permeability to moisture of glasses, ceramics, and metals, however, is very low and is orders of magnitude lower than for any plastic material. Hence, the only true hermetic packages are those made of metals, ceramics, and glasses. The common feature of hermetic packages is the use of a lid or a cap to seal in the semiconductor device mounted on a suitable substrate. The leads entering the package also need to be hermetically sealed.

11.7   Die Attachment Techniques

To provide electrical connections between the chip pads and package, different bonding techniques are used. These can be classified as follows.

Wire Bonding

Wire bonding (see Fig. 11.10) is a method used to connect a fine wire between an on-chip pad and a substrate pad. This substrate may simply be the ceramic base of a package or another chip. The common materials used are gold and aluminum. The main advantage of wire bonding technology is its low cost; but it cannot provide large I/O counts, and it needs large bond pads to make connections. The connections have relatively poor electrical performance.

Tape-Automated Bonding

In tape-automated bonding (TAB) technology, a chip with its attached metal films is placed on a multilayer polymer tape. The interconnections are patterned on a multilayer polymer tape. The tape is positioned above the "bare die" so that the metal tracks (on the polymer tape) correspond to the bonding sites on the die (Fig. 11.11). TAB technology provides several advantages over wire bonding technology. It requires a smaller bonding pad, smaller on-chip bonding pitch, and a decrease in the quantity of gold used for bonding.[30] It has better electrical performance, lower labor costs, higher I/O counts and lighter weight, greater densities, and the chip can be attached in a face-up or face-down configuration. TAB technology includes time and cost of designing and fabricating the tape and the capital expense of the

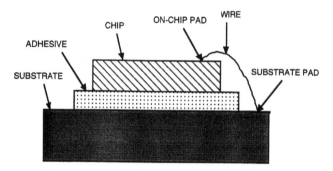

FIGURE 11.10   Wire bonding assembly.

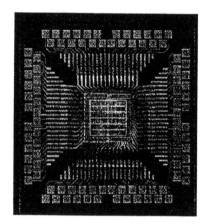

FIGURE 11.11   Tape-automated bonded die.

TAB bonding equipment. In addition, each die must have its own tape patterned for its bonding configuration. Thus, TAB technology has typically been limited to high-volume applications.

Solder Bump Bonding

Solder bumps are small spheres of solder (solder balls) that are bonded to contact areas or pads of semiconductor devices and subsequently used for face-down bonding. The length of the electrical connections between the chip and the substrate can be minimized by placing solder bumps on the die, flipping the die over, aligning the solder bumps with the contact pads on the substrate, and re-flowing the solder balls in a furnace to establish the bonding between the die and the substrate[31] (Fig. 11.12). This technology provides electrical connections with minute parasitic inductances and capacitances. In addition, the contact pads are distributed over the entire chip surface rather than being confined to the periphery. As a result, the silicon area is used more efficiently, the maximum number of interconnects is increased, and signal interconnections are shortened. But this technique results in poor thermal conduction, difficult inspection of the solder bumps, and possible thermal expansion mismatch between the semiconductor chips and the substrate.

11.8   Package Parasitics

Typically, the electrical interconnection of a chip in a package consists of chip-to-substrate interconnect, metal runs on the substrate, and finally, pins from the package. Associated with these are the electrical resistance, inductance and capacitance — referred to as package parasitics. The electrical parasitics are

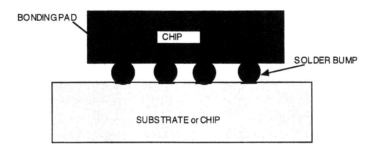

FIGURE 11.12 Flip-chip method using solder bumps.

determined by the physical parameters such as interconnect width, thickness, length, spacing, and resistivity; by the thickness of the dielectric; and by the dielectric constant.

Resistance refers to both dc and ac. The dc resistance of an interconnect is a property of its cross-sectional area, length, and material resistivity. In addition, the ac resistance depends on the frequency of the signal and is higher than the dc resistance because of the skin effect. Resistance in the power distribution path results in attenuation of input signals to the device and output signals from the device. This has the effect of increasing the path delay.

Capacitance of an interconnect is a property of its area, the thickness of the dielectric separating it from the reference potential, and the dielectric constant of the dielectric. It is convenient to consider this as two parts: capacitance with respect to ground, and capacitance with respect to other interconnections. The capacitance with respect to ground is referred to as the *load capacitance*. This is seen as part of the load by the output driver and thus can slow down the rise time of the driver. Interlead capacitance couples the voltage change on the active interconnect to the quiet interconnect.[32] This is referred to as *crosstalk*.

Inductance can be defined only if the complete current path is known. In the context of component packages, the inductance of an interconnect should be understood as part of a complete current loop. Thus, if the placement of the package in the system alters the current path in the package, the package inductance will vary. Total inductance consists of self-inductance and mutual inductance. Mutual inductance between two interconnects generates a voltage in one when there is current change in the other. Inductive effects are the leading concern in the design of power distribution paths in high-performance packages. They are manifested as "ground bounce" noise and "simultaneous switching" noise.

11.9 Package Modeling

As the complexity of devices increases, design and development efforts for packages become comparable to design and development efforts for chips. Many package design concepts must be simulated to assess their associated performance parameters.[33,34] Computer-aided design software and test chips are becoming indispensable design tools. Computer-aided design tools are extensively used to analyze the thermal, thermomechanical, mechanical, and electrical parameters of packages;[35] for example, electrical modeling extracts an equivalent electrical circuit that describes the physical structure of the package and, hence, the equivalent electrical circuit of the package can be used in circuit simulation programs to evaluate the overall performance of a packaged circuit. Until now, the equivalent electrical circuit incorporated only lumped electrical parameters; but as frequency of operation of the circuits is increasing, the distributed model of the package needs to be developed for high-frequency simulations.[36]

11.10 Packaging in Wireless Applications

Wireless applications typically involve RF, high-frequency digital, and mixed-mode circuits. Wireless packaging requires minimal electrical parasitic effects that need to be well-characterized.

In wireless applications, the trend is to integrate multiple modules on a single chip.[37] So, the thermal management of the whole chip becomes crucial. The IC package must have good thermal properties. Metal as a material shows optimal properties concerning thermal conductivity, electromagnetic shielding, mechanical and thermal stability. For thermal expansion, best match to semiconductor and ceramic material can be achieved with molybdenum, tungsten, or special composites like kovar. Ceramic materials are applied, both as parts of the package as well as for subsystem-carrying RF transmission lines. To this end, and to provide electromagnetic shielding, these materials partly have to be metallized. Aluminum nitride, beryllia, aluminum silicon carbide, and CVD diamond show best thermal conductivity and are therefore applied in high-power applications,[38,39] while alumina is well known for standard microwave applications.[40]

Integration of passive components is a major challenge in wireless packages. More and more efforts are being made to integrate passive components, power devices on a chip, with the other mixed signal circuits.[41] The size of the package becomes an issue. Micromachining technology provides a way to make miniature packages that conform to RF circuits, while providing physical and electrical shielding. Conformal packages made by applying micromachining technology provide the capability to isolate individual circuit elements and improve circuit performance by eliminating the radiation and cross-coupling between the adjacent circuits.[42,43]

At high frequencies, interconnections need to be carefully designed. Microstrip interconnects and co-planar waveguides are mostly used for microwave packaging.[44] Flip-chip packaging has tremendous potential for future RF packaging.[45]

11.11   Future Trends

Packaging is the bridge between silicon and electronics systems. Packaging design and fabrication are increasingly important to system applications. Consideration of factors affecting waveform integrity for both power and signal (i.e., timing, cross-talk, and ground bounce) will affect device layout on the chip, chip layout on the package, and interconnect.

Conventional surface mount packages will dominate in the region of low pin count and low clock frequency. Ball-grid array packages and chip-scale packages will be used for medium pin counts. Technically bare chip solutions can cover the whole area, but have a reliability versus cost tradeoff. Bare chip solutions could be very competitive with packaged solutions, as they can accomplish very high density and very good electrical performance.

Packaging needs are driven as much by market application requirements as by silicon technology. Cost drives technology tradeoffs for all market segments. As the complexity of package technology continues to increase, new materials will be needed to meet design and performance challenges. Significant engineering development will be needed for power increases at each technology generation.

An integrated design environment of physical, electrical, thermal, thermo-mechanical, chip, package, and system design needs to be evolved. Most of these integrated solutions will provide modeling and simulation capabilities that will be embodied in packaging computer aided design systems. Design tools are required to manage the complexity of packaging that is being pushed to its performance limits.

References

1. Tummalam, R. R. and Klopfenstein, A. G., *Microelectronics Packaging Handbook*, Van Nostrand Reinhold, New York, 1989.
2. Bakoglu, H. B., *Circuits, Interconnections, and Packaging for VLSI*, Addison-Wesley, New York, 1990.
3. Tummala, R. R., "Electronic Packaging in the 1990s — A Perspective from America," *IEEE Trans. Components, Hybrids, Manuf. Technol.*, vol. 14, no. 2, pp. 262, June 1991.
4. Wessely, H., "Electronic Packaging in the 1990s — A Perspective from Europe," *IEEE Trans. Components, Hybrids, Manuf. Technol.*, vol. 14, no. 2, pp. 272, June 1991.

5. Mones, A. H. and Spielberger, R. K., "Interconnecting and Packaging VLSI Chips," *Solid State Technology*, pp. 119-122, Jan. 1984.

6. Kakei, M., "Low Stress Molding Compounds for VLSI Devices," *Nikkei Microdevices*, 1984.

7. Lyman, J., "VLSI Packages are Presenting Diversified Mix," *Electronics*, pp. 67-73, Sept. 1984.

8. Bakoglu, H. B., "Packaging for High-Speed System," *IEEE International Solid-State Circuits Conference (ISSCC'98)*, pp. 100-101, San Francisco, Feb. 1988.

9. Kaupp, H. R., "Characteristics of Microstrip Transmission Line," *IEEE Trans. Computers*, EC-16, pp. 185, April 1967.

10. Trivedi, M. and Shenai, K., "Framework for Power Package Design and Optimization," *Intl. Workshop on Integrated Power Packaging (IWIPP'98)*, 1998, Chicago, IL.

11. Khandelwal, P., Trivedi, M., and Shenai, K., "Thermal Issues in LDMOSFET's Packages," *European Solid-State Device Research Conference (ESSDRC)*, 1998.

12. Manchester, K. and Bird, K., "Thermal Resistance: A Reliability Consideration," *IEEE Trans. Components, Hybrids, Manuf. Technol.*, vol. 31, no. 4, pp. 550, Dec. 1980.

13. Chu, R. C., Hwang, U. P., and Simons, R. E., "Conduction Cooling for LSI Packages, A One-Dimensional Approach," *IBM Journal of Research and Development*, vol. 26, no. 1, pp. 45-54, Jan. 1982.

14. Mohan, N., Undeland, T. M., and Robbins, W. P., *Power Electronics: Converters, Applications, and Design*, John Wiley & Sons, 1996.

15. Fukuzawa, I., "Moisture Resistance Degradation of Plastic LSIs by Reflow Soldering," *Proc. International Reliability Physics Symposium*, pp. 192, 1985.

16. Lau, J. H., *Solid Joint Reliability: Theory and Applications*, Van Nostrand Reinhold, New York, 1991.

17. Gallace, L. J. and Rosenfield, M., "Reliability of Plastic Encapsulated Integrated Circuit in Moisture Environments," *RCA Review*, vol. 45, no. 2, pp. 95-111, June 1984.

18. Lau, J. H., "Thermal Stress Analysis of SMT PQFP Packages and Interconnections," *J. Electronic Packaging, Trans. ASME*, vol. 2, pp. 111, March 1989.

19. Kawai, S., "Structure Design of Plastic IC Packages," *Proc. SEMI Tech. Symposium*, pp. 349, Nov. 1988.

20. White, M. L., "Attaining Low Moisture Levels in Hermetic Packages," *Proc. 20th International Reliability Physics Symposium*, pp. 253, 1982.

21. Sepulveda, J. L. and Siglianu, R., "BeO Packages House High-Power Components," *Microwaves & RF*, pp. 111-124, March 1998.

22. Fehr, G., Long, J., and Tippetts, A., "New Generation of High Pin Count Packages," *Proc. IEEE Custom Integrated Circuits Conference (CICC'85)*, pp. 46-49, 1985.

23. Sudo, T., "Considerations of Package Design for High Speed and High Pin Count CMOS Devices," *Proc. 39th Electronic Components and Technology Conference*, 1989.

24. Midford, T. A., Wooldridge, J. J., and Sturdivant, R. L., "The Evolution of Packages for Monolithic Microwave and Millimeter Wave Circuits," *IEEE Trans. Antennas and Propagation*, vol. 43, no. 9, pp. 983, Sept. 1995.

25. Yamaji, Y., Juso, H., Ohara, Y., Matsune, Y., Miyata, K., Sota, Y., Narai, A., and Kimura, T., "Development of Highly Reliable CSP," *Proc. 47th Electronic Components and Technology Conference*, pp. 1022-1028, 1997.

26. Su, L. S., Louis, M., and Reber, C., "Cost Analysis of Chip Scale Packaging," *Proc. 21st IEEE/CPMT Intl. Electronics Manufacturing Technology Symposium*, pp. 216-223, 1997.

27. Lyman, J., "Multichip Modules Aim at Next Generation VLSI," *Electronic Design*, pp. 33-34, March 1989.

28. Barlett, C. J., Segelken, J. M., and Teneketgen, N. A., "Multichip Packaging Design for VLSI Based Systems," *IEEE Trans. Components, Hybrids, Manuf. Technol.*, vol. 12, no. 4, pp. 647-653, Dec. 1987.

29. Spielberger, R. K., Huang, C. D., Nunne, W. H., Mones, A. H., Fett, D. C., and Hampton, F. L., "Silicon-on-Silicon Packaging," *IEEE Trans. Components, Hybrids, Manuf. Technol.*, vol. 7, no. 2, pp. 193-196, June 1984.

30. Andrews, W., "High Density Gate Arrays Tax Utility, Packaging, and Testing," *Computer Design*, pp. 43-47, Aug. 1988.

31. Fujimoto, H., "Bonding of Ultrafine Terminal-Pitch LSI by Micron Bump Bonding Method," *Proc. IMC*, p. 115, 1992.

32. Dang, R. L. M. and Shigyo, N., "Coupling Capacitance for Two-Dimensional Wires," *IEEE Electron Device Letters*, vol. EDL-2, pp. 196-197, Aug. 1981.

33. Jackson, R. W., "A Circuit Topology for Microwave Modeling of Plastic Surface Mount Packages," *IEEE Trans. Microwave Theory and Techniques*, vol. 44, pp. 1140-1146, 1997.

34. Gupta, R., Allstot, D. J., and Meixner, R., "Parasitic-Aware Design and Optimization of CMOS RF Integrated Circuits," *IEEE MTT-S International Symposium*, vol. 3, pp. 1867-1870, 1998.

35. Perugupalli, P., Xu, Y., and Shenai, K., "Measurement of Thermal and Packaging Limitations in LDMOSFETs for RFIC Applications," *IEEE Instrumentation and Measurement Technology Conference (IMTC)*, pp. 160-164, May 1998.

36. Raid, S. M., Su, W., Salma, I., Riad, A. E., Rachlin, M., Baker, W., and Perdue, J., "Plastic Packaging Modeling and Characterization at RF/Microwave Frequencies," *Proc. 3rd International Symposium on Advanced Packaging Materials*, p. 147, Mar. 1997.

37. Bugeau, J. L., Heitkamp, K. M., and Kellerman, D., "Aluminum Silicon Carbide for High Performance Microwave Packages," *IEEE MTT-S International Symposium*, vol. 3, pp. 1575-1578, 1995.

38. Gomes-Casseres, M. and Fabis, P. M., "Thermally Enhanced Plastic Package Using Diamond for Microwave Applications," *IEEE MTT-S International Symposium*, vol. 1, pp. 227-230, 1996.

39. Kemerley, R. T., "Emerging Microwave Packaging Technologies," *IEEE 3rd Tropical Meeting on Electrical Performance of Electronic Packaging*, pp. 139-242, 1994.

40. Smith, C. R., "Power Module Technology for the 21st Century," *Proc. IEEE National Aerospace and Electronics Conference*, vol. 1, pp. 106-113, 1995.

41. Drayton, R. F. and Katehi, L.P.B., "Micromachined Conformal Packages for Microwave and Millimeter-Wave Applications," *IEEE MTT-S International Symposium*, vol. 3, pp. 1387-1390, 1995.

42. Drayton, R. F., Henderson, R. M., and Katehi, L.P.B., "Advanced Monolithic Packaging Concepts for High Performance Circuits and Antennas," *IEEE MTT-S International Symposium*, vol. 3, pp. 1615-1618, 1996.

43. Wein, D. S., "Advanced Ceramic Packaging for Microwave and Millimeter Wave Applications," *IEEE Trans. Antennas and Propagation*, vol. 43, pp. 940-948, Sept. 1995.

44. Krems, T., Haydll, W. H., Massler, H., and Rudiger, J., "Advances of Flip Chip Technology in Millimeter-Wave Packaging," *IEEE MTT-S International Symposium*, vol. 2, pp. 987-990, 1997.

12

Multichip Module Technologies

Victor Boyadzhyan
Jet Propulsion Laboratory

John Choma, Jr.
University of Southern California

12.1 Introduction

From the pioneering days to its current renaissance, the electronics industry has become the largest and most pervasive manufacturing industry in the developed world. Electronic products have the hallmark of innovation, creativity, and cost competitiveness in the world market place. The way the electronics are packaged, in particular, has progressed rapidly in response to customers' demands in general for diverse functions, cost, performances, and robustness of different products. For practicing engineers, there is a need to access the current state of knowledge in design and manufacturing tradeoffs.

Thus arises a need for electronics technology-based knowledge to optimize critical electronic design parameters such as speed, density, and temperature, resulting in performance well beyond PC board design capabilities. By removing discrete component packages and using more densely packed interconnects, electronic circuit speeds increase. The design challenge is to select the appropriate packaging technology, and to manage any resulting thermal problems.

The expanding market for high-density electronic circuit layouts calls for multi-chip modules (MCMs) to be able to meet the requirements of fine track and gap dimensions in signal layers, the retention of accurately defined geometry in multilayers, and high conductivity to minimize losses. Multi-chip module technologies fill this gap very nicely. This chapter provides engineers/scientists with an overview of existing MCM technologies and briefly explains similarities and differences of existing MCM technologies. The text is reinforced with practical pictorial examples, omitting extensive development of theory and details of proofs.

The simplest definition of a multi-chip module (MCM) is that of a single electronic package containing more than one integrated circuit (IC) die.[1] An MCM combines high-performance ICs with a custom-designed common substrate structure that provides mechanical support for the chips and multiple layers of conductors to interconnect them.

One advantage of this arrangement is that it takes better advantage of the performance of the ICs than it does interconnecting individually packaged ICs because the interconnect length is much shorter. The really unique feature of MCMs is the complex substrate structure that is fabricated using multilayer ceramics, polymers, silicon, metals, glass ceramics, laminates, etc. Thus, MCMs are not really new. They have been in existence since the first multi-chip hybrid circuit was fabricated. Conventional PWBs utilizing chip-on-board (COB), a technique where ICs are mounted and wire-bonded directly to the board, have also existed for some time. However, if packaging efficiency (also called silicon density), defined as the percentage of area on an interconnecting substrate that is occupied by silicon ICs, is the guideline used to define an MCM, then many hybrid and COB structures with less than 30% silicon density do not qualify as MCMs. In combination with packaging efficiency, a minimum of four conductive layers and 100 I/O leads has also been suggested as criteria for MCM classification.[1]

A formal definition of MCMs has been established by the Institute for Interconnecting and Packaging Electronic Circuits (IPC). They defined three primary categories of MCMs: MCM-L, MCM-C, and MCM-D.

It is important to note that these are simple definitions. Consequently, many IC packaging schemes, which technically do not meet the criteria of any of the three simple definitions, may incorrectly be referred to as MCMs. However, when these simple definitions are combined with the concept of packaging efficiency, chip population, and I/O density, there is less confusion about what really constitutes an MCM. The fundamental (or basic) intent of MCM technology is to provide an extremely dense conductor matrix for the interconnection of bare IC chips. Consequently, some companies have designated their MCM products as high-density interconnect (HDI) modules.

12.2   Multi-Chip Module Technologies

From the above definitions, it should be obvious that MCM-Cs are descended from classical hybrid technology, and MCM-Ls are essentially highly sophisticated printed circuit boards, a technology that has been around for over 40 years. On the other hand, MCM-Ds are the result of manufacturing technologies that draw heavily from the semiconductor industry.

MCM-L

Modules constructed of plastic laminate-based dielectrics and copper conductors utilizing advanced forms of printed wiring board (PWB) technologies to form the interconnects and vias are commonly called "laminate MCMs," or MCM-Ls.[2]

Advantages	
Economic	Ability to fabricate circuits on large panels with a multiplicity of identical patterns. Reduces manufacturing cost. Quick response to volume orders.

Disadvantages
Technological | More limited in interconnect density relative to advanced MCM-C and MCM-D technologies. Copper slugs and cutouts are used in MCM-Ls for direct heat transfer. This degrades interconnection density.

MCM-L development has involved evolutionary technological advances to shrink the dimensions of interconnect lines and vias. From a cost perspective, it is desirable to use conventional PWB technologies for MCM-L fabrication. This is becoming more difficult as the need for multi-chip modules with higher interconnect density continues.

As MCM technologies are being considered for high-volume consumer products applications, a focus on containing the cost of high-density MCM-Ls is becoming critical.

The most useful characteristic in assessing the relative potential of MCM-L technology is interconnection density,[3,4] which is given by:

$$\text{Packaging efficiency (\%)} = \text{Silicon chip area/Package area} \quad (12.1)$$

The above formula measures how much of the surface of the board can be used for chip mounting pads versus how much must be avoided because of interconnect traces and holes/pads.

MCM-C

These are modules constructed on co-fired ceramic or glass-ceramic substrates using thick-film (screen printing) technologies to form the conductor patterns using fireable metals. The term "co-fired" implies that the conductors and ceramic are heated at the same time. These are also called thick-film MCMs.

Ceramic technology for MCMs can be divided into four major categories

- Thick-film hybrid process
- High-temperature co-fired alumina process (HTCC)
- Low-temperature co-fired ceramic/glass based process (LTCC)
- High T_c aluminum nitride co-fired substrate (AIN)

Thick-film hybrid technology produces by the successive deposition of conductors, dielectric, and/or resistor patterns onto a base substrate.[5] The thick-film material, in the form of a paste, is screenprinted onto the underlying layer, then dried and fired. The metallurgy chosen for a particular hybrid construction depends on a number of factors, including cost sensitivity, conductivity requirements, solderability, wire bondability, and more. A comparative summary of typical ceramic interconnect properties is compiled in Table 12.1.

TABLE 12.1 A Comparative Summary of Typical Ceramic Interconnect Properties

Item	Thick Film	HTCC	LTCC
Line width (μm)	125	100	100
Via diameter (μm)	250	125	175
Ave. no. conductor layers	1–6	1–75	1–75
Conductor res. (mohm/sq)	2–100	8–12	3–20
ε (dielectric)	5–9	9–10	5–8
CTE	4–7.5	6	3–8
T_c Dielectric (W/mC)	2	15–20	1–2
Relative cost (low volume)	Medium	High	High
Tooling costs	Low	High	High
Capital outlay	Low	High	Medium

12.3   Materials for HTCC Aluminum Packages

Metal conductors of tungsten and molybdenum are used for compatibility in co-firing to temperatures of 1600°C. Materials compatibility during co-firing dictates that raw materials of alumina with glass used for densification and any conductor metal powders (W, Mo) must be designed to closely match onset, rate, and volume shrinkage; promote adhesion; and minimize thermal expansion mismatch between conductor and dielectric.

Processing HTCC Ceramics

The raw materials used in fabrication of aluminum substrates include aluminum oxide, glass, binder, plasticizer, and solvent. Materials specifications are used to control alumina particle size, surface area, impurity, and agglomeration. Glass frit is controlled through composition, glass transition and softening point, particle size, and surface area. Molecular weight, group chemistry, and viscosity controls are used for the binder and plasticizer.

Metal Powder and Paste

A thick-film paste often uses metal powder, glass powder, organic resins, and solvents. Compositions are varied to control screening properties, metal shrinkage, and conductivity. Paste fabrication begins with batch mixing, dispersion, and deagglomeration, which are completed on a three-roll mill.

Thick-Film Metallization

The greensheet is cast, dried, stripped from the carrier film, and blanked into defect-free sheets, typically 200 mm^2. The greensheet is then processed through punching, screening, and inspection operations.

HTCC in Summary

- Electrical performance characteristics include 50-ohm impedance, low conductor resistance, ability to integrate passive components such as capacitors and inductors, the ability to achieve high wiring density (ease of increasing the number of wiring at low cost), the ability to support high-speed simultaneous switching drivers, and the ease of supporting multiple reference voltages.
- Inherent thermal performance characteristics superior to MCM-L and MCM-D.
- Time-demonstrated reliability.

12.4   LTCC Substrates

The use of glass and glass-ceramics in electronic packaging goes back to the invention of semiconductors. Glasses are used for sealing T-O type packages and CERDIPs, as crossover and inter-level dielectrics in hybrid substrates. The success of co-fired alumina substrates spurred the development of the multilayer glass-ceramic substrates. These advantages derive from the higher electrical conductivity lines of copper, silver, or gold; the lower dielectric constant of the glass ceramic; and the closer CTE match of the substrate to silicon.

Two approaches have been used to obtain glass-ceramic compositions suitable for fabricating self-supporting substrates.[6–8] In the first approach, fine powder of a suitable glass-composition is used that has the ability to sinter well in the glassy state and simultaneously crystallize to become a glass-ceramic. More commonly, mixtures of ceramic powders are used, such as alumina and a suitable glass in nearly equal proportions, to obtain densely sintered substrates.[9,10] Because many glass and ceramic powders can be used to obtain densely sintered glass-ceramic, the actual choice is often made on the basis of other desirable attributes in the resulting glass-ceramic — such as low dielectric constant for lowering the signal propagation delay and coefficient of thermal expansion (CTE) closely matched to the CTE of silicon to improve the reliability of solder interconnections. Unlike the case of a crystallizing glass, the mixed glass and ceramic approach allows for a much wider choice of materials.

TABLE 12.2 AlN Properties Comparison

Item	ALN	HTCC	LTCC	BeO	Si	Cu
Thermal conductivity (W/mK)	175	25	2	260	150	394
Density (g/cm)		3.3	3.9	2.6	2.8	8.9
Dielectric constant (Mhz)	8.9	9.5	5.0	6.7		
Dissipation factor	0.0004	0.0004	0.0002	0.0004		
Bending strength (MaP)	320	420	210	220		

Source: From Ref. 2.

12.5 Aluminum Nitride

Aluminum nitride products are used in a variety of commercial and military applications.

Thermal management with solutions such as AlN can provide superior cooling to ensure reliable system operation. AlN packages typically offer a thermal conductivity of 150 to 200 W/mK, a level which can be compared with many metals or other high thermal conductive materials such as berillia (BeO) or silicon carbide (SiC). AlN has a thermal coefficient of expansion of 4.4 ppm, which is better matched to silicon than to alumina or plastics. Table 12.2 provides a comparison of AlN properties.

12.6 Materials for Multi-Layered AlN Packages

Aluminum nitride is a synthetic compound manufactured by two processes: the carbothermal reduction of alumina (Eq. 12.2) and/or direct nitridation of aluminum metal (Eq. 12.3):

$$Al_2O_3 + 3C + N_2 \rightarrow 2AlN + 3CO \qquad (12.2)$$

$$2Al + N_2 \rightarrow 2AlN \qquad (12.3)$$

MCM-D

Modules are formed by the deposition of thin-film metals and dielectrics, which may be polymers or inorganic dielectrics. These are commonly called thin-film MCMs.

Here, the focus will be on materials to fabricate the high-density MCM-D interconnect. The materials of construction can be categorized as the thin-film dielectric, the substrate, and the conductor metallization.

12.7 Thin-Film Dielectrics

Dielectrics for the thin-film packaging are polymeric and inorganic. Here, we will try to be brief and informative about those categories. Thin-film packages have evolved to a much greater extent with polymeric materials. The capability offered by polymers includes a lower dielectric constant, the ability to form thicker layers with higher speeds, and lower cost of deposition. Polymer dielectrics have been used as insulating layers in recent microelectronics packaging.

12.8 Carrier Substrates

The thin-film substrate must have a flat and polished surface in order to build upon. The substrate should be inert to the process chemicals, gas atmospheres, and temperatures used during the fabrication of the interconnect. Mechanical properties are particularly important because the substrate must be strong enough to withstand handling, thermal cycling, and shock. The substrate must also meet certain CTE constraints because it is in contact with very large silicon chips on one side and with the package on the other side.[11,12] Thermal conductivity is another important aspect when heat-generating, closely spaced chips need that heat conducting medium. It is informative to state that high-density, large-area processing has generated interest in glass as a carrier material.

Metallic substrates have been used to optimize the thermal and mechanical requirements while minimizing substrate raw material and processing costs. Metallic sandwiches such as Cu/Mo/Cu can be tailored to control CTE and thermal properties. 5%Cu/Mo/5%Cu is reported to have a thermal conductivity (TC) of 135 W/mK, a CTE of 5.1 ppm, an as-manufactured surface finish of 0.813 μm, and a camber of 0.0005 in/in.

12.9 Conductor Metallization

In MCM, fabrication materials chosen will depend on the design, electrical requirements, and process chosen to fabricate the MCM. It is important to note that the most critical requirements for conductor metallization are conductivity and reliability.

Aluminum

Aluminum is a low-cost material that has adequate conductivity and can be deposited and patterned by typical IC techniques. It is resistant to oxidation. It can be sputtered or evaporated, but cannot be electroplated.

Copper

Copper has significant conductivity over aluminum and is more electromigration-resistant. It can be deposited by sputtering, evaporation, electroplating, or electroless plating. Copper rapidly oxidizes, forming a variety of oxides that can have poor adhesion to polymer dielectrics and copper itself.

Gold

Gold is used in thin-film structures to minimize via contact resistance problems caused by oxidation of Cu and Al. Gold can be deposited by sputtering, evaporation, electroplating, or electroless plating. Cost is high with excellent resistivity characteristic. Adhesion is poor and so it requires a layer (50 to 200 nm) of Ti or Ti/W.

12.10 Choosing Substrate Technologies and Assembly Techniques

The MCM designer has the freedom of choosing/identifying substrate and assembly technologies[13–15] from many sources.[16,17] If you are about to start a design and are looking for guidelines, finding a good source of information could be Internet access. In addition to designing to meet specific performance requirements, it is also necessary to consider ease of manufacture. For example, Maxtek publishes a set of design guidelines that inform the MCM designer of preferred assembly materials and processes, process flows, and layout guidelines.

Under Substrate Technologies and Assembly techniques it will be very informative to look at some pictorial examples, a primary source of which is Maxtek. The pictorial examples, followed by a brief description of technology or assembly technique shown, will serve as a quick guideline for someone who

Substrate Technologies	Assembly Techniques
Chip-on-board	Surface mount
Chip-on-flex	Chip-and-wire
Thick-film ceramic	Mixed technology
Cofired ceramic	Special module
Thin-film ceramic	

would like to get a feel of what technologies are viable in the MCM technology domain. Here, it is important to mention that a lot of current space electronic flight projects use MCM technologies for their final deliverables. Project Cassini, for example, used MCM hybrids in telecommunication subassemblies. On this note, take a look at some of the examples of MCM technologies currently on the market.

Chip-on-Board

Chip-on-board substrate technology (Fig. 12.1) has low set-up and production costs and utilizes Rigid FR-406, GETEK, BT Epoxy, or other resin boards.[3] Assembly techniques used are direct die attach/wire bonding techniques, combined with surface-mount technologies.

Chip-on-Flex

Chip-on-Flex substrate technologies[18,19] (Fig. 12.2) are excellent for high-frequency, space-constrained circuit implementation. In creating this particular technology, the manufacturer needs Kapton or an equivalent flex-circuit base material with board stiffeners. Here, the die can be wire-bonded, with "glob-top" protection, while other discretes can be surface mounted. Integral inductors can be incorporated (e.g., for load matching).

Thick-Film Ceramic

This technology is the most versatile technology, with low-to-medium production costs. The 1-GHz attenuator above demonstrates the versatility of thick film on ceramic substrate technology (Fig. 12.3), which utilizes both standard and custom ICs, printed resistors, and capacitors actively trimmed to 0.25% with extremely stable capacitors formed between top plate and ground plane on the other side of substrate. Thick-film thermistor here senses overheating resistor and protects the remainder of the circuit.

FIGURE 12.1 Chip-on-board.

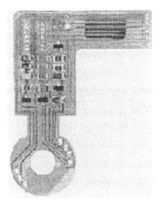

FIGURE 12.2 Chip-on-flex.

Co-Fired Ceramic

Co-fired ceramic MCM substrate technologies (low- or high-temperature co-fired multilayer ceramic) (Fig. 12.4) are particularly suited for high-density digital arrays. Despite its high set-up and tooling costs, up to 14 co-fired ceramic layers are available from this particular manufacturer. In this technology, many package styles are available, including DIP, pin-grid array, and flat pack.

Thin-Film Ceramic

For thin-film ceramic technologies (see Fig. 12.5), here the outlined technology features include:

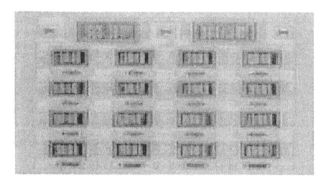

FIGURE 12.3 Thick-film ceramic.

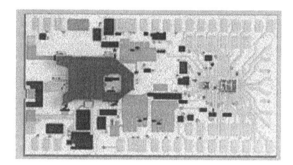

FIGURE 12.4 Co-fired ceramic.

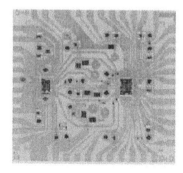

FIGURE 12.5 Thin-film ceramic.

- High-performance MCMs, offset by high set-up and tooling costs
- Alumina or BeO (as shown here) substrate
- Both sides of substrate can be used, the back side typically being a ground plane, with access through plated-through holes
- Chip-and-wire assembly with epoxied capacitors
- Many packaging styles available, including Kovar or ceramic package and lid
- Primarily used for high-frequency circuits requiring thin-film inductors or controlled impedance lines

For quick reference, some specifications and MCM technology information are summarized in Tables 12.3 and 12.4.

Other thick-film components are available, such as thermistors and spark gaps. In many applications, both sides of a substrate can be used for printed components.

12.11 Assembly Techniques

Surface-Mount Assembly

The surface-mount assembly technique can be categorized under lowest-cost, fastest turnaround assembly method, using pre-packaged components soldered to glass-epoxy board, flex circuit, or thick-film ceramic substrate. Many package styles are available, such as SIP and DIP. Pins may be attached as in-line leads, 90° leads, etc.[19]

TABLE 12.3 Thick-Film Specifications

Printed Component	Sheet Resistivity per Square	Typical Values	Comments
Conductor, Gold	3–5 mohm		Lowest resistivity, 5-mil min. line/space
Etched thick-film	3–5 mohm		2 mil min. line/space
Conductor, Pd-Ag	<50 mohm		Lowest cost, solderable, 10-mil min. line/space
Resistor	3 ohm–1 Mohm		<±100 ppm/°C, laser trimmable to ±0.25%
Capacitor		10, 50, 100, 1500	Untrimmed ±30%, may be actively trimmed dielectric constant
Inductor		4–100 nH	Untrimmed ±10%

Source: Ref. 19.

TABLE 12.4 MCM Substrate Technologies

Technology	Material	Line/Space (mils min.)	Dielectric Constant	R	C	L
Chip-on-board	Glass-epoxy (FR-4), polyimides	4	4.3-5.0 4.0-4.6	Y	Y	
Chip-on-flex	Kapton or equiv. with stiffeners	3 inner, 5 outer	3.2-3.9	Y	Y	
Thick-film ceramic	Alumina	5	9.26.3	Y	Y	Y
	BeO	5		Y	Y	
Multilayer ceramic	Lo-fire alumina	5	5.8-9.0	Y	Y	
	Hi-fire alumina	5	9.0-9.6	Y	Y	
Thin-film ceramic	Alumina	1	9.9	Y	Y	Y
	BeO	1	6.3			
Etched thick-film	Alumina	2	9.2	Y	Y	Y

Chip-and-Wire Assembly

In order to minimize interconnect electronic circuit parasitics, a high-density layout is one of the assembly techniques recommended as a solution. The highlight of this technique is epoxy attachment/wire bonding of integrated circuits and components (e.g., capacitors) to a glass-epoxy board (chip-on-board). Of course, another way is attachment to a thick- or thin-film ceramic substrate. Currently, many package styles are available and can be listed as follows:[19,20]

- Epoxy seal, using a B-stage ceramic lid epoxies to the substrate (quasi-hermetic seal)
- Encapsulation, in which bare semiconductor die are covered with a "glob top" (low-cost seal)
- Metal (typically Kovar) package with Kovar lid either welded or soldered to the package (hermetic seal)
- Leads are typically plug-in type, SIP, DIP, PGA, etc.

Mixed Technologies

Another category of assembly technique recognized as "mixed technologies" combines chip-and-wire with surface-mount assembly techniques on a single substrate, which may be a glass-epoxy board or ceramic substrate. Heat sink/heat spreaders are available in a variety of materials.

The Maxtek module shown in Fig. 12.7 includes a 4-layer, 20-mil-thick glass-epoxy board mounted to a beryllium copper heat spreader.

Selectively gold plated for wire bonding pads and pads at each end for use with elastomeric connectors,

- Three methods of IC die attach
- Epoxied directly to glass-epoxy board
- Epoxied directly to the BeCu heat spreader through a cutout in the board
- Epoxied to the head spreader, through a cutout, via a thermally conductive submount, to electrically isolate the die from the heat spreader
- Solder-mounted resistors and capacitors
- 50-ohm differential signal lines
- IC die may be "glob topped" or covered in either a ceramic or plastic lid for protection

Special Modules

Under special modules we can emphasize and highlight technologies of complex assemblies of mixed-technology substrates, flexible circuits, and/or electromechanical components. (See Fig. 12.8.) Complex assemblies of mixed-technology substrates often utilize a B-stage epoxy lid or glob top over chip-and-wire circuitry. These technologies enable designers to provide integrated solution complex

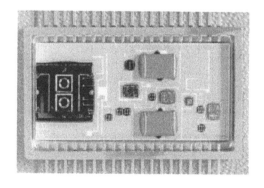

FIGURE 12.6 JPL STRV-2 project: tunneling accelerometer.

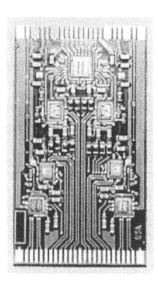

FIGURE 12.7 Example of a mixed-technology assembly technique.

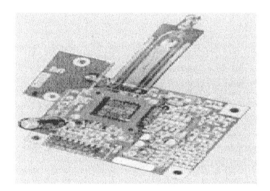

FIGURE 12.8 Example of a special module.

system problems as in a module shown on page 22 which is CRT Driver System capable of displaying 4 million pixels with 1.5-ns rise and fall times. The circuit incorporates a thin-film MCM connected by a special high-frequency connector to an FR-4 board with thick-film ceramic low-inductance load resistor and flex circuit with an integral impedance-matching inductor, connecting directly to the CRT.[19,20]

The design engineer of an MCM chip should work with the customer to partition the circuit and optimize the design to be implemented in MCM technology. Application of technologies for placement, routing, via minimization, tree searching, and layer estimation will be important to assess at this point. The general function, purpose, and configuration of the active elements, interconnects, and assembly technology should also be assessed, along with key materials and critical properties, representative manufacturing-process flows, potential failure mechanisms, qualification procedures, and design for testability.

Note that two concepts must be carefully examined for successful MCM production. An MCM design is initiated by selecting appropriate technologies from the many options available. The basic choices are for substrate technology and assembly techniques. Design tradeoffs are analyzed, and a preliminary specification is completed. Following circuit simulation, prototypes are produced and tested. When the application requires it, an ASIC can be designed to be included in the MCM.

12.12 Summary

In summary, it is customary to give an answer to the fundamental question: What can multi-chip modules do for you?… and here is the answer …

MCMs optimize critical design parameters such as speed, density, and temperature, resulting in performance well beyond PC board design capabilities. By removing discrete component packages and using more densely packed interconnects, circuit speeds increase. The design challenge is to select the appropriate packaging technology, and to manage any resulting thermal problems.[20]

MCM technologies found their way and are utilized in the wireless, fiber, and instrumentation markets and in space and military programs; and in the real world, they stand in the forefront of best merchant-market technology.

References

1. W. D. Brown, *ELEG 5273 Electronic Packaging*, University of Arkansas.
2. P. E. Garrou and I. Turlik, *Multichip Module Technology Handbook*, McGraw-Hill, 1998.
3. J. H. Reche, "High Density Multichip Interconnect for Advanced Packaging," *Proc. NEPCON West*, 1989, 1308
4. N. G. Koopman, T. C. Reiley, and P. A. Totta, "Chip and Package Interconnections," in *Microelectronics Packaging Handbook*, Van Nostrand Reinhold, New York, 1989, 361.
5. D. Suranayana et al., "Flip Chip Solder Bump Fatigue Life Enhanced by Polymer Encapsulation," *Proc. 40th ECTC*, 1990, 338
6. J. G. Aday, T. G. Tessier, H. Crews, and J. Rasul, "A Comparative Analysis of High Density PWB Technologies," *Proc. Int. MCM Conference*, Denver, 1996, 239.
7. J. G. Aday, T. G. Tessier, and H. Crews, "Selecting Flip Chip on Board Compatible High Density PWB Technologies," *Int. J. Microcircuits and Electronic Packaging*, vol. 18, No. 4, 1995, 319.
8. Y. Tsukada, S. Tsuchida and Y. Mashimoto, "Surface Laminar Circuitry Packaging," *Proc. 42nd ECTC*, 1992, 22.
9. M. Moser and T. G. Tessier, "High Density PCBs for Enhanced SMT and Bare Chip Assembly Applications," *Proc. Int. MCM Conference*, Denver, 1995, 543.
10. E. Enomoto, M. Assai, Y. Sakaguchi, and C. Ohashi, "High Density Printed Wiring Boards Using Advanced Fully Additive Processing," *Proc. IPC*, 1989, 1.
11. C. Sullivan, R. Funer, R. Rust, and M. Witty, "Low Cost MCM-L for Vehicle Application," *Proc. Int. MCM Conf.*, Denver, 1996, 142.
12. W. Schmidt, "A Revolutionary Answer to Today's and Future Interconnect Challenge," *Proc. Printed Circuit World Convention VI*, San Francisco, CA, 1994, T12-1.
13. D. P. Seraphim, D. E. Barr, W. T. Chen, G. P. Schmitt, and R. R. Tummala, "Printed Circuit Board Packaging," in *Microelectronics Packaging Handbook*, Van Nostrand Reinhold, New York, 1989, 853.
14. M. J. Begay, and R. Cantwell, "MCM-L Cost Model and Application Case Study," *Proc. Int. MCM Conference*, Denver, 1994, 332.
15. Compositech Ltd., 120 Ricefield Lane, Hauppauge, NY 11788-2071.
16. H. Holden, "Comparing Costs for Various PWB Build Up Technologies," *Proc. Int. MCM Conference*, Denver, 1996, 15.
17. J. Diekman and M. Mirhej, "Nonwoven Aramid Papers: A New PWB Reinforcement Technology," *Proc. IEPS*, 1990, 123.
18. J. Fjelstad, T. DiStefano, and K. Karavakis, "Multilayer Flexible Circuits with Area Array Interconnections for High Performance Electronics," *Proc. 2nd Int. FLEXCON*, 1995, 110.
19. Maxtek, Web site reference: http://www.maxtek.com.
20. J. J. Licari and L. R. Enlow, *Hybrid Microcircuit Technology Handbook*, July 1988, 25.

13

Channel Hot Electron Degradation-Delay in MOS Transistors Due to Deuterium Anneal

Isik C. Kizilyalli
Lucent Bell Laboratories

Karl Hess
University of Illinois at Urbana-Champaign

Joseph W. Lyding
University of Illinois at Urbana-Champaign

13.1 Introduction

Hydrogen-related degradation by hot electrons in MOS transistors has been long known and is well documented.[1] It has recently been discovered that the degradation exhibits a giant isotope effect if hydrogen is substituted by deuterium.[2–4] The isotope effect can delay the channel hot-electron degradation by factors of 10 to 100 and, with the current definition of lifetime, even much beyond that. It therefore must be an effect different to the known kinetic isotope effect and the standard changes in reaction velocity of a factor of three or so when hydrogen is substituted by deuterium.

Deuterium is a stable and abundantly available isotope of hydrogen. It is contained at a level larger than 10^{-4} in all natural water sources. Its mass is roughly twice that of hydrogen, while all its electronic energy levels and the related chemistry are identical to that of hydrogen.

The difference in channel hot-electron degradation and lifetime must therefore be due to the mass difference. There are several possible explanations of the giant isotope effect in degradation.[4] The most probable explanation at low supply voltages ($V_{DD} < 3.3$ V) is the one first advanced in Ref. 5. This explanation concerns the dynamics of hydrogen (deuterium) desorption from an Si-H (Si-D) bond at the silicon–silicon dioxide interface under extreme non-equilibrium conditions — which is the central cause of the particular degradation discussed here. There are other forms of degradation known that show no, or a much lesser, isotope effect and are not discussed here. According to Ref. 5, the heated electrons (with a typical energy of several electron volts) excite local vibrations of the Si–H bond. These vibrations have a very long lifetime of the order of nanoseconds or longer because the vibrational energy of the Si–H is mismatched to the bulk vibrations in both silicon and silicon dioxide. Therefore, a high probability exists that other hot electrons will collide with the Si–H and cause further vibrational excitation until, finally, desorption is accomplished. One vibrational mode of Si–D, on the other hand, virtually matches a bulk silicon vibrational energy. Therefore, the local Si–D vibrations are short lived and it is much less likely that Si–D is excited so much that deuterium will desorb. The basic science of these processes has meanwhile been investigated (e.g., by scanning tunneling microscopy) and is consistent with the above description.[4] Similar desorption processes are known from photochemistry, where the energy is provided by photons, and the new aspect here is only the energy supply by the channel electrons and the extent of the isotope effect due to the special interface properties. While these considerations apply only under extreme non-equilibrium conditions, there are also in-equilibrium isotopic differences between the Si–H and Si–D bond, again due to their vibrational differences.

The question that appears under the initial equilibrium conditions, before the degradation occurs, is the following. Since many processing steps in MOS technology introduce hydrogen in one form or another, and since hydrogen densities in the silicon dioxide will be of the order of 10^{18} cm^{-3} whether desired or not, how can the hydrogen be effectively replaced by deuterium? An elementary proof[6] shows that the equilibrium population of the silicon bond by H or D also depends on the vibrational properties. Because of the higher vibrational energy of some of the Si-H vibrational modes (compared to deuterium), hydrogen is less likely to saturate the bond. In fact, if H and D are present at the same density at the interface, then (around the usual anneal temperature of 425°C) the deuterium is about 10 times more likely to populate the silicon bond than hydrogen.[6]

From these facts, it is evident that a relatively simple substitution of hydrogen by deuterium can lead to very beneficial delays of hot-electron degradation. In the following, we first describe the necessity to anneal the silicon–silicon dioxide interface with H or D; then we describe the advantages of D for hot electron degradation and the introduction of D by post-metal anneal procedures. Finally, we describe confirmations and extensions to more complex introductions of D.

13.2 Post-Metal Forming Gas Anneals in Integrated Circuits

Low-temperature post-metal anneals (350–450°C) in hydrogen ambients have been successfully used in MOS fabrication technologies to passivate the silicon dangling bonds and consequently to reduce the Si/SiO$_2$ interface trap charge density.[7-11] This treatment is imperative from a fabrication standpoint since silicon dangling bonds at the Si/SiO$_2$ interface are electrically active and lead to the reduction of channel conductance and also result in deviations from the ideal capacitance–voltage characteristics. Electron spin resonance (ESR) measurements performed in conjunction with deep-level transient spectroscopy (DLTS) and capacitance–voltage (C-V) measurements have elucidated the role of hydrogen in this defect annihilation process. The passivation process is described by the equation

$$P_b + H_2 \rightarrow P_b H + H \tag{13.1}$$

where P_bH is the passivated dangling bond. These measurements indicate that for the oxides grown on Si-<111>, the density of the interface trap states in the middle of the forbidden gap decreases from 10^{11}–10^{12} cm^{-2} eV^{-1} to about 10^{10} cm^{-2} eV^{-1} after the post-metal anneal process step. The Si-<100>/SiO$_2$ material system, which is technologically more significant, exhibits the same qualitative behavior. The necessity of post-metallization anneal processing for CMOS technologies is demonstrated in Figs. 13.1 and 13.2 where the measured NMOS threshold voltage (V_{th}) and transconductance (g sub m) distributions of a wafer annealed in forming gas (10% H$_2$) is compared with an untreated wafer. The high mean value and variation of the threshold voltage and reduced channel mobility across the untreated wafer is a clear indication of the unacceptable levels of interface trap density for CMOS circuit operation and stability. As described above, MOS transistors under bias can degrade as a result of channel hot (large kinetic energy) carriers (electrons and holes) stimulating the desorption of the hydrogen that is passivating the dangling bonds at the Si/SiO$_2$ interface. These concerns are exacerbated with the ever-ongoing scaling efforts for high-performance transistors and added dielectric/metallization (and hence plasma process damage) layers in integrated circuits.

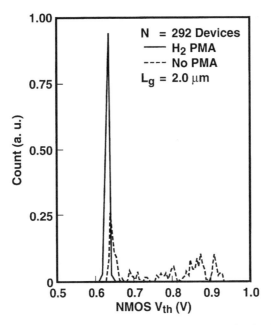

FIGURE 13.1 Histogram demonstrating the effects of hydrogen anneals on the threshold voltage V_{th} of NMOS transistors.

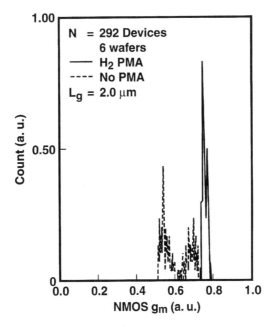

FIGURE 13.2 Same as Fig. 13.1 but for the transconductance g_m.

13.3 Impact of Hot Electron Effects on CMOS Development

The current industry practice for evaluating the intrinsic hot carrier related reliability of a MOS transistor involves two distinct accelerated electrical stress test methodologies (including certain intermediate methodologies). Although it is unlikely that the physical mechanisms responsible for gate oxide wear-out are identical in each of these cases, hydrogen appears to play some role in both modes of degradation. In the first stress configuration, the Si/SiO_2 system is degraded via large electrical fields across the gate oxide (e.g., $|V_G| >> V_D$).[12] Threshold voltage shifts in MOS capacitor structures are observed. The damage induced in this degradation mode is due to both charge trapping within the oxide and the creation of Si/SiO_2 interface trap states. Our initial studies have not identified a clear isotope effect for this mode of stress test. In this article, only the second stress configuration, namely the *channel hot carrier aging of NMOS transistors*, where we have discovered the large isotope effect, is discussed. In this mode of accelerated stress, threshold voltage instability and channel transconductance degradation in MOS transistors[13] is induced with the aid of hot carriers (electrons and holes with large kinetic energy) using the source-drain electric field. That is, the Si/SiO_2 interface is degraded by hot carriers that are traversing the device while they gain kinetic energy from the source-drain electric field. The device is biased to maximize the substrate current (e.g., $V_G \approx 1/2 V_D$). The transistor aging is accelerated by stressing the device using a drain voltage (V_D) which is much larger than the intended operating voltage. The hot carrier lifetime of the transistor is estimated by extrapolating the accelerated stress test results to operating voltage (peak substrate current specification) conditions. DC, AC, or pulsed DC waveforms are most commonly used. This mode of accelerated stress tests performed on NMOS transistors typically results in localized oxide damage, which has been identified as Si/SiO_2 interface trap states.[14,15] The asymmetry of the current–voltage characteristics under source-drain reversal indicates that the damaged region is located near the drain end of the transistor where the electric fields are the largest due to the pinch-off effect. Moreover, it has been suggested that the generation of the interface trap states is due to hot carrier-stimulated hydrogen desorption and depassivation of the silicon dangling bonds. Channel hot-carrier degradation in MOS transistors manifests itself in the form of threshold voltage (V_{th}) instability, transconductance ($g_m = dI_{DS}/dV_{GS}$) degradation, and a change in the subthreshold slope ($S_l = d \ln I_{DS} /dV_{GS}$ at $V_{GS} < V_{th}$) over time.

Various technological advances have been made to address the problem of MOS transistor degradation due to channel hot carriers. Most significant and lasting progress in fabrication technology to alleviate the channel hot carrier problem has been the development of lightly doped source-drain (LDD) and gate sidewall spacer processes.[15–17] The lightly doped drain region is used to reduce the strength of the electric field at the gate-end of the drain. Such advances have been integrated in all submicron CMOS technologies at the cost of added process complexity and intricate device design.[18] Unfortunately, processing requirements for good short channel behavior and high performance, namely, shallow source-drain junctions, reduced overlap capacitance, and low source-access resistance, are all at odds with hot carrier immune device design.

A reasonable argument (now it appears that this was merely wishful thinking) had been that the hot carrier degradation effect can be scaled away by reducing the supply voltage (constant field scaling). However, this does not appear to be the case since device feature size scaling accompanies supply voltage scaling in VLSI CMOS technologies to achieve improved performance and increased packing density. Gate oxide thickness is also reduced to maintain the device current density at low supply voltage operation. Takeda et al. have observed device degradation in a no-LDD NMOS transistor structure with gate lengths of 0.3 μm at 2.5 V.[19] Chung et al. observed hot carrier degradation for a 0.15-μm gate length transistor (with no LDD regions) at 1.8 V.[20] Current state-of-the-art high-performance 1.5- and 1.8-V sub-0.18-μm CMOS technology with good quality gate oxide exhibits hot carrier degradation effects and requires precise drain engineering.[21,22] Circuit solutions to alleviate hot carrier degradation effects in transistors, although proposed, involve further circuit design and layout complications.[23] Hot carrier-induced transistor degradation will continue to be a major roadblock for satisfying market demand for high-performance CMOS circuits such as the Lucent-DSP shown in Fig. 13.3 with transistor gate lengths phase-shifted down to 0.1 μm and operated at a large range of supply voltages (1.0 to 1.8 V).

13.4 The Hydrogen/Deuterium Isotope Effect and CMOS Manufacturing

As noted above, the major reliability limitation for the miniaturization CMOS technologies is the NMOS transistor hot carrier lifetime. A breakthrough in integrated circuit manufacturing is needed where reliable

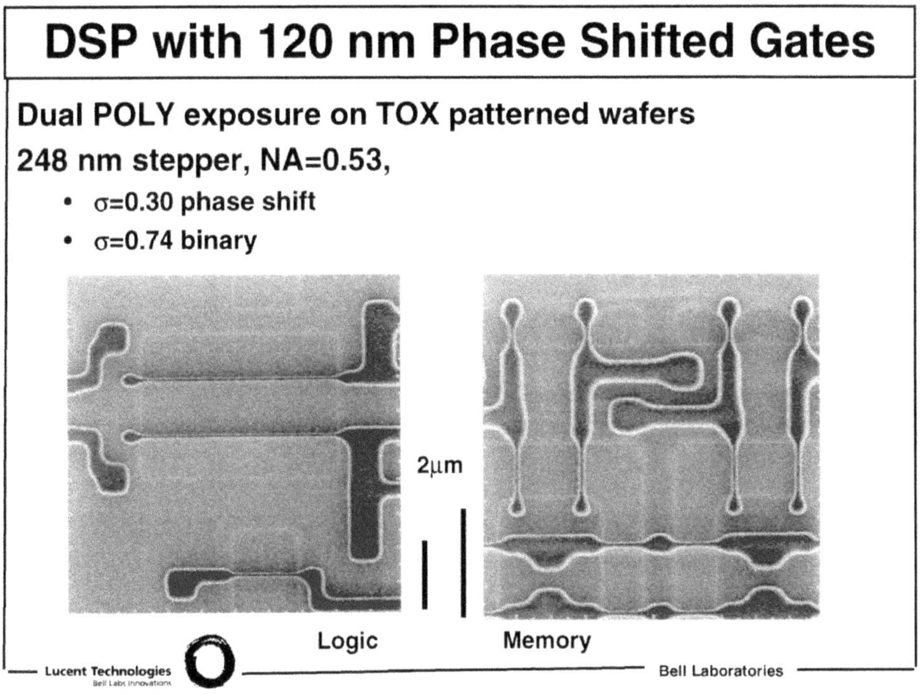

FIGURE 13.3 High-performance CMOS circuit with 0.1-μm gate length operated in a range of 1.0 V to 1.8 V of supply voltages.

CMOS scaling is enabled by eliminating the undesirable effects of channel hot carriers. The giant hydrogen/deuterium isotope effect that has been discovered and reported by Lyding, Hess, and Kizilyalli[2,3] in NMOS transistors is the extremely significant increase in time-dependent channel hot carrier transistor (reliability) lifetime in devices that have been annealed with deuterium instead of hydrogen, as shown in Figs. 13.4 and 13.5. In the fabrication sequence, deuterium is introduced, instead of hydrogen, to the Si/SiO$_2$ interface via a low-temperature (400–450°C) post-metallization anneal process. Figure 13.6 shows the transfer characteristics of uncapped NMOS and PMOS transistors annealed in deuterium and hydrogen ambients at 400°C and 1 hour. Prior to hot electron stress, transistors annealed in either ambient are electrically identical. This results in indistinguishable device function prior to hot carrier stress. Indistinguishable device function prior to hot carrier stress is also demonstrated in two experiments shown in Table 13.1. This is an expected result since the chemistry of deuterium and hydrogen is virtually identical. These results prove that deuterium and hydrogen are equally effective in reducing the interface trap charge density which results in equivalent device function. Deuterium can be substituted for hydrogen (at least in post-metal anneal processes) in semiconductor manufacturing.

Although devices annealed either in deuterium or hydrogen appear to be identical in pre-stress electrical tests, they exhibit markedly different degradation dynamics. The observed improvement in the degradation rates (lifetimes) in the transistors is a result of the large difference in the desorption rates of the two isotopes, as described in the introduction and in detail in Refs. 4 and 5.

The large hydrogen/deuterium isotope effect in NMOS transistors has been subsequently observed and verified by other laboratories.[24–27] Studies also indicate that transistors annealed in deuterium are much more resilient against plasma process-induced damage (as quantified by Si/SiO$_2$ interface trap generation and gate oxide leakage).[28] Furthermore, stability of PMOS transistors;[29] hydrogenated (deuterated) amorphous silicon-based solar cells and TFTs;[30–32] hydrogen (deuterium) terminated, porous-silicon light-emitting devices;[33] and ultra-thin oxides for non-volatile memory devices[34] have been found to improve with the isotopic substitution against degradation due to light and field exposure.

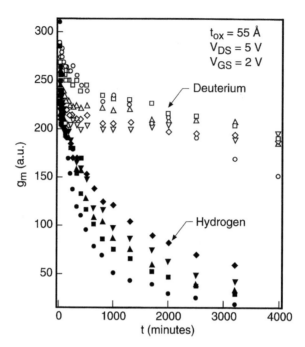

FIGURE 13.4 Transconductance channel hot-electron accelerated stress degradation with hydrogen and deuterium anneal and a significant degradation delay due to the presence of deuterium.

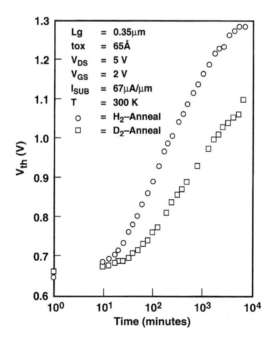

FIGURE 13.5 Same as Fig. 13.4 but for the threshold voltage.

For reasons outlined above, there is a strong motivation to introduce the deuterium anneal process to CMOS manufacturing. However, two further obstacles need to be removed for transfer of process to the factory floor. First, all modern CMOS technologies require a minimum of three levels of dielectric/metal interconnect process. Anneal processes that are found to be effective for improving channel

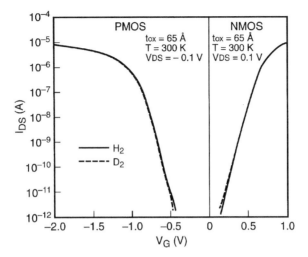

FIGURE 13.6 Subthreshold current I_{DS} as a function of gate voltage V_G for both hydrogen and deuterium anneal before degradation. No difference is shown within the experimental accuracy.

TABLE 13.1 Anneal Process and Threshold Voltage

Process	$V_{th,N}$	$\sigma\text{-}V_{th,N}$	$V_{th,P}$	$\sigma\text{-}V_{th,P}$
10%-H_2	0.51V	2.7 mV	0.983 V	2.8 mV
50%-D_2	0.51V	2.4 mV	0.986 V	4.5 mV
100%-D_2	0.51V	1.7 mV	0.987 V	1.1 mV

hot carrier reliability in one-level of metal/dielectric structures (e.g., 400°C for 0.5 hr and 10% D_2/90% N_2) may be ineffective for multi-level metal/dielectric structures. The deuterium anneal process needs to be (and in some cases has been) optimized for multi-level interconnect. Second, the benefits of the deuterium anneal should be still evident subsequent to the final SiN cap wafer passivation process.

The test vehicle used for the experiments to surmount these challenges is a development version of Lucent Technologies 0.35 μm 3.3 V transistors with a 65 Å gate oxide, very shallow arsenic implanted MDD regions, TEOS spacers, and three dielectric (doped and undoped plasma enhanced TEOS) and metal levels (Ti/TiN/AlCuSi/TiN).[35] The deuterium (5–100% D_2) anneal process was performed after the third layer of metal had been patterned. The deuterium anneal temperatures vary between 400 to 450°C and anneal times of 1/2 to 5 hours are considered. Accelerated hot carrier DC stress experiments are performed on NMOS transistors at peak substrate current conditions. In Fig. 13.7, the time-dependent deviation of V_{th} is shown with the deuterium anneal conditions as a parameter. Table 13.2 summarizes the results of all stress experiments ($V_{DS} = 5$ V and $V_{GS} = 2$ V), assuming a degradation criteria of $\delta V_{th} = 200$ mV. The degradation dynamics of S_i and $I_{D,SAT}$ are plotted in Figs. 13.8 and 13.9. Degradation in the transistor I–V characteristics is accompanied by an increase in the interface trap density (D_{it}) as extracted (Fig. 13.10) from charge-pump current measurements. Figure 13.11 shows hot electron degradation lifetime versus substrate current with 20% g_m (transconductance) degradation as the lifetime criteria. For a peak substrate current specification of $I_{SUB} = 2000$ nA/μm, the extrapolated lifetime for the hydrogen and deuterium annealed device is 0.06 years and 4 years, respectively. The hot carrier lifetime (reliability) of transistors increases continuously and dramatically with increases in deuterium anneal times and temperatures. Negligible improvement is observed for short (1/2 hour) and low concentration (<10% D_2) anneal conditions. The 400°C, 2-hour process results in a four-fold improvement in lifetime over the standard hydrogen process, while the 450°C, 3-hour deuterium anneal process yields nearly a factor of 50 improvement. Hence, whenever the deuterium content in the wafer is increased, large corresponding improvements in transistor lifetimes are measured. For longer (450°C, and 5-hour)

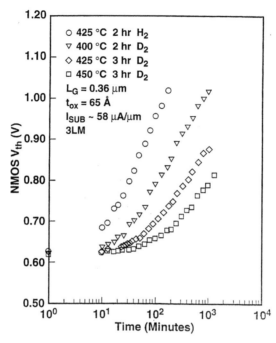

FIGURE 13.7 Degradation improvements by deuterium anneal after three levels of metallization for various temperatures and times. Only one hydrogen curve is shown; annealing with hydrogen gives identical results over wide ranges of temperature and time.

TABLE 13.2 Relative Hot Carrier Reliability (Lifetime) Improvement and D_2 Anneals

Temperature	Time	Ambient	N_2 Pre-anneal	Lifetime
425°C	2 hr	10%H_2/90%N_2	No	1.0
400°C	1/2 hr	5%D_2/95%N_2	No	1.0
400°C	1 hr	5%D_2/95%N_2	No	1.0
400°C	1 hr	10%D_2/90%N_2	No	1.0
400°C	2 hr	100%D_2	Yes	3.8
400°C	3 hr	100%D_2	No	5.5
425°C	3 hr	100%D_2	No	12.5
450°C	2 hr	100%D_2	Yes	36.0
450°C	3 hr	20%D_2/80%N_2	No	37.5
450°C	3 hr	100%D_2	No	62.5
450°C	5 hr	100%D_2	No	80.0

anneals, the lifetime improvement asymptotically reaches a factor of about 80 to 100 over the standard process, corresponding to similar findings in basic experiments using scanning tunneling microscopy.[4] In Fig. 13.12, it is demonstrated that the benefits of the deuterium anneal are still observed even if the post-metal anneal is followed by an SiN caps process. Similar findings for higher levels of metallization were made by other groups (see Ref. 27). For ultimate stability against further processing, and to avoid the complicated diffusion through many layers of metallization, it may be necessary to introduce the deuterium in a layer close to the device that can act as a reservoir and is activated in any temperature increase (anneal). A convincing proof of this possibility has been given.[36]

Figure 13.13 shows a hydrogen and deuterium profile as measured by surface ion mass spectroscopy for a successfully treated transistor. A deuterium peak concentration at the interface is a typical necessity for successful anneal. The absolute concentrations may vary according to experimental conditions.

These experiments prove that one can substitute deuterium for hydrogen in a CMOS manufacturing process with no penalty, yet with a 50 to 100-fold improvement in channel hot carrier lifetime. For

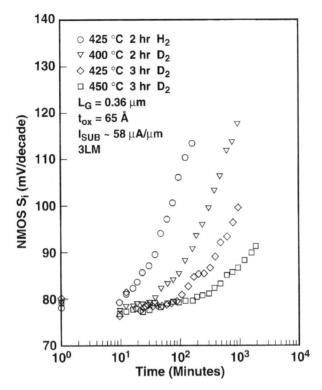

FIGURE 13.8 Degradation dynamics as in Fig. 13.7 but for S_I.

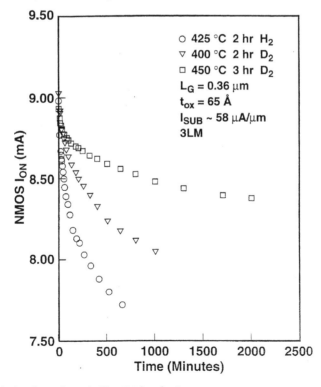

FIGURE 13.9 Degradation dynamics as in Fig. 13.7 but for I_{ON}.

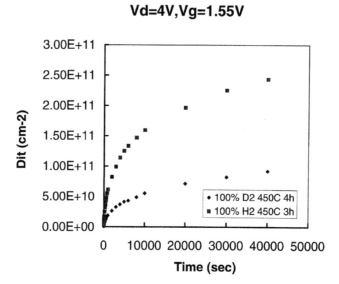

FIGURE 13.10 Interface trap density D_{it} extracted from charge-pump current measurements vs. stress time with hydrogen and deuterium anneal as indicated.

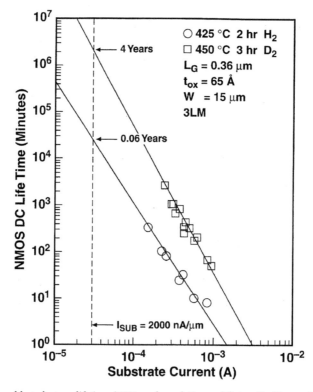

FIGURE 13.11 Channel hot electron lifetime (20% g_m degradation as lifetime limit) vs. substrate current for both hydrogen and deuterium anneal with extrapolations to actual operating conditions (dashed vertical line).

completeness, other transistor structures with varying design considerations have been evaluated and summarized elsewhere.[37,38] Transistor parameters that were explored include: (1) gate oxide thickness t_{ox} = 50–115 Å, (2) LDD implant species arsenic and phosphorus, (3) gate stack structure of n+-polysilicon

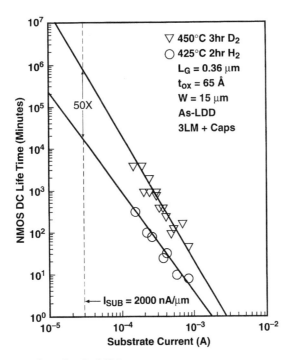

FIGURE 13.12   As in Fig. 13.11 but after final SiN cap process.

and polycide (n$^+$-polysilicon/WSi), and (4) an experimental 0.25-μm 3.3-V CMOS[39] process with 4 levels of metal. In all cases, the large isotope effect is observed. Clearly, the hydrogen/deuterium isotope effect is a general property of the semiconductor device wear-out.

It is important to correlate the observed improved hot carrier reliability to the location and quantity of deuterium in the wafer. Secondary ion mass spectroscopy (SIMS) analysis through the first interlevel oxide and silicon was performed on two uncapped (it is well known that Si$_x$N$_y$ is a barrier for deuterium) samples as shown in Fig. 13.13. The first wafer was annealed in forming gas (10% molecular hydrogen), while the other sample was annealed in forming gas comprising 10% molecular deuterium. A Cameca IMS-2f system with oxygen primary beam 60 μm^2 was used for analysis. ^{18}O$^+$ was monitored to locate the SiO$_2$/Si interface. The ^{2}D$^+$ concentration is inferred from the difference between the ^{2}H$^+$ profiles for wafers annealed in deuterium and hydrogen. Deuterium was not detected under large areas of (200 times 200 μm^2) polysilicon in wafers that were annealed in deuterium and were not capped with Si$_x$N$_y$. This indicates the finite lateral diffusion length of deuterium in the transistor gate oxide and channel region. Deuterium is detected in the interlevel oxide at concentrations of 10^{19} cm^{-3} and was found to accumulate at Si/SiO$_2$ interfaces with a surface concentration of 10^{14} cm^{-2}. This SIMS study suggests that deuterium diffuses rapidly through the interlevel oxides and the gate sidewall spacers to passivate the interface states in the transistor channel region. However, the exact lateral spread (reach) of diffused deuterium in the transistor-channel and gate-oxide region is not certain.

The question still remains regarding the purity, specification limits, as well as cost for implementing deuterium gas in semiconductor manufacturing. Deuterium is a stable isotope of hydrogen and is present as D$_2$O. Deuterium gas is produced by electrolysis of pure D$_2$O. Tritium is a radioactive isotope of hydrogen and has a lifetime of 12.3 years. Because of the nature of the electrolysis, the molar concentration of tritium in the gas will be essentially the same as the feed heavy water with the tritium gas being in the form of DT rather than pure T$_2$. If the tritium content in the feed water is 50 nCi/kg, the expected concentration in the gas would be about 45 pCi/L (1.65 Bq/L). The tritium content in heavy water varies from 5 nCi/kg (virgin heavy water) to 50 Ci/kg (heavy water used in nuclear reactors). The limit of tritium in deuterium gas suitable for CMOS manufacturing is estimated as follows. When

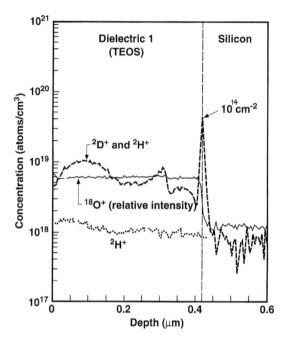

FIGURE 13.13 Typical results for hydrogen and deuterium concentrations measured after anneals by secondary ion mass spectroscopy (SIMS).

10^{14} deuterium atoms are placed in a single chip, 3×10^{-9} Bq/L of tritium are also incorporated. This implies that one tritium decay event would occur approximately every 370 days, much below the rate of other radioactive events occurring in chip technology and operation. In addition, this is only a β-decay with usually negligible consequences. Since it is very difficult to measure tritium gas at these low concentrations, specifications should be placed on the tritium content for the heavy water used in the electrolytic production process (which is straightforward). A suggested upper limit could be 6000 nCi/kg that results in 1 tritium event/month per chip.[40] The substitution of deuterium for hydrogen adds 0.1% to the total wafer cost.

13.5 Summary

It has been demonstrated that the replacement of hydrogen by deuterium in CMOS technology can lead to significant delays in channel hot electron degradation. Increases in hot electron lifetime of a factor of 10 to 100 and beyond have been shown by simple post-metal anneals in deuterium atmosphere for several levels of metallization. Deuterium has also been proven beneficial for the reliability of other devices such as deuterated amorphous thin-film silicon devices of various kinds. Since deuterium and hydrogen have the same electronic energy levels, deuterium- and hydrogen-treated devices are indistinguishable in terms of their normal pre-stress electronic characteristics. Deuterium only delays degradation due to its higher mass and different vibrational properties.

Acknowledgments

K. H. and J. W. Lyding acknowledge financial support from the Office of Naval Research under the MURI program.

References

1. Takeda, E., Yang, C. Y., and Miura-Hamada, A., *Hot Carrier Effects in MOS Devices*, Academic Press, 1995, 73.
2. Lyding, J. W., Hess, K., and Kizilyalli, I. C., "Reduction of Hot Electron Degradation in MOS Transistors By Deuterium Processing," *Appl. Phys. Lett.*, 68, 2526-2528, 1996.
3. Kizilyalli, I. C., Lyding, J. W., and Hess, K., "Deuterium Post Metal Annealing of MOSFETs for Improved Hot Carrier Reliability," *IEEE Electron Device Letters*, 18, 81-83, 1997.
4. Hess, K., Kizilyalli, I. C., and Lyding, J. W., "Giant Isotope Effect in Hot Electron Degradation of Metal Oxide Silicon Devices," *IEEE Trans. Electron Devices,* 45, 406-416, 1998.
5. Vande Walle, C. G. and Jackson, W. B., "Comment on Reduction of Hot Electron Degradation in MOS Transistors by Deuterium Processing," *Appl. Phys. Lett.*, 69, 2441, 1996.
6. Ipatova, I. P., Chikalova-Luzia, O. P., and Hess, K., "Effect of Localized Vibrations on the Si Surface Concentrations of H and D," *J. Appl. Phys.*, 83, 814-819, 1998.
7. Cheroff, G., Fang, F., and Hochberg, H., "Effect of Low Temperature Annealing on the Surface Conductivity of Si in the Si–SiO$_2$ — Al System," *IBM Journal*, 8, 416-421, 1964.
8. Balk, P., "Effects of Hydrogen Annealing on Silicon Surfaces," *Electrochemical Society Spring Meeting*, San Francisco, Abstract No. 109, pp. 237-240, 1965.
9. Johnson, N. M., Biegelsen, D. K., and Moyer, M. D., "Low-Temperature Annealing and Hydrogenation of Defects at the Si-SiO$_2$ Interface," *J. Vac. Sci. Technolog.*, 19, 390-394, 1981.
10. Grove, A. S., *Physics and Technology of Semiconductor Devices*, New York, John Wiley & Sons, 1967.
11. Nicollian E. H. and Brews, J. R., *MOS (Metal Oxide Semiconductor) Physics and Technology*, New York, John Wiley & Sons, 1982.
12. Ning, T. H., Osburn, C. M., and Yu, H. N., "Emission Probability of Hot Electrons from Silicon to Silicon Dioxide," *J. Appl. Phys.*, 48, 286-293, 1977.
13. Abbas, S. A. and Dockerty, R. C., Hot Carrier Instability in IGFET's," *Appl. Phys. Lett.*, 27, 147-148, 1975.
14. Hu, C., Tam, S. C., Hsu, F., Ko, P., Chan, T., and Terrill, K. W., "Hot-Electron-Induced MOSFET Degradation: Model, Monitor, and Improvement," *IEEE Trans. Electron Devices*, ED-32, 375-385, 1985.
15. Doyle, B., Bourcerie, M., Marchetaux, J., and Boudou, A., "Interface State Creation and Charge Trapping in the Medium-to-High Gate Voltage Range During Hot-Carrier Stressing of N-MOS Transistors," *IEEE Trans. Electron Devices*, ED-37, 744-754, 1990.
16. Tsang, P. J., Ogura, S., Walker, W. W., Shepard, J. F., and Critchlow, D. L., "Fabrication of High-Performance LDDFET's with Oxide Sidewall-Spacer Technology," *IEEE Trans. Electron Devices*, ED-29, 590-596, 1982.
17. Chatterjee, P. K., Shah, A. H., Lin, Y., Hunter, W. R., Walker, E. A., Rhodes, C. C., and Bruncke, W. C., "Enhanced Performance 4K×1 high-speed SRAM Using Optically Defined Submicrometer Devices in Selected Circuits," *IEEE Trans. Electron Devices*, ED-29, 700-706, 1982.
18. Chen, M. L., Leung, C. W., Cochran, W. T., Juengling, W., Dzuiba, C., and Yang, T., "Suppression of Hot Carrier Effects in Deep Submicrometer CMOS Technology," *IEEE Trans. Electron Devices*, ED-35, 2210-2220, 1988.
19. Takeda, E., Suzuki, N., and Hagiwara, T., "Device Performance Degradation due to Hot-Carrier Injection at Energies Below the Si-SiO$_2$ Energy Barrier," *IEEE International Electron Device Meeting Tech. Digest*, 396-399, 1983.
20. Chung, J. E., Jeng, M., Moon, J. E., Ko, P., and Hu, C., "Low-Voltage Hot-Electron Currents and Degradation in Deep-Submicrometer MOSFET's," *IEEE Trans. Electron Devices*, ED-37, 1651-1657, 1990.
21. Rodder, M., Aur, S., and Chen, I. C., "A Scaled 1.8 V, 0.18 µm Gate Length CMOS Technology: Device Design and Reliability Considerations," *IEEE IEDM Technical Digest*, 415-418, 1995.

22. Hargrove, M., Crowder, S., Nowak, E., Logan, R., Han, L., Ng, H., Ray, A., Sinitsky, D., Smeys, P., Guarin, F., Oberschmidt, J., Crabbe, E., Yee, D., and Su, L., "High-Performance sub-0.08 μm CMOS with Dual Gate Oxide and 9.7 ps Inverter Delay," *IEDM Tech. Digest*, 627-630, 1998.

23. Leblebici, Y. and Kang, S. M., in *Hot-carrier Reliability of MOS VLSI Circuits*, Boston, Kluwer, 1993.

24. Devine, R. A., Autran, J. L., Warren, W. L., Vanheusdan, K. L., and Rostaing, J. C., "Interfacial Hardness Enhancement in Deuterium Annealed 0.25 {mu m} Channel Metal Oxide Semiconductor Transistors," *Appl. Phys. Lett.*, 70, 2999-3001, 1997.

25. Aur, S., Grider, T., McNeil, V., Holloway, T., and Eklund, R., "Effects of Advanced Processes on Hot Carrier Reliability," *IEEE Proc. International Physics Symposium (IRPS)*, 1998.

26. Clark, W. F., Ference, T. G., Hook, T., Watson, K., Mittl, S., and Burnham, J., "Process Stability of Deuterium-Annealed MOSFET's," *IEEE Electron Device Lett.*, 20, 48-50, 1999.

27. Lee, J., Epstein, Y., Berti, A., Huber, J., Hess, K., and Lyding, J. W., "The Effect of Deuterium Passivation at Different Steps of CMOS Processing on Lifetime Improvements of CMOS Transistors," *IEEE Trans. Electron Devices*, submitted.

28. Krishnan, S., Rangan, S., Hattangady, S., Xing, G., Brennan, K., Rodder, M., and Ashok, S., "Assessment of Charge-Induced Damage to Ultra-Thin Gate MOSFETs," *IEEE IEDM Tech. Digest*, 445-448, 1997.

29. Li, E., Rosenbaum, E., Tao, J., and Fang, P., "CMOS Hot Carrier Lifetime Improvement from Deuterium Anneal," *56th Device Research Conference*, Santa Barbara, CA, 1998.

30. Sugiyama, S., Yang, J., and Guha, S., "Improved Stability Against Light Exposure in Amorphous Deuterated Silicon Alloy Solar Cell," *Appl. Phys. Lett.,* 70, 378-380, 1997.

31. Wei, J. H. and Lee, S. C., "Improved Stability of Deuterated Amorphous Silicon Thin Film Transistors," *J. Appl. Phys.*, 85, 543-550, 1999.

32. Wei, J. H., Sun, M. S., and Lee, S. C., "A Possible Mechanism for Improved Light-Induced Degradation in Deuterated Amorphous-Silicon Alloy," *Appl. Phys. Lett.*, 71, 1498-1450, 1997.

33. Matsumoto, T., Masumoto, Y., Nakagawa, T., Hashimoto, M., Ueno, K., and Koshida, N., "Electroluminescence from Deuterium Terminated Porous Silicon," *Jpn. J. Appl. Phys.*, 36, L1089-1091, 1997.

34. Kim, H. and Hwang, H., "High Quality Ultrathin Gate Oxide Prepared by Oxidation in D_2O," *Appl. Phys. Lett.*, 74, 709-710, 1999.

35. Kizilyalli, I. C., Lytle, S., Jones, B. R., Martin, E., Shive, S., Brooks, A., Thoma, M., Schanzer, R., Sniegowski, J., Wroge, D., Key, R., Kearney, J., and Stiles, K. "A Very High Performance and Manufacturable 3.3 V 0.35 μm CMOS Technology for ASICs," *IEEE Custom Integrated Circuits Conference (CICC) Tech. Digest*, pp. 31-34, 1996.

36. Ference, T. et al., *IEEE Trans. ED46*, pp. 747-753, 1999.

37. Kizilyalli, I. C., Abeln, G., Chen, Z., Lee, J., Weber, G., Kotzias, B., Chetlur, S., Lyding, J., and Hess, K., "Improvement of Hot Carrier Reliability with Deuterium Anneals for Manufacturing Multi-Level Metal/Dielectric MOS Systems," *IEEE Electron Device Lett.*, 19, 444-446, 1998.

38. Kizilyalli, I. C., Weber, G., Chen, Z., Abeln, G., Schofield, M., Lyding, J., and Hess, K., "Multi-level metal CMOS manufacturing with deuterium for improved hot carrier reliability," *IEEE IEDM Tech. Digest*, 935-938, 1998.

39. Kizilyalli, I. C., Huang, R., Hwang, D., Vaidya, H., and Thoma, M., "A Merged 2.5 V and 3.3 V 0.25-μm CMOS Technology for ASICs," *IEEE CICC Tech. Digest*, 159-162, 1998.

40. Machachek, R. F., Ontario Hydro (Toronto), private communication.

14

Materials

Stephen I. Long
*University of California
at Santa Barbara*

14.1 Introduction

Very-high-speed digital integrated circuit design is a multidisciplinary challenge. First, there are several IC technologies available for very-high-speed applications. Each of these claims to offer unique benefits to the user. In order to choose the most appropriate or cost-effective technology for a particular application or system, the designer must understand the materials, the devices, the limitations imposed by process on yields, and the thermal limitations due to power dissipation.

Second, very-high-speed digital ICs present design challenges if the inherent performance of the devices is to be retained. At the upper limits of speed, there are no digital circuits, only analog. Circuit design techniques formerly thought to be exclusively in the domain of analog IC design are effective in optimizing digital IC designs for highest performance.

Finally, system integration when using the highest-speed technologies presents an additional challenge. Interconnections, clock and power distribution both on-chip and off-chip require much care and often restrict the achievable performance of an IC in a system.

The entire scope of very-high-speed digital design is much too vast to present in a single tutorial chapter. Therefore, we must focus the coverage in order to provide some useful tools for the designer. We will focus primarily on compound semiconductor technologies in order to restrict the scope. Silicon IC design tutorials can be found in other chapters in this handbook. This chapter gives a brief introduction to compound semiconductor materials in order to justify the use of non-silicon materials for the highest-speed applications. The transport properties of several materials are compared. Second, a technology-independent description of device operation for high-speed or high-frequency applications will be given in Chapter 15. The charge control methodology provides insight and connects the basic material properties and device geometry with performance. Chapter 16 describes the design basics of very-high-speed ICs. Static design methods are illustrated with compound semiconductor circuit examples, but are based on generic principles such as noise margin. The transient design methods emphasize analog circuit techniques and can be applied to any technology.

Finally, Chapter 17 describes typical circuit design approaches using FET and bipolar device technologies and presents applications of current interest.

14.2 Compound Semiconductor Materials

The compound semiconductor family is composed of the group III and group V elements shown in Table 14.1. Each semiconductor is formed from at least one group III and one group V element. Group IV elements such as C, Si, and Ge are used as dopants, as are several group II and VI elements such as Be or Mg for p-type and Te and Se for n-type. Binary semiconductors such as GaAs and InP can be grown in large single-crystal ingot form using the liquid-encapsulated Czochralski method[1] and are the materials of choice for substrates. At the present time, GaAs wafers with a diameter of 100 and 150 mm are most widely used. InP is still limited to 75 mm diameter.

TABLE 14.1 Column III, IV, and V Elements Associated with Compound Semiconductors

B	C	N
Al	Si	P
Ga	Ge	As
In	Sn	Sb

Three or four elements are often mixed together when grown as thin *epitaxial* films on top of the binary substrates. The alloys thus formed allow electronic and structural properties such as bandgap and lattice constant to be varied as needed for device purposes. Junctions between different semiconductors can be used to further control charge transport as discussed in Section 14.4.

14.3 Why III-V Semiconductors?

The main motivation for using the III-V compound semiconductors for device applications is found in their electronic properties when compared with those of the dominant semiconductor material, silicon. Figure 14.1 is a plot of steady-state *electron velocity* of several n-type semiconductors versus electric field. From this graph, we see that at low electric fields the slope of the curves (*mobility*) is higher than that of silicon. High mobility means that the resistivity will be less for III-V n-type materials, and it may be easier to achieve lower access resistance. *Access resistance* is the series resistance between the device contacts and the internal active region. An example would be the base resistance of a bipolar transistor. Lower resistance will reduce some of the fundamental device time constants to be described in Chapter 15 that often dominate device high-frequency performance. Figure 14.1 also shows that the peak electron velocity is higher for III-V materials, and the peak velocity can be achieved at much lower electric fields. High velocity reduces *transit time*, the time required for a charge carrier to travel from its source to its destination, and improves device high-frequency performance, also discussed in Chapter 15. Achieving this high velocity at lower electric fields means that the devices will reach their peak performance at lower voltages, which is useful for low-power, high-speed applications. Mobilities and peak velocities of several semiconductors are compared in Table 14.2.

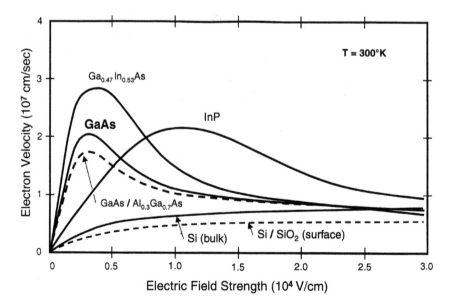

FIGURE 14.1 Electron velocity versus electric field for several n-type semiconductors.

TABLE 14.2 Comparison of Mobilities and Peak Velocities of Several n- and p-type Semiconductors

Semiconductor	E_G (eV)	ε_r	Electron Mobility (cm²/V-s)	Hole Mobility (cm²/V-s)	Peak Electron Velocity (cm/s)
Si (bulk)	1.12	11.7	1450	450	N.A.
Ge	0.66	15.8	3900	1900	N.A.
InP	1.35 D	12.4	4600	150	2.1×10^7
GaAs	1.42 D	13.1	8500	400	2×10^7
$Ga_{0.47}In_{0.53}As$	0.78 D	13.9	11,000	200	2.7×10^7
InAs	0.35 D	14.6	22,600	460	4×10^7
$Al_{0.3}Ga_{0.7}As$	1.80 D	12.2	1000	100	—
AlAs	2.17	10.1	280	—	—
$Al_{0.48}In_{0.52}As$	1.92 D	12.3	800	100	—

Note: In bandgap energy column, the symbol "D" indicates direct bandgap; otherwise, it is indirect bandgap. T = 300 K and "weak doping" limit.

On the other hand, as also shown in Table 14.2, p-type III-V semiconductors have rather poor hole mobility when compared with elemental semiconductor materials such as silicon or germanium. Holes also reach their peak velocities at much higher electric fields than electrons. Therefore, p-type III-V materials needed for the base of a bipolar transistor, for example, are used, but their thickness must be extremely small to avoid degradation in transit time. Lateral distances must also be small to avoid excessive series resistance. CMOS-like complementary FET technologies have also been developed,[2] but their performance has been limited by the poorer speed of the p-channel devices.

14.4 Heterojunctions

In the past, most semiconductor devices were composed of a single semiconductor element, such as silicon or gallium arsenide, and employed n- and p-type doping to control charge transport. Figure 14.2(a) illustrates an energy band diagram of a semiconductor with uniform composition that is in an applied electric field. Electrons will drift downhill and holes will drift uphill in the applied electric field. The electrons and/or holes could be produced by doping or by ionization due to light. In a *heterogeneous* semiconductor as shown in Fig. 14.2(b), the bandgap can be graded from wide bandgap on the left to narrow on the right by varying the composition. In this case, even without an applied electric field, a built-in quasi-electric field is produced by the bandgap variation that will transport both holes and electrons in the *same* direction.

The abrupt *heterojunction* formed by an atomically abrupt transition between AlGaAs and GaAs, shown in the energy band diagram of Fig. 14.3, creates discontinuities in the valence and conduction bands. The conduction band energy discontinuity is labeled ΔE_C and the valence band discontinuity, ΔE_V. Their sum equals the energy bandgap difference between the two materials. The potential energy steps caused by these discontinuities are used as barriers to electrons or holes. The relative sizes of these potential barriers depend on the composition of the semiconductor materials on each side of the heterojunction. In this example, an electron barrier in the conduction band is used to confine carriers into a narrow potential energy well with triangular shape. Quantum well structures such as these are used

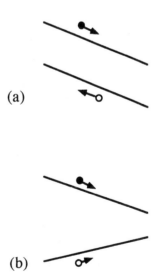

(a)

(b)

FIGURE 14.2 (a) Homogeneous semiconductor in uniform electric field, and (b) Heterogeneous semiconductor with graded energy gap. No applied electric field.

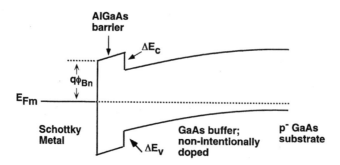

FIGURE 14.3 Energy band diagram of an abrupt heterojunction.

to improve device performance through two-dimensional charge transport channels, similar to the role played by the inversion layer in MOS devices. The structure and operation of heterojunctions in FETs and BJTs will be described in Chapter 15.

The overall principle of the use of heterojunctions is summarized in a *Central Design Principle*:

"Heterostructures use energy gap variations in addition to electric fields as forces acting on holes and electrons to control their distribution and flow."[3,4]

The energy barriers can control motion of charge both across the heterojunction and in the plane of the heterojunction. In addition, heterojunctions are most widely used in light-emitting devices, since the compositional differences also lead to either stepped or graded index of refraction, which can be used to confine, refract, and reflect light. The barriers also control the transport of holes and electrons in the light-generating regions.

Figure 14.4 shows a plot of bandgap versus lattice constant for many of the III-V semiconductors.[3] Consider GaAs as an example. GaAs and AlAs have the same lattice constant (approximately 0.56 nm) but different bandgaps (1.4 and 2.2 eV, respectively). An alloy semiconductor, AlGaAs, can be grown epitaxially on a GaAs substrate wafer using standard growth techniques. The composition can be selected by the Al-to-Ga ratio, giving a bandgap that can be chosen across the entire range from GaAs to AlAs. Since both lattice constants are essentially the same, very low lattice mismatch can be achieved for any

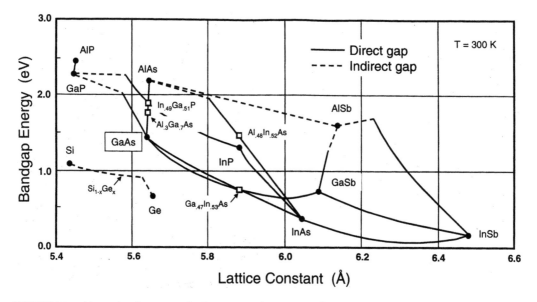

FIGURE 14.4 Energy bandgap versus lattice constant for compound semiconductor materials.

composition of $Al_xGa_{1-x}As$. Lattice matching permits low defect density, high-quality materials to be grown that have good electronic and optical properties. It quickly becomes apparent from Fig. 14.4, however, that a requirement for lattice matching to the substrate greatly restricts the combinations of materials available to the device designer. For electron devices, the low mismatch GaAs/AlAs alloys, GaSb/AlSb alloys, $Al_{.48}In_{.52}As/InP/Ga_{.47}In_{.53}As$, and $GaAs/In_{.49}Ga_{.51}As$ combinations alone are available. Efforts to utilize combinations such as GaP on Si or GaAs on Ge that lattice match have been generally unsuccessful because of problems with interface structure, polarization, and autodoping.

For several years, lattice matching was considered to be a necessary condition if mobility-damaging defects were to be avoided. This barrier was later broken when it was discovered that high-quality semiconductor materials could still be obtained although lattice-mismatched if the thickness of the mismatched layer is sufficiently small.[5,6] This technique, called *pseudomorphic* growth, opened another dimension in III-V device technology, and allowed device structures to be optimized over a wider range of bandgap for better electron or hole dynamics and optical properties.

Two of the pseudomorphic systems that have been very successful in high-performance millimeter-wave FETs are the InAlAs/InGaAs/GaAs and InAlAs/InGaAs/InP systems. The $In_xGa_{1-x}As$ layer is responsible for the high electron mobility and velocity which both improve as the In concentration x is increased. Up to x = 0.25 for GaAs substrates and x = 0.80 for InP substrates have been demonstrated and result in great performance enhancements when compared with lattice-matched combinations.

References

1. Ware, R., Higgins, W., O'Hearn, K., and Tiernan, M., Growth and Properties of Very Large Crystals of Semi-Insulating Gallium Arsenide, presented at *18th IEEE GaAs IC Symp.*, Orlando, FL, 54, 1996.
2. Abrokwah, J. K., Huang, J. H., Ooms, W., Shurboff, C., Hallmark, J. A. et al., A Manufacturable Complementary GaAs Process, presented at *IEEE GaAs IC Symposium,* San Jose, CA, 127, 1993.
3. Kroemer, H., Heterostructures for Everything: Device Principles of the 1980s?, *Japanese J. Appl. Phys.*, 20, 9, 1981.
4. Kroemer, H., Heterostructure Bipolar Transistors and Integrated Circuits, *Proc. IEEE*, 70, 13, 1982.
5. Matthews, J. W. and Blakeslee, A. E., Defects in Epitaxial Multilayers. III. Preparation of Almost Perfect Layers, *J. Crystal Growth*, 32, 265, 1976.
6. Matthews, J. W. and Blakeslee, A. E., Coherent Strain in Epitaxially Grown Films, *J. Crystal Growth*, 27, 118, 1974.

15

Compound Semiconductor Devices for Digital Circuits

Donald B. Estreich
Hewlett-Packard Company

15.1 Introduction

An *active device* is an electron device, such as a transistor, capable of delivering power amplification by converting dc bias power into time-varying signal power. It delivers a greater energy to its load than if the device were absent. The *charge control* framework[1-3] discussed below presents a unified understanding of the operation of all electron devices and simplifies the comparison of the several active devices used in digital integrated circuits.

15.2 Unifying Principle for Active Devices: Charge Control Principle

Consider a generic electron device as represented in Fig. 15.1. It consists of three electrodes encompassing a charge *transport region*. The transport region is capable of supporting charge flow (electrons as shown in the figure) between an *emitting electrode* and a *collecting electrode*. A third electrode, called the *control electrode*, is used to establish the electron concentration within the transport region. Placing a *control charge*, Q_C, on the control electrode establishes a *controlled charge*, denoted as $-Q$, in the transport region. The operation of active devices depends on the *charge control principle:*[1]

> Each charge placed upon the control electrode can at most introduce an equal and opposite charge in the transport region between the emitting and collecting electrode.

At most, we have the relationship, $|-Q| = |Q_C|$. Any parasitic coupling of the control charge to charge on the other electrodes, or remote parts of the device, will decrease the controlled charge in the transport region, that is $|-Q| < |Q_C|$ more generally. For example, charge coupling between the control

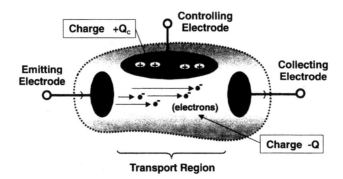

FIGURE 15.1 Generic charge control device consisting of three electrodes embedded around a charge transport region.

electrode and the collecting electrode forms a feedback or output capacitance, say C_o. Time variation of Q_C leads to the modulation of the current flow between emitting and collecting electrodes.

The generic structure in Fig. 15.1 could represent any one of a number of active devices (e.g., vacuum tubes, unipolar transistors, bipolar transistors, photoconductors, etc.). Hence, charge control analysis is very broad in scope, since it applies to all electronic transistors.

Starting from the charge control principle, we associate two characteristic time constants with an active device, thereby leading to a first-order description of its behavior. Application of a potential difference between the emitting and collecting electrodes, say V_{CC}, establishes an electric field in the transport region. Electrons in the transport region respond to the electric field and move across this region with a *transit time* τ_r. The transit time* is the first of the two important characteristic times used in charge control modeling. With charge $-Q$ in the transit region, the static (dc) current I_o between emitting and collecting electrodes is

$$I_o = -Q/\tau_r = Q_c/\tau_r \qquad (15.1)$$

A simple interpretation of τ_r is as follows: τ_r is equal to the length l of the transport region, divided by the average velocity of transit (i.e., $\tau_r = l/\langle v \rangle$). From this perspective, a charge of $-Q$ (coulombs) is swept out of the collecting electrode every τ_r seconds.

Now consider Fig. 15.2, showing the common-emitting electrode connection of the active device of Fig. 15.1 connected to input and output (i.e., load) resistances, say R_{in} and R_L, respectively. The second characteristic time of importance can now be defined: It is the *lifetime time constant*, and we denote it by τ. It is a measure of how long a charge placed on the control electrode will remain on the control terminal. The lifetime time constant is established in one of several ways, depending on the physics of the active device and/or its connection. The controlling charge may "leak away" by (1) discharging through the external resistor R_{in} as typically happens with FET devices, (2) recombining with intermixed oppositely charged carriers within the device (e.g., base recombination in a bipolar transistor), or (3) discharging through an internal shunt leakage path within the device. The dc current flowing to replenish the lost control charge is given by

$$I_{in} = -Q/\tau = Q_c/\tau \qquad (15.2)$$

The *static (dc) current gain* G_I of a device is defined as the current delivered to the output, divided by the current replenishing the control charge during the same time period. Where in τ seconds charge $-Q$ is both lost and replenished, charge Q_c times the ratio τ/τ_r has been supplied to the output resistor R_L. In symbols, the static current gain is

---

*The transit time τ_r is best interpreted as an average transit time per carrier (electron). We note that $1/\tau_r$ is common to all devices — it is related to a device's ultimate capability to process information.

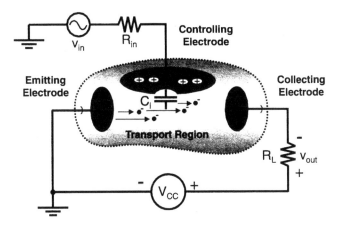

FIGURE 15.2 Generic charge control device of Fig. 15.1 connected to input and output resistors, R_{in} and R_L, respectively, with bias voltage and input signal applied.

$$G_I = I_o/I_{in} = \tau/\tau_r \tag{15.3}$$

provided $|-Q| = |Q_C|$ holds.

In the dynamic case, the process of *small-signal amplification* consists of an incremental variation of the control charge Q_c directly resulting in an incremental change in the controlled charge, $-Q$. The resulting variation in output current flowing in the load resistor translates into a time-varying voltage v_o. The charge control formalism holds just as well for large-signal situations. In the large-signal case, the changes in control charge are no longer small incremental changes. Charge control analysis under large charge variations is less accurate due to the simplicity of the model, but still very useful for approximate switching calculations in digital circuits.

An important dynamic parameter is the *input capacitance* C_i of the active device. Capacitance C_i is a measure of the work required to introduce a charge carrier in the transport region. Capacitance C_i is given by the change in charge Q for a corresponding change in input voltage v_{in}. It is desirable to maximize C_i in an active device. The *transconductance* g_m is calculated from

$$g_m = \left(\frac{\partial I_o}{\partial v_{in}} \right)_{v_o} = \left(\frac{\partial I_o}{\partial Q} \right) \cdot \left(\frac{\partial Q}{\partial v_{in}} \right) \tag{15.4}$$

The first partial derivative on the right-hand side of Eq. 15.4 is simply $(1/\tau_r)$, and the second partial derivative is C_i. Hence, the transconductance g_m is the ratio

$$g_m = \frac{C_i}{\tau_i} \tag{15.5}$$

A physical interpretation of g_m is the ratio of the work required to introduce a charge carrier to the average transit time of a charge carrier in the transport region. The transconductance is one of the most commonly used device parameters in circuit design and analysis.

In addition to C_i, another capacitance, say C_o, is introduced and associated with the collecting electrode. Capacitance C_o accounts for charge on the collecting electrode coupled to either static charge in the transport region or charge on the control electrode. A non-zero C_o indicates that the coupling between the controlling electrode and the charge in transit is less than unity (i.e., $|-Q| < |Q_C|$).

For small-signal analysis the capacitance parameters are usually taken at fixed numbers evaluated about the device's bias state. When using charge control in the large-signal case, the capacitance parameters

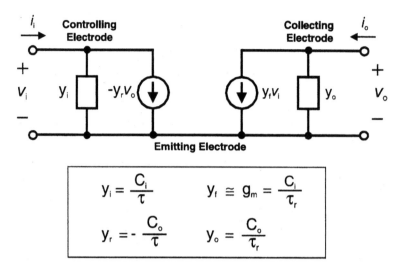

FIGURE 15.3 Two-port, small-signal, admittance charge control model with the emitting electrode selected as the common terminal to both input and output.

must include the voltage dependencies. For example, the input capacitance C_i can be strongly dependent upon the control electrode to emitting electrode and collecting electrode potentials. Hence, during the change in bias state within a device, the magnitude of the capacitance C_i is time varying. This variation can dramatically affect the switching speed of the active device. Parametric dependencies on the instantaneous bias state of the device are at the heart of accurate modeling of large-signal or switching behavior of active devices.

We introduce the *small-signal admittance charge control model* shown in Fig. 15.3. This model uses the emitting electrode as the common terminal in a two-port connection. The transconductance g_m is the magnitude of the real part of the *forward admittance* y_f and is represented as a voltage-controlled current source positioned from collecting-to-emitting electrode. The *input admittance*, denoted by y_i, is equivalent to (C_i/τ), where τ is the control charge lifetime time constant. Parameter y_i can be expressed in the form $(g_i + sC_i)$ where $s = j\omega$. An *output admittance*, similarly denoted by y_o, is given by (C_o/τ_r) where τ_r is the transit time and, in general $y_o = (g_o + sC_o)$. Finally, the *output-to-input feedback admittance* y_r is included using a voltage-controlled current source at the input. Often, y_r is small enough to approximate as zero (the model is then said to be *unilateral*).

Consider the frequency dependence of the *dynamic (ac) current gain* G_i. The low-frequency current gain is interpreted as follows: an incremental charge q_c is introduced on the control electrode with lifetime τ. This produces a corresponding incremental charge $-q$ in the transport region. Charge $-q$ is swept across the transport region every transit time τ_r seconds. In time τ, charge $-q$ crosses the transit region τ/τ_r times, which is identically equal to the low-frequency current gain.

The lifetime τ associated with the control electrode arises from charge "leaking off" the controlling electrode. This is modeled as an RC time constant at the input of the equivalent circuit shown in Fig. 15.4(a) with τ equal to $R_{eq}C_i$. R_{eq} is the equivalent resistance presented to capacitor C_i. That is, R_{eq} is determined by the parallel combination of $1/g_i$ and any external resistance at the input. The *break frequency* ω_B associated with the control electrode is

$$\omega_B = \frac{1}{\tau} = \frac{1}{R_{eq}C_i} \tag{15.6}$$

When the charge on the control electrode varies at a rate ω less than ω_B, G_i is given by τ/τ_r because charge "leaks off" the controlling electrode faster than $1/\omega$. Alternatively, when ω is greater than ω_B, G_i

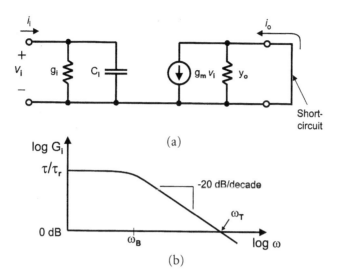

(a)

(b)

FIGURE 15.4 (a) Small-signal admittance model with output short-circuited, and (b) magnitude of the small-signal current gain G_i plotted as a function of frequency. The unity current gain crossover (i.e., $G_i = 1$) defines the parameter f_T (or ω_T).

decreases with increasing ω because the applied signal charge varies upon the control electrode more rapidly than $1/\tau$. In this case, G_i is inversely proportional to ω, that is,

$$G_i = \frac{1}{\omega \tau_r} = \frac{\omega_T}{\omega} \tag{15.7}$$

where ω_T is the common-emitter *unity current gain frequency*. At $\omega = \omega_T$ ($= 2\pi f_T$), the ac current gain equals unity, as illustrated in Fig. 15.4(b).

Consider the *current gain-bandwidth product* $G_i \Delta f$. A purely capacitive input impedance cannot define a bandwidth. However, a finite real impedance always appears at the input terminal in any practical application. Let R_i be the *effective input resistance* of the device (i.e., R_i will be equal to $(1/g_i)$ in parallel with the external resistance R_{in}). Since the input current is equal to q_c/τ and the output current is equal to q/τ_r, the current gain-bandwidth product becomes

$$G_i \cdot \Delta f = \frac{q/\tau_r}{q_c/\tau} \frac{\omega}{2\pi} \tag{15.8}$$

For $\omega \gg \omega_B$, at $\tau = 1/\omega$, and assuming $|q_c| = |-q|$,

$$G_i \cdot \Delta f = \frac{1}{2\pi \tau_r} = \frac{\omega_T}{2\pi} = f_T \tag{15.9}$$

f_T (or ω_T) is a widely quoted parameter used to compare or "benchmark" active devices. Sometimes, f_T (or ω_T) is interpreted as a measure of the maximum speed a device can drive a replica of itself. It is easy to compute and historically has been easy to measure with bridges and later using S-parameters. However, f_T does have interpretative limitations because it is defined as current into a short-circuit output. Hence, it ignores input resistance and output capacitance effects upon actual circuit performance.

Likewise, voltage and power gain expressions can be derived. It is necessary to define the output impedance before either can be quantified. Let R_o be the *effective output resistance* at the output terminal of the active device. Assuming both the input and output RC time constants to be identical (i.e., $R_i C_i = R_o C_o$), the *voltage gain* G_v can be expressed in terms of G_i as

$$G_v = G_i \frac{R_o}{R_i} = G_i \frac{C_i}{C_o} \tag{15.10}$$

where R_o is the parallel equivalent output resistance from all resistances at the output node.

The *power gain* G_p is computed from the product of $G_i \cdot G_v$ along with the *power gain-bandwidth product*. These results are listed in Table 15.1 as summarized from Johnson and Rose.[1] These simple expressions are valid for all devices as interpreted from the charge control perspective. They provide for a first-order comparison, in terms of a few simple parameters, among the active devices commonly available. From an examination of Table 15.1, it is evident that maximizing C_i and minimizing τ_r leads to higher transconductance, higher parametric gains, and greater frequency response. This is an important observation in understanding how to improve upon the performance of any active device.

Whereas f_T has limitations, the frequency at which the maximum power gain extrapolates to unity, denoted by ω_{max}, is often a more useful indicator of device performance. The primary limitation of ω_{max} is that it is very difficult to calculate and is usually extrapolated from S-parameter measurements in which the extrapolation is approximate at best.

15.3 Comparing Unipolar and Bipolar Transistors

Unipolar transistors are active devices that operate using only a single charge carrier type, usually electrons, in their transport region. *Field-effect transistors* fall into the unipolar classification. In contrast, *bipolar transistors* depend on positive and negative charged carriers (i.e., both majority and minority carriers) within the transport region. A fundamental difference arises from the relative locations of the control

TABLE 15.1   Charge Control Relations for All Active Devices

Parameter	Symbol	Expression
Transconductance	g_m	$\dfrac{C_i}{\tau_r} \Leftrightarrow \omega_T C_i$
Current amplification	G_i	$\dfrac{1}{\omega \tau_r} \Leftrightarrow \dfrac{\omega_T}{\omega}$
Voltage amplification	G_v	$\dfrac{1}{\omega \tau_r} \dfrac{C_i}{C_o} \Leftrightarrow \dfrac{\omega_T}{\omega} \dfrac{C_i}{C_o}$
Power amplification	$G_p = G_i G_v$	$\dfrac{1}{\omega^2 \tau_r^2} \dfrac{C_i}{C_o} \Leftrightarrow \dfrac{\omega_T^2}{\omega^2} \dfrac{C_i}{C_o}$
Current gain-bandwidth product	$G_i \cdot \Delta f$	$\dfrac{1}{\tau_r} \Leftrightarrow \omega_T$
Voltage gain-bandwidth product	$G_v \cdot \Delta f$	$\dfrac{1}{\tau_r} \dfrac{C_i}{C_o} \Leftrightarrow \omega_T \dfrac{C_i}{C_o}$
Power gain-bandwidth product	$G_p \cdot \Delta f^2$	$\dfrac{1}{\tau_r^2} \dfrac{C_i}{C_o} \Leftrightarrow \omega_T^2 \dfrac{C_i}{C_o}$

Note: Table assumes $R_i C_i = R_o C_o$. (After Johnson and Rose (Ref. 1), March 1959. © 1959 IEEE, reproduced with permission of IEEE.)

electrode and transport region — in unipolar devices, they are physically separated, whereas in bipolar devices, they are merged into the same physical region (i.e., base region). Before reviewing the physical operation of each, transport in semiconductors is briefly reviewed.

Charge Transport in Semiconductors[4-6]

Bulk semiconducting materials are useful because their conductivity can be controlled over many orders of magnitude by changing the doping level. Both electrons and holes[4] can conduct current in semiconductors. In integrated circuits metal, semiconductor, and insulator layers are used together in precisely positioned shapes and thicknesses to form useful device and circuit functions.

Fig. 14.1 illustrates the behavior of electron velocity as a function of local electric field strength for several important semiconducting materials. Two characteristic regions of behavior can be identified: a *linear* or *ohmic* region at low electric fields, and a *velocity-saturated* region at high fields. At low fields, current transport is proportional to the carrier's mobility. Mobility is a measure of how easily carriers move through a material.[4] At high fields, carriers saturate in velocity; hence, current levels will correspondingly saturate in active devices. The data in Fig. 14.1 assume low doping levels (i.e., $N_x < 10^{15}$ cm^{-3}). The dashed curve represents transport in a GaAs quantum well formed adjacent to an $Al_{0.3}Ga_{0.7}As$ layer — in this case, *interface scattering* lowers the mobility. A similar situation is found for transport in silicon at a semiconductor–oxide interface such as found in metal-oxide-semiconductor (MOS) devices.

Several general conclusions can be extracted from this data:

1. Compound semiconductors generally have higher electron mobilities than silicon.
2. At high fields (say $E > 20,000$ V/cm), saturated electron velocities tend to converge to values close to 1×10^7 cm/s.
3. Many compound semiconductors show a transition region between low and high electric field strengths with a *negative differential mobility* due to electron transfer from the Γ ($\mathbf{k} = 0$) valley to conduction band valleys with higher effective masses (this gives rise to the *Gunn Effect*[7]).

Hole mobilities are much lower than electron mobilities in all semiconductors. Saturated velocities of holes are also lower at higher electric fields. This is why n-channel field-effect transistors have higher performance than p-channel field-effect transistors, and why npn bipolar transistors have higher performance than pnp bipolar transistors. Table 14.2 compares electron and hole mobilities for several semiconducting materials.

Field-Effect (Unipolar) Transistor[8-10]

Fig. 15.5(a) shows a conceptual view of an n-channel field-effect transistor (FET). As shown, the n-type channel is a homogeneous semiconducting material of thickness b, with electrons supporting the drain-to-source current. A p-type channel would rely on mobile holes for current transport and all voltage polarities would be exactly reversed from those shown in Fig. 15.5(a). The control charge on the gate region (of length L and width W) establishes the number of conduction electrons per unit area in the channel by electrostatic attraction or exclusion. The cross-section on the FET channel in Fig. 15.5(b) shows a *depletion layer*, a region void of electrons, as an intermediary agent between the control charge and the controlled charge. This depletion region is present in *junction* FET and *Schottky barrier junction* (i.e., *metal-semiconductor junction*) MESFET structures.

In all FET structures, the gate is physically separated from the channel. By physically separating the control charge from the controlled charge, the gate-to-channel impedance can be very large at low frequencies. The gate impedance is predominantly capacitive and, typically, very low gate leakage currents are observed in high-quality FETs. This is a distinguishing feature of the FET — its high input impedance is desirable for many circuit applications.

The channel, positioned between the source and drain ohmic contacts, forms a resistor whose resistance is modulated by the applied gate-to-channel voltage. We know the gate potential controls the channel charge by the charge control relation. Time variation of the gate potential translates into a corresponding

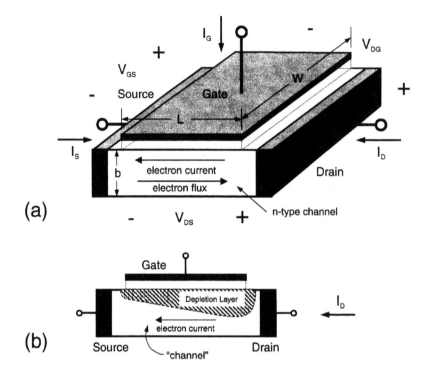

FIGURE 15.5 (a) Conceptual view of a field-effect transistor with the channel sandwiched between source and drain ohmic contacts and a gate control electrode in close proximity; and (b) cross-sectional view of the FET with a depletion layer shown such as would be present in a compound semiconductor MESFET.

time variation of the drain current (and the source current also). Therefore, transconductance g_m is the natural parameter to describe the FET from this viewpoint.

Fig. 15.6(a) shows the I_D-V_{DS} characteristic of the n-channel FET in the common-source connection with constant electron mobility and a long channel assumed. Two distinct operating regions appear in Fig. 15.6(a) — the *linear* (i.e., *non-saturated*) region, and the *saturated* region, separated by the dashed parabola. The origin of current saturation corresponds to the onset of *channel pinch-off* due to carrier exclusion at the drain end of the channel. Pinch-off occurs when the drain voltage is positive enough to

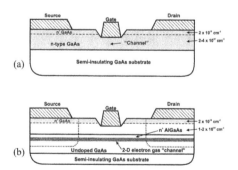

FIGURE 15.6 (a) Field-effect transistor drain current (I_D) versus drain-to-source voltage (V_{DS}) characteristic with the gate-to-source voltage (V_{GS}) as a stepped parameter; (b) I_D versus V_{GS} "transfer curve" for a constant V_{DS} in the saturated region of operation, revealing its "square-law" behavior; (c) transconductance g_m versus V_{GS} for a constant V_{DS} in saturated region of operation corresponding to the transfer curve in (b). These curves assume constant mobility, no velocity saturation, and the "long-channel FET approximation."

deplete the channel completely of electrons at the drain end; this corresponds to a gate-to-source voltage equal to the *pinch-off voltage*, denoted as $-V_p$ in Figs. 15.6(b) and (c). For constant V_{DS} in the saturated region, the I_D vs. V_{GS} *transfer curve* approximates "square law" behavior; that is,

$$I_D = I_{D,sat} = I_{DSS}\left[1 - \frac{V_{GS}}{(-V_P)}\right]^2 \quad \text{for } -V_P \le V_{GS} \le \varphi \tag{15.11}$$

where I_{DSS} is the drain current when $V_{GS} = 0$, and φ is a built-in potential associated with the gate-to-channel junction or interface (e.g., a metal-semiconductor Schottky barrier as in the MESFET). The symbol $I_{D,sat}$ denotes the drain current in the saturated region of operation. Transconductance g_m is linear with V_{GS} for the saturation transfer characteristic of Eq. (15.11) and is approximated by

$$g_m = \frac{\partial I_D}{\partial V_{GS}} \cong 2\frac{D_{DSS}}{V_P}\left[1 - \frac{V_{GS}}{(-V_P)}\right] \quad \text{for } -V_P \le V_{GS} \le \varphi \tag{15.12}$$

Equations 15.11 and 15.12 are plotted in Figs. 15.6(b) and (c), respectively.

Bipolar Junction Transistors (Homojunction and Heterojunction)[7-11]

In the *bipolar junction transistor* (BJT), both the control charge and the controlled charge occupy the same region (i.e., the *base region*). A control charge is injected into the base region (i.e., this is the base current flowing in the base terminal), causing the emitter-to-base junction's potential barrier to be lowered. Barrier lowering results in majority carrier diffusion across the emitter-to-base junction. Electrons diffuse into the base and holes into the emitter in the npn BJT shown in Fig. 15.7. By controlling the emitter-to-base junction's physical structure, the dominant carrier diffusion across this n-p junction should be injection into the base region. For our npn transistor, the dominant carrier transport is electron diffusion into the base region where the electrons are minority carriers. They transit the base region, of base width W_b, by both diffusion and drift. When collected at the collector-to-base junction, they establish the collector current I_C. The base width must be short to minimize recombination in the base region (this is reflected in the current gain parameter commonly used with BJT and HBT devices).

In *homojunction* BJT devices, the emitter and base regions have the same bandgap energy. The respective carrier injection levels are set by the ratio of the emitter-to-base doping levels. For high emitter efficiency, that is, the number of carriers diffusing into the base being much greater than the number of carriers simultaneously diffusing into the emitter, the emitter must be much more heavily doped than the base region. This places a limit on the maximum doping level allowed in the base of the homojunction BJT, thereby leading to higher base resistance than the device designer would normally desire.[10] In contrast, the *heterojunction bipolar transistor* (HBT) uses different semiconducting materials in the base and emitter regions to achieve high emitter efficiency. A wider bandgap emitter material allows for high emitter efficiency while allowing for higher base doping levels which in turn lowers the parasitic base resistance. An example of a wider bandgap emitter transistor is shown in Fig. 15.8. In this example, the emitter is AlGaAs whereas the base and collector are formed with GaAs. Figure 15.8 shows the band diagram under normal operation with the emitter–base junction forward-biased and the collector–base junction reverse-biased. The discontinuity in the valence band edge at the emitter–base heterojunction is the origin of the reduced diffusion into the emitter region. The injection ratio determining the emitter efficiency depends exponentially on this discontinuity. If ΔE_g is the valence band discontinuity, the injection ratio is proportional to the exponential of ΔE_g normalized to the thermal energy kT:[10a]

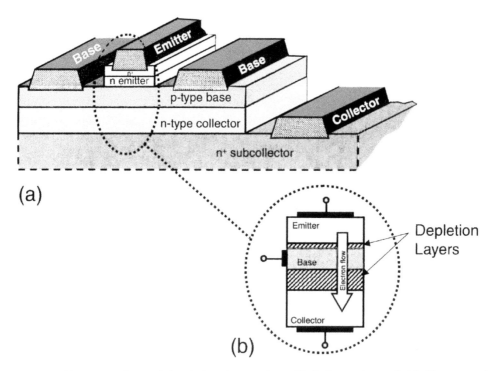

FIGURE 15.7 (a) Conceptual view of a bipolar junction transistor with the base region sandwiched between emitter and collector regions. Structure is representative of a compound semiconductor heterojunction bipolar transistor. (b) Simplified cross-sectional view of a vertically structured BJT device with primary electron flow represented by large arrow.

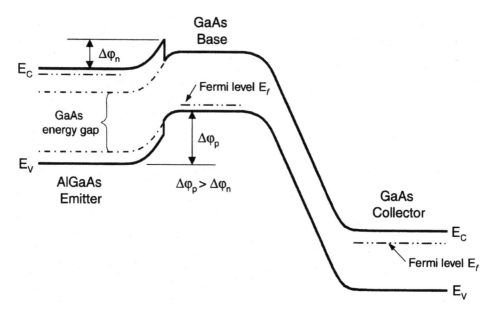

FIGURE 15.8 The bandgap diagram for an HBT AlGaAs/GaAs device with the wider bandgap for the AlGaAs emitter (solid line) compared with a homojunction GaAs BJT emitter (dot-dash line). The double dot-dashed line represents the Fermi level in each region.

$$\frac{J_n}{J_p} \propto \exp\left(-\Delta E_g / kT\right) \tag{15.13}$$

For example, ΔE_g equal to 8kT gives an exponential factor of approximately 8000, thereby leading to an emitter efficiency of nearly unity, as desired. The use of the emitter-base band discontinuity is a very efficient way to hold high emitter efficiencies.

In bipolar devices, the collector current I_C is given by the exponential of the *base-emitter forward voltage* V_{BE} normalized to the *thermal voltage* kT/q

$$I_C - I_S \exp\left(qV_{BE} / kT\right) \tag{15.14}$$

The saturation current I_S is given by a quantity that depends on the structure of the device; it is inversely proportional to the base doping charge Q_{BASE} and proportional to the device's area A, namely

$$I_S = \frac{qADn_i^2}{Q_{BASE}} \tag{15.15}$$

where the other symbols have their usual meanings (D is the minority carrier *diffusion constant* in the base, n_i is the *intrinsic carrier concentration* of the semiconductor, and q is the *electron's charge*).

A typical collector current versus collector-emitter voltage characteristic, for several increasing values of (forward-biased) emitter-base voltages, is shown in Fig. 15.9(a). Note the similarity to Fig. 15.6(a),

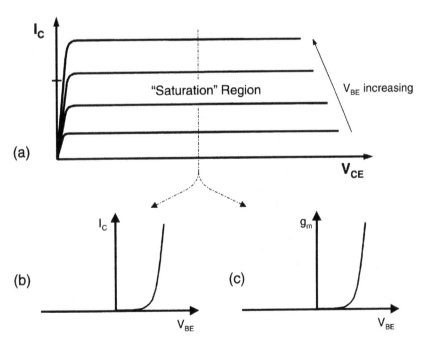

FIGURE 15.9 (a) Collector current (I_C) versus collector-to-emitter voltage (V_{CE}) characteristic curves with the base-to-emitter voltage (V_{BE}) as stepped parameter; (b) I_C versus V_{BE} "transfer curve" for a constant V_{CE} in saturated region of operation shows exponential behavior; and (c) transconductance g_m versus V_{BE} for a constant V_{CE} in the saturated region of operation corresponding to the transfer curve in (b).

with the BJT having a quicker turn-on for low V_{CE} values compared with the softer knee for the FET. The transconductance of the BJT and HBT is found by taking the derivative of Eq. 15.14, thus

$$g_m = \frac{\partial I_C}{\partial V_{BE}} = \frac{qI_S}{kT} \exp\left(qV_{BE}/kT\right) \tag{15.16}$$

Both I_C and g_m are of exponential form, as observed in Fig. 15.13; Eqs. 15.14 and 15.16 are plotted in Figs. 15.9(b) and (c), respectively. The transconductance of the BJT/HBT is generally much larger than that of the best FET devices (this can be verified by comparing Eq. (15.12) with Eq. (15.16) with typical parameter values inserted). This has significant circuit design advantages for the BJT/HBT devices over the FET devices because high transconductance is needed for high current drive to charge load capacitance in digital circuits. In general, higher g_m values allow a designer to use feedback to a greater extent in design and this provides for greater tolerance to process variations.

Comparing Parameters

Table 15.2 compares some of the more important features and parameters of the BJT/HBT device with the FET device. For reference, a common-source FET configuration is compared with a common-emitter BJT/HBT configuration. One of the most striking differences is the input impedance parameter. A FET has a high input impedance at low to mid-range frequencies because it essentially is a capacitor. As the frequency increases, the magnitude of the input impedance decreases as $1/\omega$ because a capacitive reactance varies as $\left|C_{gs}/\omega\right|$. The BJT/HBT emitter-base is a forward-biased pn junction, which is inherently a low impedance structure because of the lowered potential barrier to carriers. The BJT/HBT input is also capacitive (i.e., a large diffusion capacitance due to stored charge), but a large conductance (or small resistance) appears in parallel assuring a low input impedance even at low frequencies.

BJT/HBT devices are known for their higher transconductance g_m, which is proportional to collector current. An FET's g_m is proportional to the saturated velocity v_{sat} and its input capacitance C_{gs}. Thus, device structure and material parameters set the performance of the FET whereas thermodynamics play the key role in establishing the magnitude of g_m in a BJT/HBT.

TABLE 15.2 Comparing Electrical Parameters for BJT/HBT vs. FET

Parameter	BJT/HBT	FET
Input impedance Z	Low Z due to forward-biased junction; large diffusion capacitance C_{be}	High Z due to reverse biased junction or insulator; small depletion layer capacitance C_{gs}
Turn-on Voltage	Forward voltage V_{BE} highly repeatable; set by thermodynamics	Pinch-off voltage V_p not very repeatable; set by device design
Transconductance	High g_m [= $I_C/(kT/q)$]	Low g_m [$\cong v_{sat}C_{gs}$]
Current gain	β (or h_{FE}) = 50 to 150; β is important due to low input impedance	Not meaningful at low frequencies and falls as $1/\omega$ at high frequencies
Unity current gain cutoff frequency f_T	$f_T = g_m/2\pi C_{BE}$ is usually lower than for FETs	$f_T = g_m/2\pi C_{gs}$ (= $v_{sat}/2\pi L_g$) higher for FETs
Maximum frequency of oscillation f_{max}	$f_{max} = [f_T/(8\pi r_b \cdot C_{bc})]^{1/2}$	$f_{max} = f_T [r_{ds}/R_{in}]^{1/2}$
Feedback capacitance	C_{bc} large because of large collector junction	Usually C_{gd} is much smaller than C_{bc}
1/f Noise	Low in BJT/HBT	Very high 1/f noise corner frequency
Thermal behavior	Thermal runaway and second breakdown	No thermal runaway
Other		Backgating is problem in semi-insulating substrates

Thermodynamics also establishes the magnitude of the *turn-on voltage* (this follows simply from Eq. 15.14) in the BJT/HBT device. For digital circuits, turn-on voltage (or *threshold voltage*) is important in terms of repeatability and consistency for circuit robustness. The BJT/HBT is clearly superior to the FET in this regard because doping concentration and physical structure establish an FET's turn-on voltage. In general, these variables are less controllable. However, the forward turn-on voltage in the AlGaAs/GaAs HBT is higher (~1.4 V) because of the band discontinuity at the emitter–base heterojunction. For InP-based HBTs, the forward turn-on voltage is lower (~0.8 V) than that of the AlGaAs/GaAs HBT and comparable to the approximate 0.7 V found in silicon BJTs. This is important in digital circuits because reducing the signal swing allows for faster circuit speed and lowers power dissipation by allowing for reduced power supply voltages.

For BJT/HBT devices, *current gain* (often given the symbol of β or h_{FF}) is a meaningful and important parameter. Good BJT devices inject little current into the emitter and, hence, operate with low base current levels. The current gain is defined as the collector current divided by the base current and is therefore a measure of the quality of the device (i.e., traps and defects, both surface and bulk, degrade the current gain due to higher recombination currents). At low to mid-range frequencies, current gain is not especially meaningful for the high input impedance FET device because of the capacitive input.

The intrinsic gain of an HBT is higher because of its higher *Early voltage* V_A. The Early voltage is a measure of the intrinsic output conductance of a device. In the HBT, the change in the collector voltage has very little effect on the modulation of the collector current. This is true because the band discontinuity dominates the establishment of the current collected at the collector–base junction. A figure of merit is the *intrinsic voltage gain* of an active device, given by the product $g_m V_A$, and the HBT has the highest values compared to silicon BJTs and compound semiconductor FETs.

It is important to have a dynamic figure of merit or parameter to assess the usefulness of an active device for high-speed operation. Both the unity current gain cutoff frequency f_T and maximum frequency of oscillation f_{max} have been discussed in the charge control section above. Both of these figures of merit are used because they are simple and can generally be correlated to circuit speed. The higher the value of both parameters, the better the high-speed circuit performance. This is not the whole story because in digital circuits other factors such as output node-to-substrate capacitance, external load capacitances, and interconnect resistance also play an important role in determining the speed of a circuit.

Generally, $1/f$ noise is much higher in FET devices than in the BJT/HBT devices. This is usually of more importance in analog applications and oscillators however. Thermal behavior in high-speed devices is important as designers push circuit performance. Bipolar devices are more susceptible to thermal runaway than FETs because of the positive feedback associated with a forward-biased junction (i.e., a smaller forward voltage is required to maintain the same current at higher temperatures). This is not true in the FET; in fact, FETs generally have negative feedback under common biases used in digital circuits. Both GaAs and InP have poorer thermal conductivity than silicon, with GaAs being about one-third of silicon and InP being about one-half of silicon.

Finally, circuits built on GaAs or InP semi-insulating substrates are susceptible to *backgating*. Backgating is similar to the backgate-bias effects in MOS transistors, only it is not as predictable or repeatable as the well-known *backgate-bias effect* is in silicon MOSFETs on silicon lightly doped substrates. Interconnect traces with negatively applied voltages and located adjacent to devices can change their threshold voltage (or turn-on voltage). It turns out that HBT devices do not suffer from backgating, and this is one of their advantages. Of course, semi-insulating substrates are nearly ideal for microstrip transmission lines on top of the substrates because of their very low loss. Silicon substrates are much more lossy in comparison and this is a decided advantage in GaAs and InP substrates.

15.4 Typical Device Structures

In this section, a few typical device structures are described. We begin with FET structures and then follow with HBT structures. There are many variants on these devices and the reader is referred to the literature for more information.[10,11,13-16]

FET Structures

In the silicon VLSI world, the MOSFET (*metal-oxide-semiconductor field-effect transistor*) dominates. This device forms a channel at the oxide–semiconductor interface upon applying a voltage to the gate to attract carriers to this interface.[17] The thin layer of mobile carriers forms a two-dimensional sheet of carriers. One of the limitations with the MOSFET is that the oxide–semiconductor interface scatters the carriers in the channel and degrades the performance of the MOSFET. This is evident in Fig. 14.1 where the lower electron velocity at the Si-SiO$_2$ interface is compared with electron velocities in compound semi-conductors. For many years, device physicists have looked for device structures and materials which increase electron velocity. FET structures using compound semiconductors have led to much faster devices such as the MESFET and the HEMT.

The MESFET (*metal-semiconductor FET*) uses a thin doped channel (almost always n-type because electrons are much more mobile in semiconductors) with a reverse-biased Schottky barrier for the gate control.[9] The cross-section of a typical MESFET is shown in Fig. 15.10(a). A recessed gate is used along with a highly doped n$^+$ layer at the surface to reduce the series resistance at both the source and drain connections. The gate length and electron velocity in the channel dominate in determining the high-speed performance of a MESFET. Much work has gone into developing processes that form shorter gate structures. For digital devices, lower breakdown voltages are permissible, and therefore shorter gate lengths and higher channel doping are more compatible with such devices. For a given semiconductor material, a device's breakdown voltage BV_{GD} times its unity current gain cutoff frequency f_T is a constant. Therefore, it is possible to tradeoff BV_{GD} for f_T in device design. A high f_T is required in high-speed digital circuits because devices with a high f_T over their logic swing will have a high g_m/C ratio for large-signal operation. A high g_m/C ratio translates into a device's ability to drive load capacitances.

It is also desirable to maximize the charge in the channel per unit gate area. This allows for higher currents per unit gate width and greater ability to drive large capacitive loads. The higher current per

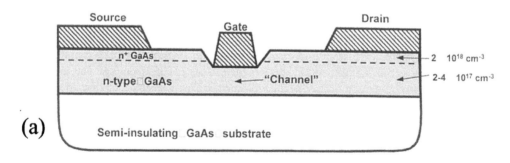

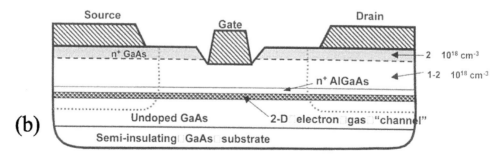

FIGURE 15.10   Typical FET cross-sections for (a) GaAs MESFET device with doped channel, and (b) AlGaAs/GaAs HEMT device with single quantum well containing and two-dimensional electron gas.

unit gate width also favors greater IC layout density. In the MESFET, the doping level of the channel sets this limit. MESFET channels are usually ion-implanted and the added lattice damage further reduces the electron mobility.

To achieve still higher currents per gate width and even higher figures of merit (such as f_T and f_{max}), the HEMT (*high electron mobility transistor*) structure has evolved.[10,16] The HEMT is similar to the MESFET except that the doped channel is replaced with a *two-dimensional quantum well* containing electrons (sometimes referred to as a *2-D electron gas*). The quantum well is formed by a discontinuity in conduction band edges between two different semiconductors (such as AlGaAs and GaAs in Fig. 14.3). From Fig. 14.4 we see that GaAs and $Al_{0.3}Ga_{0.7}As$ have nearly identical lattice constants but with somewhat different bandgaps. One compound semiconductor can be grown (i.e., using *molecular beam epitaxy* or *metalo-organic chemical vapor deposition* techniques) on a different compound semiconductor if the lattice constants are identical. Another example is $Ga_{0.47}In_{0.53}As$ and InP, where they are lattice matched. The difference in conduction band edge alignment leads to the formation of a quantum well. The greater the edge misalignment, the deeper the quantum well can be, and generally the greater the number of carriers the quantum well can hold. The charge per unit area that a quantum well can hold directly translates into greater current per unit gate width. Thus, the information in Fig. 14.4 can be used to *bandgap engineer* different materials that can be combined in lattice matched layers.

A major advantage of the quantum well comes from being able to use semiconductors that have higher electron velocity and mobility than the substrate material (e.g., GaAs) and also avoid charge impurity scattering in the quantum well by locating the donor atoms outside the quantum well itself. Figure 15.10(b) shows a HEMT cross-section where the dopant atoms are positioned in the wider bandgap AlGaAs layer. When these donors ionize, electrons spill into the quantum well because of its lower energy. Higher electron mobility is possible because the ionized donors are not located in the quantum well layer. A recessed gate is placed over the quantum well, usually on a semiconductor layer such as the AlGaAs layer in Fig. 15.10(b), allowing modulation of the charge in the quantum well.

There are only a few lattice-matched structures possible. However, semiconductor layers for which the lattice constants are not matched are possible if the layers are thin enough (of the order of a few nanometers). Molecular beam epitaxy and MOCVD make it possible to grow layers of a few atomic layers. Such structures are called *pseudomorphic HEMT* (PHEMT) devices.[10,16] This gives more flexibility in selecting quantum well layers which hold greater charge and have higher electron velocities and mobilities. The highest performance levels are achieved with pseudomorphic HEMT devices.

FET Performance

All currently used FET structures are n-channel because hole velocities are very low compared with electron velocities. Typical gate lengths range from 0.5 microns down to about 0.1 microns for the fastest devices. The most critical fabrication step in producing these structures is the gate recess width and depth.

The GaAs MESFET (ca. 1968) was the first compound semiconductor FET structure and is still used today because of its simplicity and low cost of manufacture. GaAs MESFET devices have f_T values in the 20 GHz to 50 GHz range corresponding to gate lengths of 0.5 microns down to 0.2 microns, and g_m values of the order of 200 to 400 mS/mm, respectively. These devices will typically have I_{DSS} values of 200 to 400 mA/mm, where parameter I_{DSS} is the common-source, drain current with zero gate voltage applied in a saturated state of operation.

In comparison, the first HEMT used an AlGaAs/GaAs material structure. These devices are higher performance than the GaAs MESFET (e.g., given an identical gate length, the AlGaAs/GaAs HEMT has an f_T about 50% to 100% higher, depending on the details of the device structure and quality of material). Correspondingly higher currents are achieved in the AlGaAs/GaAs HEMT devices.

Higher performance still is achieved using InP based HEMTs. For example, the $In_{0.53}Ga_{0.47}As/In_{0.52}Al_{0.48}As$ on InP lattice-matched HEMT have reported f_T numbers greater than 250 GHz with gate lengths of the order of 0.1 microns. Furthermore, such devices have I_{DSS} values approaching 1000 mA/mm and very high transconductances of greater than 1500 mS/mm.[16,18] These devices do have low breakdown voltages of the order of 1 or 2 V because of the small bandgap of InGaAs. Changing the stoichiometric ratios to $In_{0.15}Ga_{0.85}As/In_{0.15}Al_{0.30}As$ on a GaAs substrate produces a pseudomorphic HEMT structure. The $In_{0.15}Ga_{0.85}As$ is a strained layer when grown on GaAs. The use of strained layers gives the device designer more flexibility in accessing a wider variety of quantum wells depths and electronic properties.

Heterojunction Bipolar Structures

Practical heterojunction bipolar transistors (HBT) devices[13,15] are still evolving. Molecular beam epitaxy (MBE) is used to grow the doped layers making up the vertical semiconductor structure in the HBT. In fact, HBT structures were not really practical until the advent of MBE, although the idea behind the HBT goes back to around 1950 (Shockley). The vastly superior compositional control and layer thickness control with MBE is what made HEMTs and HBTs possible. The first HBT devices used an AlGaAs/GaAs junction with the wider bandgap AlGaAs layer forming the emitter region. Compound semiconductor HBT devices are typically mesa structures, as opposed to the more nearly planar structures used in silicon bipolar technology, because top surface contacts must be made to the collector, base, and emitter regions. Molecular beam epitaxy grows the stack of layers over the entire wafer, whereas, in silicon VLSI processes, selective implantations and oxide masking localize the doped regions. Hence, etching down to the respective layers allows for contact to the base and collector regions. An example of such a mesa HBT structure[14] is shown in Fig. 15.11. The HBT shown uses an InGaP emitter primarily for improved reliability over the AlGaAs emitter and a carbon-doped p[+] base GaAs layer.

Recently, InP-based HBTs[15] have emerged as candidates for use in high-speed circuits. The two dominant heterojunctions are InP/InGaAs and AlInAs/InGaAs in InP devices. The small but significant bandgap difference between AlInAs directly on InP greatly limits its usefulness. InP-based HBT device structures are similar to those of GaAs-based devices and the reader is referred to Chapter 5 of Jalali and Pearton[10] for specific InP HBT devices. Generally, InP has advantages of lower surface recombination (higher current gain results), better electron transport, lower forward turn-on voltage, and higher substrate thermal conductivity.

HBT Performance

Typical current gain values in production-worthy HBT devices range from 50 at the low range to 150 at the high range. Cutoff frequency f_T values are usually quoted under the best (i.e., peak) bias conditions. For this reason f_T values must be carefully interpreted because in digital circuits, the bias state varies widely over the entire switching swing. For this reason, probably an averaged f_T value would be better, but it is difficult to determine. Typical f_T values for HBT processes in manufacturing (say 1998) are in

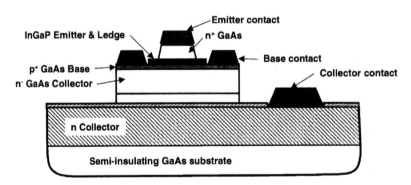

FIGURE 15.11   Cross-section of an HBT device with carbon-doped p[+] base and an InGaP emitter.[14] Note the commonly used mesa structure, where selective layer etching is required to form contacts to the base and collector regions.

the 50 to 150 GHz range. For example, for the HBT example in Fig. 15.11 with a 2 μm × 2 μm emitter f_T is approximately 65 GHz at a current density of 0.6 mA/μm^2 and its dc current gain is around 50. Of course, higher values for f_T have been reported for R&D or laboratory devices. In HBT devices, the parameter f_{max} is often lower than its f_T value (e.g., for the device in Fig. 15.11, f_{max} is about 75 GHz). Base resistance (refer to Table 15.2 for equation) is the dominant limiting factor in setting f_{max}. The best HBT devices have f_{max} values only slightly higher than their f_T values. In comparison, MESFET and HEMT devices typically have higher f_{max}/f_T ratios, although in digital circuits this may be of little importance.

Where the HBT really excels is in being able to generate much higher values of transconductance. This is a clear advantage in driving larger loading capacitances found in large integrated circuits. Biasing the HBT in the current range corresponding to the highest transconductance is essential to take advantage of the intrinsically higher transconductance.

References

1. Johnson, E. O. and Rose, A., Simple General Analysis of Amplifier Devices with Emitter, Control, and Collector Functions, *Proc. IRE*, 47, 407, 1959.
2. Cherry, E. M. and Hooper, D. E., *Amplifying Devices and Low-Pass Amplifier Design*, John Wiley & Sons, New York, 1968, Chap. 2 and 5.
3. Beaufoy, R. and Sparkes, J. J., The Junction Transistor as a Charge-Controlled Device, *ATE Journal*, 13, 310, 1957.
4. Shockley, W., *Electrons and Holes in Semiconductors*, Van Nostrand, New York, 1950.
5. Ferry, D. K., *Semiconductors*, Macmillan, New York, 1991.
6. Lundstrom, M., *Fundamentals of Carrier Transport*, Addison-Wesley, Reading, MA, 1990.
7. Sze, S. M., *Physics of Semiconductor Devices*, second ed., John Wiley & Sons, New York, 1981.
8. Yang, E. S., *Fundamentals of Semiconductor Devices*, McGraw-Hill, New York, 1978, Chap. 7.
9. Hollis, M. A. and Murphy, R. A., Homogeneous Field-Effect Transistors, *High-Speed Semiconductor Devices*, Sze, S. M., Ed., Wiley-Interscience, New York, 1990.
10. Pearton, S. J. and Shah, N. J., Heterostructure Field-Effect Transistors, *High-Speed Semiconductor Devices*, Sze, S. M., Ed., Wiley-Interscience, 1990, Chap. 5.
10a. Kroemer, H., Heterostructure Bipolar Transistors and Integrated Circuits, *Proc. IEEE*, 15, 13, 1982.
11. Muller, R. S. and Kamins, T. I., *Device Electronics for Integrated Circuits*, second ed., John Wiley & Sons, New York, 1986, Chap. 6 and 7.
12. Gray, P. E., Dewitt, D., Boothroyd, A. R., and Gibbons, J. F., *Physical Electronics and Circuit Models of Transistors*, John Wiley & Sons, New York, 1964, Chap. 7.
13. Asbeck, P. M., Bipolar Transistors, *High-Speed Semiconductor Devices*, S. M. Sze, Ed., John Wiley & Sons, New York, 1990, Chap. 6.
14. Low, T. S. et al., Migration from an AlGaAs to an InGaP Emitter HBT IC Process for Improved Reliability, presented at *IEEE GaAs IC Symposium Technical Digest*, Atlanta, GA, 153, 1998.
15. Jalali, B. and Pearson, S. J., *InP HBTs Growth, Processing and Applications*, Artech House, Boston, 1995.
16. Nguyen, L. G., Larson, L. E., and Mishra, U. K., Ultra-High-Speed Modulation-Doped Field-Effect Transistors: A Tutorial Review, *Proc. IEEE*, 80, 494, 1992.
17. Brews, J. R., The Submicron MOSFET, *High-Speed Semiconductor Devices*, Sze, S. M., Ed., Wiley-Interscience, New York, 1990, Chap. 3.
18. Nguyen, L. D., Brown, A. S., Thompson, M. A., and Jelloian, L. M., 50-nm Self-Aligned-Gate Pseudomorphic AlInAs/GaInAs High Electron Mobility Transistors, *IEEE Trans. Elect. Dev.*, 39, 2007, 1992.

16

Logic Design Principles and Examples

Stephen I. Long
*University of California
at Santa Barbara*

16.1 Introduction

The logic circuits used in high-speed compound semiconductor digital ICs must satisfy the same essential conditions for design robustness and performance as digital ICs fabricated in other technologies. The static or dc design of a logic cell must guarantee adequate voltage and/or current gain to restore the signal levels in a chain of similar cells. A minimum *noise margin* must be provided for tolerance against process variation, temperature, and induced noise from ground bounce, crosstalk, and EMI so that functional circuits and systems are produced with good electrical yield. Propagation delays must be determined as a function of loading and power dissipation.

Compound semiconductor designs emphasize speed, so logic voltage swings are generally low, τ_r low so that transconductances and f_T are high, and device access resistances are made as low as possible in order to minimize the lifetime time constant τ. This combination makes circuit performance very sensitive to parasitic R, L, and C, especially when the highest operation frequency is desired. The following sections will describe the techniques that can be used for static and dynamic design of high-speed logic.

16.2 Static Logic Design

A basic requirement for any logic family is that it must be capable of *restoring* the logic voltage or current swing. This means that the voltage or current gain with loading must exceed 1 over part of the transfer characteristic. Figure 16.1 shows a typical V_{out} vs. V_{in} *dc transfer characteristic* for a static ratioed logic inverter as is shown in the schematic diagram of Fig. 16.2. It can be seen that a chain of such inverters will restore steady-state logic voltage levels to V_{OL} and V_{OH} because the high-gain transition region around $V_{in} = V_{TH}$ will result in voltage amplification. Even if the voltage swing is very small, if centered on the *inverter threshold* voltage V_{TH}, defined as the intersection between the transfer characteristic and the $V_{in} = V_{out}$ line, the voltage will be amplified to the full swing again by each successive stage.

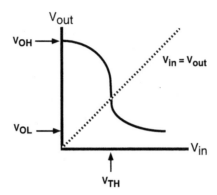

FIGURE 16.1   Typical voltage transfer characteristic for the logic inverter shown in Fig. 16.2.

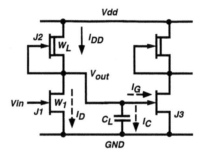

FIGURE 16.2   Schematic diagram of a direct-coupled FET logic (DCFL) inverter.

Ratioed logic implies that the *logic high and low voltages* V_{OH} and V_{OL} shown in Fig. 16.1 are a function of the widths W_1 and W_L of the FETs in the circuit shown in Fig. 16.2. In III-V technologies, this circuit is implemented with either MESFETs or HEMTs. The circuit in Fig. 16.2 is called *Direct Coupled FET Logic* or DCFL.

The logic levels of *non-ratioed* logic are independent of device widths. Non-ratioed logic typically occurs when the switching transistors do not conduct any static current. This is typical of logic families such as static CMOS or its GaAs equivalent *CGaAs*[2] which make use of complementary devices. Dynamic logic circuits such as precharged logic[1] and pass transistor logic[2,3] also do not require static current in pull-down chains. Such circuits have been used with GaAs FETs in order to reduce static power dissipation. They have not been used, however, for the highest speed applications.

Direct-Coupled FET Logic

DCFL is the most widely used logic family for the high-complexity, low-power applications that will be discussed in Chapter 17. The operation of DCFL shown in Fig. 16.2 is easily explained using a load line analysis. Currents are indicated by arrows in this figure. Solid arrows correspond to currents that are nearly constant. Dashed arrows represent currents that depend on the state of the output of the inverter. Figure 16.3 presents an I_D–V_{DS} characteristic of the enhancement mode (normally-off with threshold voltage $V_T > 0$) transistor J1. A family of characteristic curves is drawn representing several V_{GS} values. In this circuit, $V_{GS} = V_{in}$ and $V_{DS} = V_{out}$.

A load line representing the I–V characteristic of the active load ($V_{GS} = 0$) depletion-mode (normally-on with $V_T < 0$) transistor J2 is also superimposed on this drawing. Note that the logic low level V_{OL} is determined by the intersection of the two curves when $I_D = I_{DD}$. Load line 1 corresponds to a load device with narrow width; load line 2 for a wider device. It is evident that the narrow, weaker device will provide a lower V_{OL} value and thus will increase the logic swing. However, the weaker device will also have less

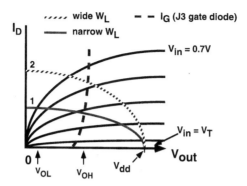

FIGURE 16.3 Drain current versus drain-source voltage characteristic of J1. The active load, J2, is also shown superimposed over the J1 characteristics as a load line. Two load lines corresponding to wide and narrow J2 widths are shown. In addition, the gate current I_G of J3 versus V_{out} limits the logic high voltage.

current available to drive any load capacitance, so the inverter with load line 1 will therefore be slower than the one with load line 2. There is therefore a tradeoff between speed and logic swing. So far, the analysis of this circuit is the same as that of an analogous nMOS E/D inverter.

In the case of DCFL logic inverters implemented with GaAs-based FETs, the Schottky barrier gate electrode of the next stage will limit the maximum value of V_{OH} to the forward voltage drop across the gate-source diode. This is shown by the gate diode I_G–V_{GS} characteristic also superimposed on Fig. 16.3. V_{OH} is given by the point of intersection between the load current I_{DD} and the gate current I_G, because a logic high output requires that the switch transistor J1 is off. V_{OH} will therefore also depend on the load transistor current. Effort must be made not to overdrive the gate since excess gate current will flow through the internal parasitic source resistance of the driven device J3, degrading V_{OL} of this next stage.

Source-Coupled FET Logic

A second widely used type of logic circuit — source-coupled FET Logic or SCFL is shown in Fig. 16.4. SCFL, or its bipolar counterpart, ECL, is widely used for very high-speed applications, which will be discussed in Chapter 17. The core of the circuit consists of a differential amplifier, J_1 and J_2, a current source J_3, and pull-up resistors R_L on the drains. The differential topology is beneficial for rejection of common-mode noise. The static design procedure can be illustrated again by a load-line analysis. A maximum current I_{CS} can flow through either J_1 or J_2.

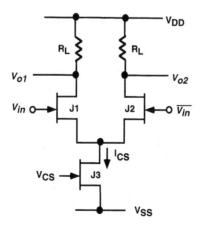

FIGURE 16.4 Schematic of differential pair J1,J2 used as a source-coupled FET logic (SCFL) cell.

Figure 16.5 shows the I_D–V_{DS} characteristic of J_1 for example. The maximum current I_{CS} is shown by a dotted line. The output voltage V_{O1} is either V_{DD} or $V_{DD} - I_{CS}R_L$; therefore, the maximum differential voltage swing, $\Delta V = 2\,I_{CS}R_L$, is determined by the choice of R_L. Next, the width of J_1 should be selected so that the change in V_{GS} needed to produce the voltage drop $I_{CS}R_L$ at the drain is less than $I_{CS}R_L$. This will ensure that the voltage gain is greater than 1 (needed to compensate for the source followers described below) and that the device is biased in its saturation region or cutoff at all times. The latter requirement is necessary if the maximum speed is to be obtained from the SCFL stage, since device capacitances are minimized in saturation and cutoff.

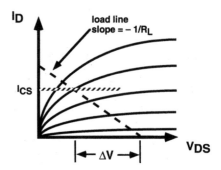

FIGURE 16.5 Load-line analysis of the SCFL inverter cell.

Source followers are frequently used on the output of an SCFL stage or at the inputs of the next stage. Figure 16.6(a) shows the schematic diagram of the follower circuit. The follower can serve two functions: level shifting and buffering of capacitive loads. When used as a level shifter, a negative or positive voltage offset can be obtained between input and output. The only requirement is that the V_{GS} of the source follower must be larger than the FET threshold voltage. If the source follower is at the output of an SCFL cell, it can be used as a buffer to reduce the sensitivity of delay to load capacitance or fanout.

The voltage gain of a source follower is always less than 1. This can be illustrated by another load-line analysis. Figure 16.6(b) presents the I_{D1}–V_{DS1} characteristic of the source follower FET, J1. A constant V_{GS1} is applied for every curve plotted in the figure. The load line (dashed line) of a depletion-mode, active current source J2, is also superimposed. In this circuit, the output voltage is $V_{out} = V_{DD} - V_{DS1}$. The V_{out} is determined by the intersection of the load line with the I_{D1} characteristic curves. The current of the pull-down current source is selected according to the amount of level shifting needed. A high current will result in a greater amount of level shift than a small load current. If the devices have high output resistance, and are accordingly very flat, very little change in V_{GS1} will be required to change V_{out} over the

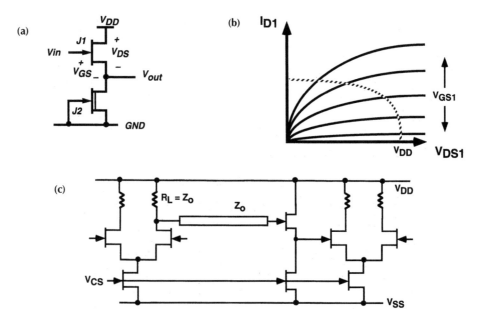

FIGURE 16.6 (a) Schematic of source follower, (b) load-line analysis of source follower, and (c) source follower buffer between SCFL stages.

full range from V_{OL} to V_{DD}. If V_{GS1} remains nearly constant, then V_{out} follows V_{in}, hence the name of the circuit. Since the input voltage to the source follower stage is $V_{in} = V_{GS1} + V_{out}$, a small change in V_{GS1} would produce an incremental voltage gain close to unity. If the output resistance is low, then the characteristic curves will slope upward and a larger range of V_{GS1} will be necessary to traverse the output voltage range. This condition would produce a low voltage gain. Small signal analysis shows that

$$A_v = \frac{1}{1 + 1 / \left[g_{m1} \left(r_{ds1} \| r_{ds2} \right) \right]} \tag{16.1}$$

The buffering effect of the source follower is accomplished by reducing the capacitive loading on the drain nodes of the differential amplifier because the input impedance of a source follower is high. Since the output tries to follow the input, the change in V_{GS} will be less than that required by a common source stage. Therefore, the input capacitance is dominated by C_{GD}, typically quite small for compound semiconductor FETs biased in saturation. The effective small-signal input capacitance is

$$C_G = C_{GD} + C_{GS} \left(1 - A_v \right) \tag{16.2}$$

where $A_v = dV_{out}/dV_{in}$ is the incremental voltage gain.

The source follower also provides a low output impedance, whose real part is approximately $1/g_m$ at low frequency. The current available to charge and discharge the load capacitance can be adjusted by the width ratio of J1 and J2. If the load is capacitive, V_{out} will be delayed from V_{in}. This will cause V_{GS1} to temporarily increase, providing excess current I_{D1} to charge the load capacitance. Ideally, for equal rise and fall times, the peak current available from J1 should equal the steady-state current of J2.

Source followers can also be used at the input of an SCFL stage to provide level shifting as shown in Fig. 16.6(c). In this case, the drain resistors, R_L, should be chosen to provide the proper termination resistance for the on-chip interconnect transmission line. These resistors provide a reverse termination to absorb signals that reflect from the high input impedance of the source follower. Alternatively, the drain resistors can be located at the gate of the source follower, thereby providing a shunt termination at the destination end of the interconnect. This practice results in good signal integrity, but because the practical values of characteristic impedance are less than 100 Ω, the current swing in the differential amplifier core must be large. This will increase power dissipation per stage.

SCFL logic structures generally employ more than one level of differential pairs so that complex logic functions (XOR, latch, and flip-flop) that require multiple gates to implement in logic families such as DCFL can be implemented in one stage. More details on SCFL gate structures and examples of their usage will be given in Chapter 17.

Static and Dynamic Noise Margin and Noise Sources

Noise margin is a measure of the ability of a logic circuit to provide proper functionality in the presence of noise.[4] There are many different definitions of noise margin, but a simple and intuitive one for the static or dc noise margin is illustrated by Fig. 16.8. Here, the transfer characteristic from Fig. 16.1 is plotted again. In order to evaluate the ability of a chain of such inverters to reject noise, a loop consisting of two identical inverters is considered. This might be representative of the positive feedback core of a bistable latch. Because the inverters are connected in a loop, an infinite chain of inverters is represented. The transfer characteristic of inverter 2 in Fig. 16.7 is plotted in gray lines. For inverter 2, $V_{out\ 1} = V_{in2}$ and $V_{out2} = V_{in1}$. Therefore, the axes are reversed for the characteristic plotted for inverter 2. If a series noise source V_N were placed within the loop as shown, the maximum static noise voltage allowed will be represented by the *maximum width* of the loops formed by the transfer characteristic.[4] These widths, labeled V_{NL} and V_{NH}, respectively, for the low and high noise margins, are shown in the figure. If the voltage V_N exceeds V_{NL} or V_{NH}, the latch will be set into the opposite state and will remain there until

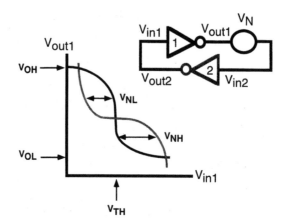

FIGURE 16.7 Voltage transfer characteristics of an inverter pair connected in a loop. Noise margins are shown as V_{NH} and V_{NL}.

reset. This would constitute a logic failure. Therefore, we must insist that any viable logic circuit provide noise margins well in excess of ambient noise levels in the circuit.

The static noise margin defined above utilized a dc voltage source V_N in series with logic inverters to represent static noise. This source might represent a static offset voltage caused by IR drop along IC power and ground distribution networks. The DCFL inverter, for example, would experience a shift in V_{TH} that is directly proportional to a ground voltage offset. This shift would skew the noise margins. The smallest noise margin would determine the circuit electrical yield. The layout of the power and ground distribution networks must consider this problem. The width of power and ground buses on-chip must be sufficient to guarantee a maximum IR drop that does not compromise circuit operation. It is important to note that this width is frequently much greater than what might be required by electromigration limits. It is essential that the designer consider IR drop in the layout. Some digital IC processes allow the topmost metal layer to form a continuous sheet, thereby minimizing voltage drops.

The static noise voltage source V_N might also represent static threshold voltage shifts on the active devices due to statistical process variation or backgating effects. Therefore, the noise margin must be several times greater than the variance in device threshold voltages provided by the fabrication process so that electrical yields will not be compromised.[5]

The above definition of maximum width noise margin has assumed a steady-state condition. It does not account for transient noise sources and the delayed response of the logic circuit to noise pulses. Unfortunately, pulses of noise are quite common in digital systems. For example, the ground potential can often be modified dynamically by simultaneous switching events on the IC chip.[6] Any ground distribution bus can be modeled as a transmission line with impedance Z_0 where

$$Z_0 = \sqrt{\frac{L_o}{C_o}} \qquad (16.3)$$

Here, L_o is the equivalent series inductance per unit length and C_o the equivalent shunt capacitance per unit length. Since the interconnect exhibits a series inductance, there will be transient voltage noise ΔV induced on the line by current transients as predicted by

$$\Delta V = L\frac{dI}{dt} \qquad (16.4)$$

This form of noise is often called *ground bounce*. The ground bounce ΔV is particularly severe when many devices are being switched synchronously, as would be the case in many applications involving flip-flops in shift registers or pipelined architectures. The high peak currents that result in such situations can generate large voltage spikes. For example, output drivers are well-known sources of noise pulses on power and ground buses unless they are carefully balanced with fully differential interconnections and are powered by power and ground pins separate from the central logic core of the IC.

Designing to minimize ground bounce requires minimization of inductance. Bakoglu[6] provides a good discussion of power distribution noise in high-speed circuits. There are several steps often used to reduce switching noise. First, it is standard practice to make extensive use of multiple ground pins on the chip to reduce bond-wire inductance and package trace inductance when conventional packaging is used. Bypass capacitance off-chip can be useful if it can be located inside the package and can have a high series resonant frequency. On-chip bypass capacitance is also helpful, especially if enough charge can be supplied from the capacitance to provide the current during clock transitions. The objective is to provide a low impedance between power and ground on-chip at the clock frequency and at odd multiples of the clock frequency. Finally, as mentioned above, high-current circuits such as clock drivers and output drivers should not share power and ground pins with other logic on-chip.

Crosstalk is another common source of noise pulses caused by electromagnetic coupling between adjacent interconnect lines. A signal propagating on a driven interconnect line can induce a crosstalk voltage and current in a coupled line. The duration of the pulse depends on the length of interconnect; the amplitude depends on the mutual inductance and capacitance between the coupled lines.[7]

In order to determine how much noise is acceptable in a logic circuit, the noise margin definition can be modified to accommodate the transient noise pulse situation. The logic circuit does not respond instantaneously to the noise pulse at the input. This delay in the response is attributed to the device and interconnect capacitances and the device current limitations which will be discussed extensively in Section 16.3. Consider the device input capacitance. Sufficient charge must be transferred during the noise pulse to the input capacitance to shift the control voltage either above or below threshold. In addition, this voltage must be maintained long enough for the output to respond if a logic upset is to occur. Therefore, a logic circuit can withstand a larger noise pulse amplitude than would be predicted from the static noise margin if the pulse width is much less than the propagation delay of the circuit. This increased noise margin for short pulses is called the dynamic noise margin (DNM). The DNM approaches the static NM if the pulse width is wide compared with the propagation delay because the circuit can charge up to the full voltage if not constrained by time.

The DNM can be predicted by simulation. Figure 16.8(a) shows the loop connection of the set-reset NOR latch similar to that which was used for the static NM definition in Fig.10.7. The inverter has been modified to become a NOR gate in this case. An input pulse train $V_1(t)$ of fixed duration but with

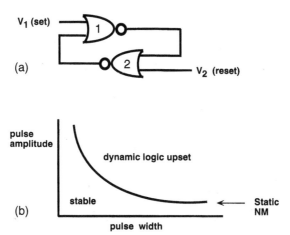

FIGURE 16.8 (a) Set-reset latch used to describe dynamic noise margin simulation, and (b) plot of the pulse amplitude applied to the set input in (a) that results in a logic upset.

gradually increasing amplitude can be applied to the set input. The latch was initialized by applying an initial condition to the reset input $V_2(t)$. The output response is observed for the input pulse train. At some input amplitude level, the output will be set into the opposite state. The latch will hold this state until it is reset again. The cross-coupled NOR latch thus becomes a logic upset detector, dynamically determining the maximum noise margin for a particular pulse width. The simulation can be repeated for other pulse widths, and a plot of the pulse amplitude that causes the latch to set for each pulse duration can be constructed, as shown in Fig. 16.8(b). Here, any amplitude or duration that falls on or above the curve will lead to a logic upset condition.

Power Dissipation

Power dissipation of a static logic circuit consists of a static and a dynamic component as shown below.

$$P_D = V_{DD}\bar{I}_{DD} + C_L \Delta V^2 \, f\eta \qquad (16.5)$$

In the case of DCFL, the current I_{DD} from the pull-up transistor J2 is relatively constant, flowing either in the pull-down (switch) device J1 or in the gate(s) of the subsequent stage(s). Taking its average value, the static power is $V_{DD}\bar{I}_{DD}$. The dynamic power $C_L \Delta V^2 f\eta$ depends on the frequency of operation f, the load capacitance C_L, and the duty factor η. η is application dependent. Since the voltage swing is rather small for the DCFL inverter under consideration (about 0.6 V for MESFETs), the dynamic power will not be significant unless the load capacitance is very large, such as in the case of clock distribution networks. The V_{DD} power supply voltage is traditionally 2 V because of compatibility with the bipolar ECL V_{TT} supply, but a V_{DD} as low as 1 V can be used for special low-power applications. Typical power dissipation per logic cell (inverter, NOR) depends on the choice of supply voltage and on I_{DD}. Power is typically determined based on speed and is usually in the range of 0.1 to 0.5 mW/gate. DCFL logic circuits are often used when the application requires high circuit density and very low power.

16.3 Transient Analysis and Design for Very-High-Speed Logic

Adequate attention must be given to static or dc design, as described in the previous section, in order to guarantee functionality under the worst-case situations. In addition, since the only reason to use the compound semiconductor devices for digital electronics at all is their speed, attention must be given to the dynamic performance as well. In this section, we will describe three methods for estimating the performance of high-speed digital logic circuit functional blocks. Each of these methods has its strengths and weaknesses.

The most effective methods for guiding the design are those that provide insight that helps to identify the dominant time constants that determine circuit performance. These are not necessarily the most accurate methods, but are highly useful because they allow the designer to determine what part of the circuit or device is limiting the speed. Circuit simulators are far more accurate (at least to the extent that the device models are valid), but do not provide much insight into performance limitations. Without simple analytical techniques to guide the design, performance optimization becomes a trial-and-error exercise.

Zero-Order Delay Estimate

The first technique, which uses the simple relationship between voltage and current in a capacitor,

$$I = C_L \frac{dV}{dt} \qquad (16.6)$$

is relevant when circuit performance is dominated by wiring or fan-out capacitance. This will be the case if the delay predicted by Eq. 16.6 due to the total loading capacitance, C_L, significantly exceeds the intrinsic delay of a basic inverter or logic gate. To apply this approach, determine the average current available from the driving logic circuit for charging (I_{LH}) and discharging (I_{HL}) the load capacitance. The logic swing ΔV is known, so low-to-high (t_{PLH}) and high-to-low (t_{PHL}) propagation delays can be determined from Eq. 16.6. These delays represent the time required to charge or discharge the circuit output to 50% of its final value. Thus, t_{PLH} is given by

$$t_{PLH} = \frac{C_L \Delta V}{2I_{LH}} \qquad (16.7)$$

where I_{LH} is the average charging current during the output transition from V_{OL} to $V_{OL} + \Delta V/2$. The net propagation delay is given by

$$t_P = \frac{t_{PLH} + t_{PHL}}{2} \qquad (16.8)$$

At this limit, where speed is dominated by the ability to drive load capacitance, we see that increasing the currents will reduce t_P. In fact, the product of power (proportional to current) and delay (inversely proportional to current) is nearly constant under this situation. Increases in power lead to reduction of delay until the interconnect distributed RC delays or electromagnetic propagation delays become comparable to t_P.

The equation also shows that small voltage swing ΔV is good for speed if the noise margin and drive current are not compromised. This means that the devices must provide high transconductance.

For example, the DCFL inverter of Fig. 16.2 can be analyzed. Figure 16.9 shows equivalent circuits that represent the low-to-high and high-to-low transitions. The current available for the low-to-high transi-

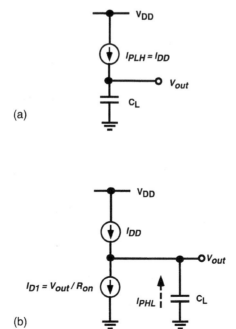

FIGURE 16.9   (a) Equivalent circuit for low-to-high transition; and (b) Equivalent circuit for high-to-low transition.

tion, I_{PLH}, shown in Fig. 16.9(a), is equal to the average pullup current, I_{DD}. If we assume that $V_{OL} = 0.1$ V and $V_{OH} = 0.7$ V, then the ΔV of interest is 0.6 V. This brings the output up to 0.4 V at $V_{50\%}$. In this range of V_{out}, the active load transistor J2 is in saturation at all times for $V_{DD} > 1$ V, so I_{DD} will be relatively constant, and all of the current will be available to charge the capacitor.

The high-to-low transition is more difficult to model in this case. V_{out} will begin at 0.7 V and discharge to 0.4 V. The discharge current through the drain of J1 is going to vary with time because the device is below saturation over this range of V_{out}. Looking at the Vin = 0.7 V characteristic curve in Fig. 16.3, we see that its I_D–V_{DS} characteristic is resistive. Let's approximate the slope by $1/R_{on}$. Also, the discharge current I_{PHL} is the difference between I_{DD} and $I_{D1} = V_{out}/R_{on}$, as shown in Fig. 16.9(b). The average current available to discharge the capacitor can be estimated by

$$I_{HL} = \frac{V_{OH} + V_{50\%}}{2R_{on}} - I_{DD} \tag{16.9}$$

Then, t_{PHL} is estimated by

$$t_{PHL} = \frac{C_L \Delta V}{2I_{HL}} \tag{16.10}$$

Time Constant Delay Methods: Elmore Delay and Risetime

Time constant delay estimation methods are very useful when the wiring capacitance is quite small or the charging current is quite high. In this situation, typical of very-high-speed SSI and MSI circuits that push the limits of the device and process technology, the circuit delays are dominated by the devices themselves. Both methods to be described rely on a large-signal equivalent circuit model of the transistors, an approximation dubious at best. But, the objective of these techniques is not absolute accuracy. That is much less important than being able to identify the dominant contributors to the delay and risetime, since more accurate but less intuitive solutions are easily available through circuit simulation. The construction of the large signal equivalent circuit requires averaging of non-linear model elements such as transconductance and certain device capacitances over the appropriate voltage swing.

The propagation delay definition described above, the delay required to reach 50% of the logic swing, must be relaxed slightly to apply methods based on linear system analysis. It was first shown by Elmore in 1948[8] and apparently rediscovered by Ashar in 1964[9] that the delay time t_D between an impulse function $\delta(0)$ applied at t = 0 to the input of a network and the centroid or "center-of-mass" of the impulse response (output) is quite close to the 50% delay. This definition of delay t_D is illustrated in Fig. 16.10. Two conditions must be satisfied in order to use this approach. First, the step response of the network is monotonic. This implies that the impulse response is purely a positive function. Monotonic step response is valid only when the circuit poles are all negative and real, or the circuit is heavily damped. Due to feedback through device capacitances, this condition is seldom completely correct. Complex poles often exist. However, strongly underdamped circuits are seldom useful for reliable logic circuits because their transient response will exhibit ringing, so efforts to compensate or damp such oscillations are needed in these cases anyway. Then, the circuit becomes heavily damped or at least dominated by a single pole and fits the above requirement more precisely.

Second, the correspondence between t_D and t_{PLH} is improved if the impulse response is symmetric in shape, as in Fig. 16.10(b). It is shown in Ref. 9 that cascaded stages with similar time constants have a tendency to approach a Gaussian-shaped distribution as the number of stages becomes large. Most logic systems require several cascaded stages, so this condition is often true as well.

Assuming that these conditions are approximately satisfied, we can make use of the fact that the impulse response of a circuit in the frequency domain is given by its transfer (or network) function F(s) in the complex frequency s = σ + jω. Then, the propagation delay, t_D, can be determined by

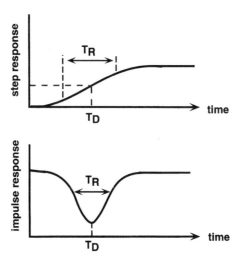

FIGURE 16.10 (a) Monotonic step response of a network; (b) corresponding impulse response. The delay t_D is defined as the centroid of the impulse response.

$$t_D = \frac{\displaystyle\int_0^\infty tf(t)dt}{\displaystyle\int_0^\infty f(t)dt} = \lim_{s\to 0}\frac{\displaystyle\int_0^\infty tf(t)e^{-st}dt}{\displaystyle\int_0^\infty f(t)e^{-st}dt} = \left[\frac{-\dfrac{d}{ds}F(s)}{F(s)}\right]_{s=0} \tag{16.11}$$

Fortunately, the integration never needs to be performed. t_D can be obtained directly from the network function F(s) as shown. But, the network function must be calculated from the large-signal equivalent circuit of the device, including all important parasitics, driving impedances, and load impedances. This is notoriously difficult if the circuit includes a large number of capacitances or inductances.

Fortunately, in most cases, circuits of interest can be subdivided into smaller networks, cascaded, and the presumed linearity of the circuits can be employed to simplify the task. In addition, the evaluation of the function at s = 0 eliminates many terms in the equations that result. In particular, Tien[10] shows that two corollaries are particularly useful in cascading circuit blocks:

1. If the network function F(s) = A(s)/B(s), then

$$t_D = \left[\frac{-\dfrac{d}{ds}A(s)}{A(s)}\right]_{s=0} + \left[\frac{\dfrac{d}{ds}B(s)}{B(s)}\right]_{s=0} \tag{16.12}$$

2. If F(s) = A(s)B(s)C(s), then

$$t_D = \left[\frac{-\dfrac{d}{ds}A(s)}{A(s)}\right]_{s=0} + \left[\frac{-\dfrac{d}{ds}B(s)}{B(s)}\right]_{s=0} + \left[\frac{-\dfrac{d}{ds}C(s)}{C(s)}\right]_{s=0} \tag{16.13}$$

This shows that the total delay is just the sum of the individual delays of each circuit block. When computing the network functions, care must be taken to include the driving point impedance of the previous stage and to represent the previous stage as a Thevenin-equivalent open-circuit voltage source. A good description and illustration of the use of this approach in the analysis of bipolar ECL and CML circuits can be found in Ref. 10.

Risetime: the standard definition of risetime is the 10 to 90% time delay of the step response of a network. While convenient for measurement, this definition is analytically unpleasant to derive for anything except simple, first-order circuits. Elmore demonstrated that the standard deviation of the impulse response could be used to estimate the risetime of a network.[8] This definition provides estimates that are close to the standard definition. The standard deviation of the impulse response can be calculated using

$$T_R^2 = 2\pi \left[\int_0^\infty t^2 f(t) dt - t_D^2 \right] \tag{16.14}$$

Since the impulse response frequently resembles the Gaussian function, the integral is easily evaluated.

Once again, the integration need not be performed. Lee[11] has pointed out that the transform techniques can also be used to obtain the Elmore risetime directly from the network function F(s).

$$T_R^2 = 2\pi \left[\frac{\dfrac{d^2}{ds^2} F(s)}{F(s)} \right]_{s=0} - 2\pi \left[\frac{\dfrac{d}{ds} F(s)}{F(s)} \right]_{s=0}^2 \tag{16.15}$$

This result can also be used to show that the risetimes of cascaded networks add as the square of the individual risetimes. If two networks are characterized by risetimes T_{R1} and T_{R2}, the total risetime $T_{R,total}$ is given by the RMS sum of the individual risetimes

$$T_{R,total}^2 = T_{R1}^2 + T_{R2}^2 \tag{16.16}$$

Time Constant Methods: Open Circuit Time Constants

The frequency domain/transform methods for finding delay and risetime are particularly valuable for design optimization because they identify dominant time constants. Once the time constants are found, the designer can make efforts to change biases, component values, or optimize the design of the transistors themselves to improve the performance through addressing the relevant bottleneck in performance. The drawback in the above technique is that a network function must be derived. This becomes tedious and time-consuming if the network is of even modest complexity. An alternate technique was developed[12,13] that also can provide reasonable estimates for delay, but with much less computational difficulty. The open-circuit time constant (OCTC) method is widely used for the analysis of the bandwidth of analog electronic circuits just for this reason. It is just as applicable for estimating the delay of very-high-speed digital circuits.

The basis for this technique again comes from the transfer or network function F(s) = Vo(s)/Vi(s). Considering transfer functions containing only poles, the function can be written as

$$F(s) = \frac{a_0}{b_n s^n + b_{n-1} s^{n-1} + \cdots + b_1 s + 1} \tag{16.17}$$

The denominator comes from the product of n factors of the form $(\tau_j s + 1)$, where τ_j is the time constant associated with the *j*-th pole in the transfer function. The b_1 coefficient can be shown to be equal to the sum

$$b_1 = \sum_{j=1}^{n} \tau_j \qquad (16.18)$$

of the time constants and b_2 the product of all the time constants. Often, the first-order term dominates the frequency response. In this case, the 3-dB bandwidth is then estimated by $\omega_{3dB} = 1/b_1$. The higher-order terms are neglected. The accuracy of this approach is good, especially when the circuit has a dominant pole. The worst error would occur when all poles have the same frequency. The error in this case is about 25%. Much worse errors can occur however if the poles are complex or if there are zeros in the transfer function as well. We will discuss this later.

Elmore has once again provided the connection we need to obtain delay and risetime estimates from the transfer function. The Elmore delay is given by

$$D = b_1 - a_1 \qquad (16.19)$$

where a_1 is the corresponding coefficient of the first-order zero (if any) in the numerator. The risetime is given by

$$T_R^2 = b_1^2 - a_1^2 + 2\left(a_2 - b_2\right) \qquad (16.20)$$

In Eq. 16.20, a_2 and b_2 correspond to the coefficients of the second-order zero and pole, respectively.

At this point, it would appear that we have gained nothing since finding that the time constants associated with the poles and zeros is well known to be difficult. Fortunately, it is possible to obtain the b_1 and b_2 coefficients directly by a much simpler method: open-circuit time constants. It has been shown that[11,12]

$$b_1 = \sum_{j=1}^{n} R_{jo} C_j = \sum_{j=1}^{n} \tau_{jo} \qquad (16.21)$$

that is, the sum of the time constants τ_{jo}, defined as the product of the effective open-circuit resistance R_{jo} across each capacitor C_j when all other capacitors are open-circuited, equals b_1. These time constants are very easy to calculate since open-circuiting all other capacitors greatly simplifies the network by decoupling many other components. Dependent sources must be considered in the calculation of the R_{jo} open-circuit resistances. Note that these open-circuit time constants are not equal to the pole time constants, but their sum gives the same result for b_1. It should also be noted that the individual OCTCs give the time constant of the network if the j-th capacitor were the only capacitor. Thus, each time constant provides information about the relative contribution of that part of the circuit to the bandwidth or the delay.[11] If one of these is much larger than the rest, this is the place to begin working on the circuit to improve its speed.

The b_2 coefficient can also be found by a similar process,[14] taking the sum of the product of time constants of all possible pairs of capacitors. For example, in a three-capacitor circuit, b_2 is given by

$$b_2 = R_{1o} C_1 R_{2s}^1 C_2 + R_{1o} C_1 R_{3s}^1 C_3 + R_{2o} C_2 R_{3s}^2 C_3 \qquad (16.22)$$

where the R_{js}^i resistance is the resistance across capacitor Cj calculated when capacitor C_i is *short*-circuited and all other capacitors are open-circuited. The superscript indicates which capacitor is to be shorted. So, R_{3s}^2 is the resistance across C_3 when C_2 is short-circuited and C_1 is open-circuited. Note that the first time constant in each product is an open-circuit time constant that has already been calculated. In

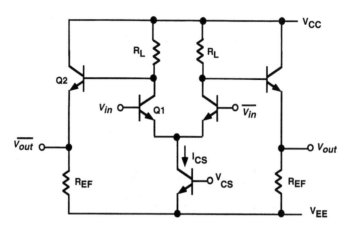

FIGURE 16.11   Schematic of basic ECL inverter.

addition, for any pair of capacitors in the network, we can find an OCTC for one and a SCTC for the other. The order of choice does not matter because

$$R_{io}C_iR_{js}^iC_j = R_{jo}C_jR_{is}^jC_i \tag{16.23}$$

so we are free to choose whichever combination minimizes the computational effort.[14]

At this stage, it would be helpful to illustrate the techniques described above with an example. An ECL inverter whose schematic is shown in Fig. 16.11 is selected for this purpose. The analysis is based on work described in more detail in Ref. 15.

The first step is to construct the large-signal equivalent circuit. We will discuss how to evaluate the large-signal component values later. Figure 16.12 shows such a model applied to the ECL inverter, where

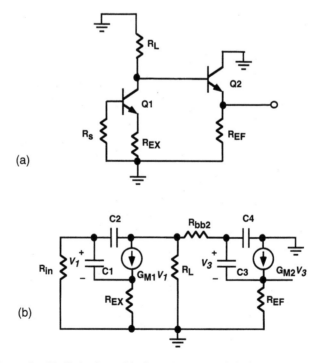

FIGURE 16.12   (a) Large-signal half-circuit model of ECL inverter; and (b) large-signal equivalent circuit of (a).

the half-circuit approximation has been used in Fig. 16.12(a) due to the inherent symmetry of differential circuits.[16] The hybrid-pi BJT model shown in Fig. 16.12(b) has been used with several simplifications. The dynamic input resistance, r_π, has been neglected because other circuit resistances are typically much smaller. The output resistance, r_o, has also been neglected for the same reason. The collector-to-substrate capacitance, C_{CS}, has been neglected because in III-V technologies, semi-insulating substrates are typically used. The capacitance to substrate is quite small compared to other device capacitances. Retained in the model are resistances R_{bb}, the extrinsic and intrinsic base resistance, and R_{EX}, the parasitic emitter resistance. Both of these are very critical for optimizing high-speed performance.

In the circuit itself, R_{IN} is the sum of the driving point resistance from the previous stage, probably an emitter follower output, and R_{bb1} of Q_1. R_L is the collector load resistor, whose value is determined by half of the output voltage swing and the dc emitter current, I_{CS}. $R_L = \Delta V / 2 I_{CS}$. The R_{EX} of the emitter follower is included in R_{EF}.

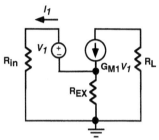

We must calculate open-circuit time constants for each of the four capacitors in the circuit. First consider C_1, the base-emitter diffusion and depletion capacitance of Q_1. C_2 is the collector-base depletion capacitance of Q_1. C_3 and C_4 are the corresponding base-emitter and base-collector capacitances of Q_2. Figure 16.13 represents the equivalent circuit schematic when $C_2 = C_3 = C_4 = 0$. A test source, V_1, is placed at the C_1 location. $R_{1o} = V_1 / I_1$ is determined by circuit analysis to be

FIGURE 16.13 Equivalent large-signal half-circuit model for calculation of R_{1o}.

$$R_{1o} = \frac{R_{IN} + R_{EX}}{1 + G_{M1} R_{EX}} \tag{16.24}$$

Table 16.1 shows the result of similar calculations for R_{2o}, R_{3o}, and R_{4o}. The b_1 coefficient (first-order estimate of t_D) can now be found from the sum of the OCTCs:

$$b_1 = R_{1o} C_1 + R_{2o} C_2 + R_{3o} C_3 + R_{4o} C_4 \tag{16.25}$$

Considering the results in Table 16.1, one can see that there are many contributors to the time constants and that it will be possible to determine the dominant terms after evaluating the model and circuit parameters.

Next, estimates must be made of the non-linear device parameters, G_{Mi} and C_i. The large signal transconductances can be estimated from

$$G_M = \frac{\Delta I_C}{\Delta V_{BE}} \tag{16.26}$$

For the half-circuit model of the differential pair, the current ΔI_C is the full value of I_{CS} since the device switches between cutoff and I_{CS}. The ΔV_{BE} corresponds to the input voltage swing needed to switch the device between cutoff and I_{CS}. This is on the order of $3V_T$ (or 75 mV) for half of a differential input. So, $G_{M1} = I_{CS}/0.075$ is the large-signal estimate for transconductance of Q_1.

The emitter follower Q_2 is biased at I_{EF} when the output is at V_{OL}. Let us assume that an identical increase in current, I_{EF}, will provide the logic swing needed on the output of the inverter to reach V_{OH}. Thus, $\Delta I_{C2} = I_{EF}$ and $R_{EF} = (V_{OH} - V_{OL})/I_{EF}$. The difference in V_{BE} at the input required to double the collector current can be calculated from

TABLE 16.1 Effective Zero-Frequency Resistances for Open-Circuit Time-Constant Calculation for the Circuit of Fig. 16.12 ($G'_{M1} = G_{M1}/(1 + G_{M1} R_{EX})$)

R_{1o}	$\dfrac{R_{IN} + R_{EX}}{1 + G_{M1} R_{EX}}$
R_{2o}	$R_{IN} + R_L + G'_{M1} R_{IN} R_L$
R_{3o}	$\dfrac{R_{bb} + R_L + R_{EF}}{1 + G_{M2} R_{EF}}$
R_{4o}	$R_{bb} + R_L$

$$\Delta V_{BE} = V_T \ln(2) = 0.7 V_T = 17.5 \, \text{mV} \tag{16.27}$$

Thus, $G_{M2} = I_{EF}/0.0175$.

C_1 and C_3 consist of the parallel combination of the *depletion (space charge) layer capacitance*, C_{be}, and the *diffusion capacitance*, C_D. C_2 and C_4 are the base-collector depletion capacitances. Depletion capacitances are voltage varying according to

$$C(V) = C(0) \left(1 - \frac{V}{\phi} \right)^{-m} \tag{16.28}$$

where $C(0)$ is the capacitance at zero bias, ϕ is the built-in voltage, and m the grading coefficient. An equivalent large-signal capacitance can be calculated by

$$C = \frac{Q_2 - Q_1}{V_2 - V_1} \tag{16.29}$$

Q_i is the charge at the initial (1) or final (2) state corresponding to the voltages V_i. $Q_2 - Q_1 = \Delta Q$ and

$$\Delta Q = \int_{V_1}^{V_2} C(V) dV \tag{16.30}$$

The large-signal diffusion capacitance can be found from

$$C_D = G_M \tau_f \tag{16.31}$$

where τ_f is the forward transit delay (τ_r) as defined in Section 15.2.

Finally, the Elmore risetime estimate requires the calculation of b_2. Since there are four capacitors in the large-signal equivalent circuit, six terms will be necessary:

$$b_2 = R_{1o} C_1 R_{2s}^1 C_2 + R_{1o} C_1 R_{3s}^1 C_3 + R_{1o} C_1 R_{4s}^1 C_4 + R_{2o} C_2 R_{3s}^2 C_3 + R_{2o} C_2 R_{4s}^2 C_4 + R_{3o} C_3 R_{4s}^3 C_4 \tag{16.32}$$

R_{2s}^1 will be calculated to illustrate the procedure. The remaining short-circuit equivalent resistances are shown in Table 16.2. Referring to Fig. 16.14, the equivalent circuit for calculation of R_{2s}^1 is shown. This is the resistance seen across C_2 when C_1 is shorted. If C_1 is shorted, $V_1 = 0$ and the dependent current source is dead. It can be seen from inspection that

$$R_{2s}^1 = R_{IN} \| R_{EX} + R_L \tag{16.33}$$

Time Constant Methods: Complications

As attractive as the time constant delay and risetime estimates are computationally, the user must beware of complications that will degrade the accuracy by a large margin. First, consider that both methods have depended on a restrictive assumption regarding monotonic risetime. In many cases, however, it is not unusual to experience complex poles. This can occur due to feedback which leads to inductive input or output impedances and emitter or source followers which also have inductive output impedance. When

TABLE 16.2 Effective Resistances for Short Circuit Time Constant Calculation for the Circuit of Fig. 16.12

R^1_{2s}	$R_{IN} \| R_{EX} + R_L$
R^1_{3s}	R_{3o}
R^1_{4s}	R_{4o}
R^2_{3s}	$\dfrac{\left(\dfrac{1}{G'_{M1}} \| R_{IN} \| R_L\right) + R_{bb} + R_{EF}}{1 + G_{M2}R_{EF}}$
R^2_{4s}	$\left(\dfrac{1}{G'_{M1}} \| R_{IN} \| R_L\right) + R_{bb}$
R^3_{4s}	$\left(R_L + R_{bb}\right) \| R_{EF}$

combined with a predominantly capacitive input impedance, complex poles will generally result unless the circuit is well damped. The time constant methods ignore the complex pole effects which can be quite significant if the poles are split and $\sigma \ll j\omega$. In this case, the circuit transient response will exhibit ringing, and time constant estimates of bandwidth, delay, and risetime will be in serious error. Of course, the ringing will show up in the circuit simulation, and if present, must be dealt with by adding damping resistances at appropriate locations.

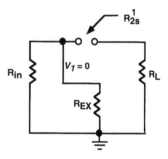

FIGURE 16.14 Equivalent circuit model for calculation of R_{2s}^1.

An additional caution must be given for circuits that include zeros. Although Elmore's equations can modify the estimates for t_D and T_R when there are zeros, the OCTC method provides no help in finding the time constants of these zeros. Zeros often occur in wideband amplifier circuits that have been modified through the addition of inductance for shunt peaking, for example. The addition of inductance, either intentionally or accidentally, can also produce complex pole pairs. Zeros are intentionally added for the optimization of speed in very-high-speed digital ICs as well; however, the large area required for the spiral inductors when compared with the area consumed by active devices tends to discourage the use of this method in all but the simplest (and fastest) designs.[11]

References

1. Yuan, J.-R. and Svensson, C., High-Speed CMOS Circuit Technique, *IEEE J. Solid-State Circuits*, 24, 62, 1989.
2. Weste, N. H. E. and Eshraghian, K., *Principles of CMOS VLSI Design – A Systems Perspective*, second ed., Addison-Wesley, Reading, MA, 1993.
3. Rabaey, J. M., *Digital Integrated Circuits: A Design Perspective*, Prentice-Hall, New York, 1996.
4. Hill, C. F., Noise Margin and Noise Immunity in Logic Circuits, *Microelectronics*, 1, 16, 1968.
5. Long, S. and Butner, S., *Gallium Arsenide Digital Integrated Circuit Design*, McGraw-Hill, New York, 1990, Chap. 3.
6. Bakoglu, H. B., *Circuits, Interconnections, and Packaging*, Addison-Wesley, Reading, MA, 1990, Chap. 7.
7. Long, S. and Butner, S., *Gallium Arsenide Digital Integrated Circuit Design*, McGraw-Hill, New York, 1990, Chap. 5.

8. Elmore, W. C., The Transient Response of Damped Linear Networks with Particular Regard to Wideband Amplifiers, *J. Appl Phys.*, 19, 55, 1948.

9. Ashar, K. G., The Method of Estimating Delay in Switching Circuits and the Fig. of Merit of a Switching Transistor, *IEEE Trans. Elect. Dev.*, ED-11, 497, 1964.

10. Tien, P. K., Propagation Delay in High Speed Silicon Bipolar and GaAs HBT Digital Circuits, *Int. J. High Speed Elect.*, 1, 101, 1990.

11. Lee, T. H., *The Design of CMOS Radio-Frequency Integrated Circuits*, Cambridge Univ. Press, Cambridge, U.K., 1998, Chap. 7.

12. Gray, P. E. and Searle, C. L., *Electronic Principles: Physics, Models, and Circuits*, John Wiley & Sons, New York, 1969, 531.

13. Gray, P. and Meyer, R., *Analysis and Design of Analog Integrated Circuits*, 3rd ed., John Wiley & Sons, New York, 1993, Chap. 7.

14. Millman, J. and Grabel, A., *Microelectronics*, second ed., McGraw-Hill, New York, 1987, 482.

15. Hurtz, G. M., Applications and Technology of the Transferred-Substrate Schottky-Collector Heterojunction Bipolar Transistor, M.S. Thesis, University of California, Santa Barbara, 1995.

16. Gray, P. and Meyer, R., *Analysis and Design of Analog Integrated Circuits*, 3rd ed., John Wiley & Sons, New York, 1993, Chap. 3.

17

Logic Design Examples

Charles E. Chang
Conexant Systems, Inc.

Meera Venkataraman
Troika Networks, Inc.

Stephen I. Long
*University of California
at Santa Barbara*

17.1 Design of MESFET and HEMT Logic Circuits

The basis of dc design, definition of logic levels, noise margin, and transfer characteristics were discussed in Chapter 16 using a DCFL and SCFL inverter as examples. In addition, methods for analysis of high-speed performance of logic circuits were presented. These techniques can be further applied to the design of GaAs MESFET, HEMT, or P-HEMT logic circuits with depletion-mode, enhancement-mode, or mixed E/D FETs. Several circuit topologies have been used for GaAs MESFETs, like direct-coupled FET logic (DCFL), source-coupled FET logic (SCFL), as well as dynamic logic families,[1] and have been extended for use with heterostructure FETs. Depending on the design requirements, whether it be high speed or low power, the designer can adjust the power-delay product by choosing the appropriate device technology and circuit topology, and making the correct design tradeoffs.

Direct-Coupled FET Logic (DCFL)

Among the numerous GaAs logic families, DCFL has emerged as the most popular logic family for high-complexity, low-power LSI/VLSI circuit applications. DCFL is a simple enhancement/depletion-mode GaAs logic family, and the circuit diagram of a DCFL inverter was shown in Fig. 16.2. DCFL is the only static ratioed GaAs logic family capable of VLSI densities due to its compactness and low power dissipation. An example demonstrating DCFL's density is Vitesse Semiconductor's 350K sea-of-gates array. The array uses a two-input DCFL NOR as the basic logic structure. The number of usable gates in the array is 175,000. A typical gate delay is specified at 95 ps with a power dissipation of 0.59 mW for a buffered two-input NOR gate with a fan-out of three, driving a wire load of 0.51 mm.[2] However, a drawback of DCFL is its low noise margin, the logic swing being approximately 600 mV. This makes the logic sensitive to changes in threshold voltage and ground bus voltage shifts.

DCFL NOR and NAND Gate

The DCFL inverter can easily be modified to perform the NOR function by placing additional enhancement-mode MESFETs in parallel as switch devices. A DCFL two-input NOR gate is shown in Fig. 17.1. If any input rises to V_{OH}, the output will drop to V_{OL}. If n inputs are high simultaneously, then V_{OL} will be decreased because the width ratio W_1/W_L in Fig. 16.2 has effectively increased by a factor of n. There is a limit to the number of devices that can be placed in parallel to form very wide NOR functions. The

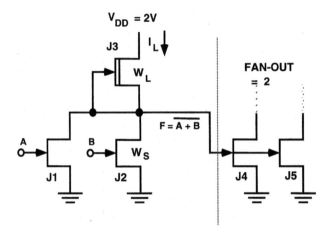

FIGURE 17.1   DCFL two-input NOR gate schematic.

drain capacitance will increase in proportion to the number of inputs, slowing down the risetime of the gate output. Also, the subthreshold current contribution from n parallel devices could become large enough to degrade V_{OH}, and therefore the noise margin. This must be evaluated at the highest operating temperature anticipated because the subthreshold current will increase exponentially with temperature according to:[3,4]

$$I_D = I_S \left[1 - \exp\left(\frac{cV_{DS}}{V_T} \right) \right] \left[\exp\left(\frac{bV_{DS}}{V_T} \right) \right] \left[\exp\left(\frac{aV_{GS}}{V_T} \right) \right] \quad (17.1)$$

The parameters a, b, and c are empirical fitting parameters. The first term arises from the diffusion component of the drain current which can be fit from the subthreshold I_D–V_{DS} characteristic at low drain voltage. The second and third terms represent thermionic emission of electrons over the channel barrier from source to drain. The parameters can be obtained by fitting the subthreshold I_D–V_{DS} and I_D–V_{GS} characteristics, respectively, measured in saturation.[5] For the reasons described above, the fan-in of the DCFL NOR is seldom greater than 4.

In addition to the subthreshold current loading, the forward voltage of the Schottky gate diode of the next stage drops with temperature at the rate of approximately –2mV/degree. Higher temperature operation will therefore reduce V_{OH} as well, due to this thermodynamic effect.

A NAND function can also be generated by placing enhancement-mode MESFETs in series rather than in parallel for the switch function. However, the low voltage swing inherent in DCFL greatly limits the application of the NAND function because V_{OL} will be increased by the second series transistor unless the widths of the series devices are increased substantially from the inverter prototype. Also, the switching threshold V_{TH} shown in Fig. 16.1 will be slightly different for each input even if width ratios are made different for the two inputs. The combination of these effects reduces the noise margin even further, making the DCFL NAND implementation generally unsuitable for VLSI applications.

Buffering DCFL Outputs

The output (drain) node of a DCFL gate sources and sinks the current required to charge and discharge the load capacitance due to wiring and fan-out. Excess propagation delay of the order of 5 ps per fan-out is typically observed for small DCFL gates. Sensitivity to wiring capacitance is even higher, such that unbuffered DCFL gates are never used to drive long interconnections unless speed is unimportant. Therefore, an output buffer is frequently used in such cases or when fan-out loading is unusually high.

The superbuffer shown in Fig. 17.2(a) is often used to improve the drive capability of DCFL. It consists of a source follower J_3 and pull-down J_4. The low-to-high transition begins when $V_{IN} = V_{OL}$. J_4 is cut off

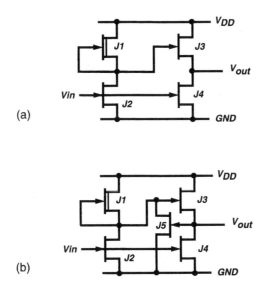

FIGURE 17.2 (a). Superbuffer schematic, and (b) modified superbuffer with clamp transistor. J5 will limit the output current when $V_{out} > 0.7$ V.

and J_3 becomes active, driving the output to V_{OH}. V_{OUT} follows the DCFL inverter output. For the output high-to-low transition, J_4 is driven into its linear region, and the output can be pulled to $V_{OL} = 0$ V in steady state. J_3 is cut off when the DCFL output (drain of J_1) switches from high to low. Since this occurs one propagation delay after the input switched from low-to-high, it is during this transition that the superbuffer can produce a current spike between V_{DD} and ground. J_4 attempts to discharge the load capacitance before the DCFL gate output has cut off J_3. Thus, superbuffers can become an on-chip noise source, so ground bus resistance and inductance must be controlled.

There is also a risk that the next stage might be overdriven with too much input current when driven by a superbuffer. This could happen because the source follower output is capable of delivering high currents when its V_{GS} is maximum. This occurs when $V_{out} = V_{OH} = 0.7$ V, limited by forward conduction of the gate diodes being driven. For a supply voltage of 2 V, a maximum $V_{GS} = 0.7$ V is easily obtained on J_3, leading to the possibility of excess static current flowing into the gates. This would degrade V_{OL} of the subsequent stage due to voltage drop across the internal source resistance. Figure 17.2(b) shows a modified superbuffer design that prevents this problem through the addition of a clamp transistor, J_5. J_5 limits the gate potential of J_3 when the output reaches V_{OH}, thus preventing the overdriving problem.

Source-Coupled FET Logic (SCFL)

SCFL is the preferred choice for very-high-speed applications. An SCFL inverter, a buffered version of the basic differential amplifier cell shown in Fig. 16.4, is shown in Fig. 17.3. The high-speed capability of SCFL stems from four properties of this logic family: small input capacitance, fast discharging time of the differential stage output nodes, good drive capability, and high F_t.

In addition to higher speed, SCFL is characterized by high functional equivalence and reduced sensitivity to threshold voltage variations.[5a] The current-mode approach used in SCFL ensures an almost constant current consumption from the power supplies and, therefore, the power supply noise is greatly reduced as compared to other logic families. The differential input signaling also improves the dc, ac, and transient characteristics of SCFL circuits.[6]

SCFL, however, has two drawbacks. First, SCFL is a low-density logic family due to the complex gate topology. Second, SCFL dissipates more power than DCFL, even with the high functional equivalence taken into account.

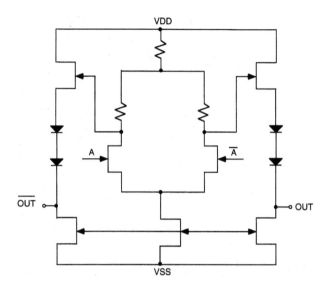

FIGURE 17.3   Schematic diagram of SCFL inverter with source follower output buffering.

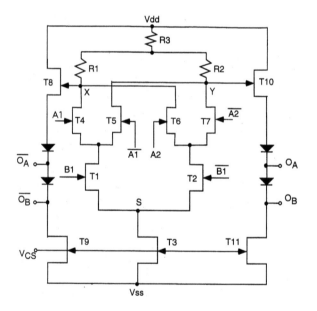

FIGURE 17.4   SCFL two-level series-gated circuit.

SCFL Two-Level Series-Gated Circuit

A circuit diagram of a two-level series-gated SCFL structure is shown in Fig. 17.4. 2-to-1 MUXs, XOR gates, and D-latches and flip-flops can be configured using this basic structure. If the A inputs are tied to the data signals and the B inputs are tied to the select signal, the resulting circuit is a 2-to-1 MUX. If the data are fed to the A1 input, the clock is connected to B and the A outputs (O_A, $\overline{O}_A$) are fed back to the A2 inputs. The resulting circuit is a D-latch as seen in Fig. 17.5. Finally, an XOR gate is created by connecting $A_1 = \overline{A}_2$, forming a new input A_{IN} and $\overline{A}_1 = A_2$ to complementary new input $\overline{A}_{IN}$.

The inputs to the two levels require different dc offsets in order for the circuit to function correctly; thus, level-shifting networks using diodes or source followers are required. Series logic such as this also requires higher supply voltages in order to keep the devices in their saturation region. This will increase the power dissipation of SCFL.

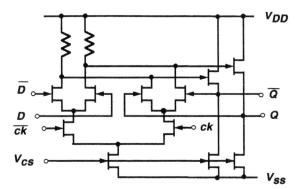

FIGURE 17.5   SCFL D latch schematic. Two cascaded latch cells with opposite clock phasing constitute a master-slave flip-flop.

The logic swing of the circuit shown in Fig. 17.4 is determined by the size of the current source T3 and the load resistors R1 and R2 (R1 = R2). Assuming T3 is in saturation, the logic swing on nodes X and Y is

$$\Delta V_{X,Y} = I_{ds3} \, R1 = I_{dss3} \, R1 \qquad (17.2)$$

where I_{dss3} is the saturation current of T3 at Vgs = 0 V. The logic high and low levels on node X ($V_{X,H}$, $V_{X,L}$) are determined from the voltage drop across R3 and Eq. 17.2.

$$V_{X,H} = V_{dd} - (I_{dss3} \, R3) \qquad (17.3)$$

$$V_{X,L} = V_{dd} - [I_{dss3} \, (R1 + R3)] \qquad (17.4)$$

The noise margin is the difference between the minimum voltage swing required on the inputs to switch the current from one branch to the other (V_{SW}) and the logic swing $\Delta V_{X,Y}$. V_{SW} is set by the ratio between the sizes of the switch transistors (T1,T2, T4-T7) and T3. For symmetry reasons, the sizes of all the switch transistors are kept the same size. Assuming the saturation drain-source current of an FET can be described by the simplified square-law equation:

$$I_{ds} = \beta W (V_{gs} - V_T)^2 \qquad (17.5)$$

where V_T is the threshold voltage, W is the FET width, and β is a process-dependent parameter. V_{SW} is calculated assuming all the current from T3 flows only through T2.

$$V_{SW} = |V_T| \sqrt{(W3/W2)} \qquad (17.6)$$

For a fixed current source size (W3), the larger the size of the switch transistors, the smaller the voltage swing required to switch the current and, hence, a larger noise margin. Although a better noise margin is desirable, it needs to be noted that the larger switch transistors means increased input capacitance and decreased speed. Depending on the design specifications, noise margin and speed need to be traded off.

Since all FETs need to be kept in the saturation region for the correct operation of an SCFL gate, level-shifting is needed between nodes A and B and the input to the next gate, in order to keep T1, T2, and T4-T7 saturated. T3 is kept in saturation if the potential at node S is higher than $V_{SS} + V_{ds,sat}$. The potential at node S is determined by the input voltages to T1 and T2. V_S settles at a potential such that the drain-source current of the conducting transistor is exactly equal to the bias current, I_{dss3}, since no current flows through the other transistor. The minimum logic high level at the output node B ($V_{OB,H}$) is

$$V_{OB,H} \geq V_{ss} + V_{ds,sat} + V_{gs} = V_{ss} + V_{ds,sat} + V_{SW} + V_{th} \tag{17.7}$$

To keep T9 and T11 in saturation, however, requires that

$$V_{OB,H} \geq V_{ss} + V_{ds,sat} + V_{SW} \tag{17.8}$$

As with the voltage on node S, the drain voltages of T1 and T2 are determined by the voltage applied to the A inputs. The saturation condition for T1 and T2 is

$$V_{OA,H} - V_{SW} - V_{th} - V_S = V_{OA,H} - V_{OB,H} \geq V_{ds,sat} \tag{17.9}$$

Equation 17.9 shows that the lower switch transistors are kept in saturation if the level-shifting difference between the A and B outputs is larger than the FET saturation voltage. Since diodes are used for level-shifting, the minimum difference between the two outputs is one diode voltage drop, V_D. If $V_{ds,sat} > V_D$, more diodes are required between the A and B outputs.

The saturation condition for the upper switch transistors, T4 to T7, is determined by the minimum voltage at nodes A and B and the drain voltage of T1 and T2.

$$V_{A,min} - (V_{OA,H} - V_{SW} - V_{th}) \geq V_{ds,sat} \tag{17.10}$$

Substituting Eq. 17.4 into Eq. 17.10 yields

$$\left(V_{dd} - I_{dss3} * (R1 + R3)\right) - \left(V_{OA,H} - V_{SW} - V_{th}\right) \geq V_{ds,sat} \tag{17.11}$$

Rewriting Eq. 17.11 using Eq. 17.8 gives the minimum power supply range

$$V_{dd} - V_{ss} \geq I_{dss3} * (R1 + R3) + 3V_{ds,sat} \tag{17.12}$$

Equation 17.11 allows the determination of the minimum amount of level-shifting required between nodes A and B to the outputs

$$V_{A,H} - V_{OA,H} \geq V_{ds,sat} + I_{dss3} * R1 - V_{th} - V_{SW} \tag{17.13}$$

Equations 17.8 to 17.13 can be used for designing the level shifters. The design parameters available in the level-shifters are the widths of the source followers (W8, W10), the current sources (W9, W11), and the diode (W_D). Assuming the current source width (W9) is fixed, the voltage drop across the diodes is partially determined by the ratio (W_D/W9). This ratio should not be made too small. Operating Schottky diodes at high current density will result in higher voltage drop, but this voltage will be partially due to the $I_D R_S$ drop across the parasitic series resistance. Since this resistance is often process dependent and difficult to reproduce, poor reproducibility of V_D will result in this case.

The ratio between the widths of the source follower and the current source (W8/W9) determines the gate-source voltage of the source follower (V_{gs8}). V_{gs8} should be kept below 0.5 V to prevent gate-source conduction.

The dc design of the two-level series-gated SCFL gate in Fig. 17.4 can be accomplished by applying Eqs. 17.2 to 17.13. Ratios between most device sizes can be determined by choosing the required noise margin and logic swing. Only W3 in the differential stage and W9 among the level-shifters are unknown at this stage. All other device sizes can be expressed in terms of these two transistor widths.

The relation between W3 and W9 can be determined only by considering transient behavior. For a given total power dissipation, the ratio between the power dissipated in the differential stage and the output buffers determines how fast the outputs are switched. If fast switching at the outputs is desired, more power needs to be allocated to the output buffers and, consequently, less power to the differential

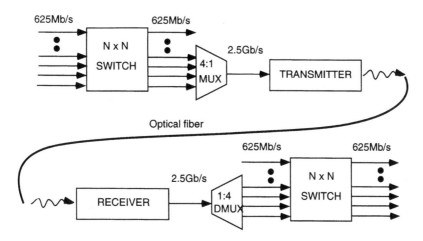

FIGURE 17.6 2.5-Gb/s optical communication system.

stage. While this allocation will ensure faster switching at the output, the switching speed of the differential stage is reduced because of the reduced current available to charge and discharge the large input capacitance of the output buffers.

Finally, it is useful to note that scaling devices to make a speed/power tradeoff is simple in SCFL. If twice as much power is allocated to a gate, all transistors and diodes are made twice as wide while all resistors are reduced by half.[6]

Advanced MESFET/HEMT Design Examples

High-Speed TDM Applications

The need for high bandwidth transmission systems continues to increase as the number of bandwidth-intensive applications in the areas of video imaging, multimedia, and data communication (such as database sharing and database warehousing) continues to grow. This has led to the development of optical communication systems with transmission bit rates, for example, of 2.5 Gb/s and 10 Gb/s. A simplified schematic of a 2.5 Gb/s communication system is shown in Fig. 17.6.

As seen in Fig. 17.6, MUXs, DMUXs, and switches capable of operating in the Gb/s range are crucial for the operation of these systems. GaAs MESFET technology has been employed extensively in the design of these high-speed circuits because of the excellent intrinsic speed performance of GaAs. SCFL is especially well suited for these circuits where high speed is of utmost importance and power dissipation is not a critical factor.

The design strategies employed in the previous subsection can now be further applied to a high-speed 4:1 MUX, as shown in Fig. 17.7. It was shown that the two-level series gated SCFL structure could be easily configured into a D-latch. The MSFF in the figure is simply a master-slave flip-flop containing two D-latches. The PSFF is a phase-shifting flip-flop that contains three D-latches and has a phase shift of 180° compared with an MSFF.

The 4:1 MUX is constructed using a tree-architecture in which two 2:1 MUXs merge two input lines each into one output operating at twice the input bit rate. The 2:1 MUX at the second stage takes the two outputs of the first stage and merges it into a single output at four times the primary input bit rate. The architecture is highly pipelined, ensuring good timing at all points in the circuit. The inherent propagation delay of the flip-flops ensures that the signals are passed through the selector only when they are stable.[6]

The interface between the two stages of 2:1 MUXs is timing-critical, and care needs to be taken to obtain the best possible phase margin at the input of the last flip-flop. To accomplish this, a delay is added between the CLK signal and the clock input to this flip-flop. The delay is usually implemented

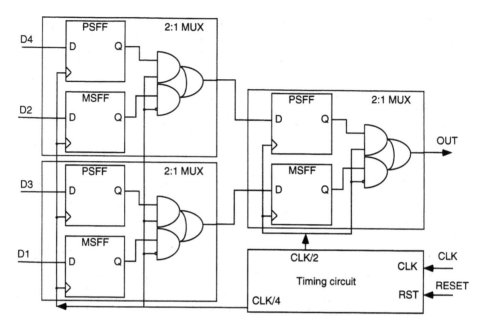

FIGURE 17.7   High-speed 4:1 multiplexer (MUX).

using logic gates because their delays are well characterized in a given process. Output jitter can be minimized if 50% duty-cycle clock signals are used. Otherwise, a retiming MSFF will be needed at the output of the 4:1 MUX.

The 4:1 MUX is a good example of an application of GaAs MESFETs with very-high-speed operation and low levels of integration. Vitesse Semiconductor has several standard products operating at the Gb/s range fabricated in GaAs using their own proprietary E/D MESFET process. For example, the 16 × 16 crosspoint switch, VSC880, has serial data rates of 2.0 Gb/s. The VS8004 4-bit MUX is a high-speed, parallel-to-serial data converter. The parallel inputs accept data at rates up to 625 Mb/s and the differential serial data output presents the data sequentially at 2.5 Gb/s, synchronous with the differential high-speed clock input.[2]

While the MESFET technologies have proven capable at 2.5 and 10 Gb/s data rates for optical fiber communication applications, higher speeds appear to require heterojunction technologies. The 40-Gb/s TDM application is the next step, but it is challenging for all present semiconductor device IC technologies. A complete 40-Gb/s system has been implemented in the laboratory with 0.1-μm InAlAs/InGaAs/InP HEMT ICs as reported in Refs. 7 and 8. Chips were fabricated that implemented multiplexers, photodiode preamplifiers, wideband dc 47-GHz amplifiers, decision circuits, demultiplexers, frequency dividers, and limiting amplifiers. The high-speed static dividers used the super-dynamic FF approach.[9]

A 0.2-μm AlGaAs/GaAs/AlGaAs HEMT quantum well device technology has also demonstrated 40-Gb/s TDM system components. A single chip has been reported that included clock recovery, data decision, and a 2:4 demultiplexer circuit.[10] The SCFL circuit approach was employed.

Very-High-Speed Dynamic Circuits

Conventional logic circuits using static DCFL or SCFL NOR gates such as those described above are limited in their maximum speed by loaded gate delays and serial propagation delays. For example, a typical DCFL NOR-implemented edge-triggered DFF has a maximum clock frequency of approximately $1/5\tau_D$ and the SCFL MSFF is faster, but it is still limited to $1/2\tau_D$ at best. Frequency divider applications that require clock frequencies above 40 GHz have occasionally employed alternative circuit approaches which are not limited in the same sense by gate delays and often use dynamic charge storage on gate

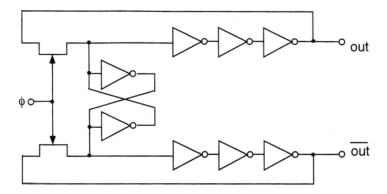

FIGURE 17.8   Dynamic frequency divider (DFD) divide-by-2 circuit. (Ref. 11, ©1989 IEEE, with permission.)

nodes for temporarily holding a logic state. These approaches have been limited to relatively simple circuit functions such as divide-by-2 or -4.

The dynamic frequency divider (DFD) technique is one of the well-known methods for increasing clock frequency closer to the limits of a device technology. For an example, Fig. 17.8 shows a DFD circuit using a single-phase clock, a cross-coupled inverter pair as a latch to reduce the minimum clock frequency, and pass transistors to gate a short chain of inverters.[11,12] These have generally used DCFL or DCFL superbuffers for the inverters. The cross-coupled inverter pair can be made small in width, since its serial delay is not in the datapath. But the series inverter chain must be designed to be very fast, generally requiring high power per inverter in order to push the power-delay product to its extreme high-speed end. Since fan-out is low, the intrinsic delays of an inverter in a given technology can be approached.

The maximum and minimum clock frequencies of this circuit can be calculated from the gate delays of the n series inverters as shown in Eqs. 17.14 and 17.15. An odd number n is required to force an inversion of the data so that the circuit will divide-by-2. Here, t_1 is the propagation delay of the pass transistor switches, J1 and J2, and t_D is the propagation delay of the DCFL inverters. The parameter "a" is the duty cycle of the clock. For a 50% clock duty cycle, the range of minimum to maximum clock frequency is about 2 to 1.

$$f_{\phi max} = \frac{1}{t_1 + nt_D} \tag{17.14}$$

$$f_{\phi min} = \frac{a}{t_1 + nt_D} \tag{17.15}$$

Clock frequencies as high as 51 GHz have been reported using this approach with a GaAs/AlGaAs P-HEMT technology.[12] The power dissipation was relatively high, 440 mW. Other DFD circuit approaches can also be found in the literature.[13-15]

A completely different approach, as shown in Fig. 17.9, utilizes an injection-locked push-pull oscillator (J1 and J2) whose free running frequency is a subharmonic of the desired input frequency.[16] FETs J3 and J4 are operating in their ohmic regions and act as variable resistors. The variation in V_{GS1} and V_{GS2} cause the oscillator to subharmonically injection-lock to the input source. Here, a divide-by-4 ratio was demonstrated with an input frequency of 75 GHz and a power dissipation of 170 mW using a 0.1-μm InP-based HEMT technology with f_T = 140 GHz and f_{max} = 240 GHz. This divider also operated in the 59–64 GHz range with only –10 dBm RF input power. The frequency range is limited by the tuning range of the oscillator. In this example, about 2 octaves of frequency variation was demonstrated.

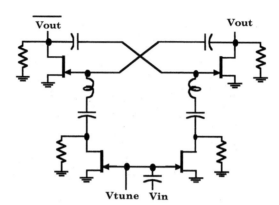

FIGURE 17.9   Injection-locked oscillator divide-by-4. (Ref. 16, ©1996 IEEE. With permission.)

Finally, efforts have also been made to beat the speed limitations of a technology by dynamic design methods while still maintaining minimum power dissipation. The quasi-dynamic FF[8,9] and quasi-differential FF[17] are examples of circuit designs emphasizing this objective. The latter has achieved 16-GHz clock frequency with approximately 2 mW of power per FF.

17.2 HBT Logic Design Examples

From a circuit topology perspective, both III-V HBTs and silicon BJTs are interchangeable, with myriad similarities and a few essential differences. The traditional logic families developed for the Si BJTs serve as the starting point for high-speed logic with III-V HBTs. During the period of intense HBT development in the 1980s and early 1990s, HBTs have implemented ECL, CML, DTL, and, I[2]L logic topologies as well as novel logic families with advanced quantum devices (such as resonant tunneling diodes[17a]) in the hopes of achieving any combination of high-speed, low-power, and high-integration level. During that time, III-V HBTs demonstrated their potential integration limits with an I[2]L 32-bit microprocessor[18] and benchmarked its high-speed ability with an ECL 30-GHz static master/slave flip-flop based frequency divider. During the same time, advances in Si based technology, especially CMOS, have demonstrated that parallel circuit algorithms implemented in a technology with slower low-power devices capable of massive integration will dominate most applications. Consequently, III-V-based technologies such as HBTs and MESFET/HEMT have been relegated to smaller but lucrative niche markets.

As HBT technology evolved into a mature production technology in the mid-1990s, it was clear that III-V HBT technology had a clear advantage in high-speed digital circuits, microwave integrated circuits, and power amplifier markets. Today, in the high-speed digital arena, III-V HBTs have found success in telecom and datacom lightwave communication circuits for SONET/ATM-based links that operate from 2.5 to 40 Gb/s. HBTs also dominate the high-speed data conversion area with Nyquist-rate ADCs capable of gigabit/gigahertz sampling rates/bandwidths, sigma-delta ADCs with very high oversampling rates, and direct digital synthesizers with gigahertz clock frequencies and ultra-low spurious outputs. In these applications, the primary requirement is ultra-high-speed performance with LSI (10 K transistors) levels of integration. Today, the dominant logic type used in HBT designs is based on non-saturating emitter coupled pairs such as ECL and current-mode logic (CML), which is the focus of this chapter.

III-V HBT for Circuit Designers

III-V HBTs and Si BJTs are inherently bipolar in nature. Thus, from a circuit point of view, both share many striking similarities and some important differences. The key differences between III-V HBT technology and Si BJT technology, as discussed below, can be traced to three essential aspects: (1) heterojunction vs. homojunction, (2) III-V material properties, and (3) substrate properties.

First, the primary advantage of a base–emitter heterojunction is that the wide bandgap emitter allows the base to be doped higher than the emitter (typically 10 to 50X in GaAs/AlGaAs HBTs) without a reduction in current gain. This translates to lower base resistance for improved f_{max} and reduces base width modulation with V_{ce} for low output conductance. Alternatively, the base can be made thinner for lower base-transit time (τ_b) and higher f_t without having R_b too high. If the base composition is also graded from high bandgap to low, an electric field can be established to sweep electrons across the base for reduced τ_b and higher f_t. With a heterojunction B-E and a homojunction B-C, the junction turn-on voltage is higher in the B-E than it is in the B-C. This results in a common-emitter I–V curve offset from the off to saturation transition. This offset is approximately 200 mV in GaAs/AlGaAs HBTs. With a highly doped base, base punch-through is not typically observed in HBTs and does not limit the f_t-breakdown voltage product as in high-performance Si BJTs and SiGe HBT with thin bases. Furthermore, if a heterojunction is placed in the base–collector junction, a larger bandgap material in the collector can increase the breakdown voltage of the device and reduce the I–V offset.

Second, III-V semiconductors typically offer higher electron mobility than Si for overall lower τ_b and collector space charge layer transit times (τ_{cscl}). Furthermore, many III-V materials exhibit velocity overshoot in the carrier drift velocity. When HBTs are designed to exploit this effect, significant reductions in τ_{cscl} can result. With short collectors, the higher electron mobility can result in ultra-high f_t; however, this can also be used to form longer collectors with still acceptable τ_{cscl}, but significantly reduced C_{bc} for high f_{max}. The higher mobility in the collector can also lead to HBTs with lower turn on resistance in the common emitter I–V curves.

Since GaAs/AlGaAs and GaAs/InGaP have wider bandgaps than Si, the turn-on voltage of the B–E ($V_{be,on}$) junction is typically on the order of 1.4 V vs. 0.9 V for advanced high-speed Si BJT. InP-based HBTs can have $V_{be,on}$ on the order of 0.7 V; however, most mature production technologies capable of LSI integration levels are based on AlGaAs/GaAs or InGaP/GaAs. The base–collector turn-on voltage is typically on the order of 1 V in GaAs-based HBTs. This allows V_{ce} to be about 600 mV lower than V_{be} without placing the device in saturation. The wide bandgap material typically results in higher breakdown voltages, so III-V HBTs typically have a high Johnson figure of merit (f_t * breakdown voltage) compared with Si- and SiGe-based bipolar transistors.

The other key material differences between III-V vs. silicon materials are the lack of a native stable oxide in III-V, the extensive use of poly-Si in silicon-based processes, and the heavy use of implants and diffusion for doping silicon devices. III-V HBTs typically use epitaxial growth techniques, and interconnect step height coverage issues limit the practical structure to one device type, so PNP transistors are not typically included in an HBT process. These key factors contribute to the differences between HBTs and BJTs in terms of fabrication.

Third, the GaAs substrate used in III-V HBTs is semi-insulating, which minimizes parasitic capacitance to ground through the substrate, unlike the resistive silicon substrate. Therefore, the substrate contact as in Si BJTs is unnecessary with III-V HBTs. In fact, the RF performance of small III-V HBT devices can be measured directly on-wafer without significant de-embedding of the probe pads below 26 GHz. For interconnects, the line capacitance is typically dominated by parallel wire-to-wire capacitance, and the loss is not limited by the resistive substrate. This allows for the formation of high-Q inductors, low-loss transmission lines, and longer interconnects that can be operated in the 10's of GHz. Although BESOI and SIMOX Si wafers are insulating, the SiO_2 layer is typically thin resulting in reduced but still significant capacitive coupling across this thin layer.[19]

Most III-V substrates have a lower thermal conductivity than bulk Si, resulting in observed self-heating effects. For a GaAs/AlGaAs HBT, this results in observed negative output conductance in the common-emitter I–V curve measured with constant I_b. The thermal time constant for GaAs/AlGaAs HBTs is on the order of microseconds. Since thermal effects cannot track above this frequency, the output conductance

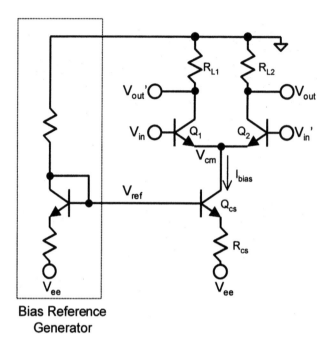

Bias Reference
Generator

FIGURE 17.10   Standard differential CML buffer with a simple reference generator.

of HBTs at RF (> 10 MHz) is low but positive. This effect does result in a small complication for HBT models based on the standard Gummel Poon BJT model.

Current-Mode Logic

The basic current-mode logic (CML) buffer/inverter cell is shown in Fig. 17.10. The CML buffer is a differential amplifier that is operated with its outputs clipped or in saturation. The differential inputs $(V_{in}$ and $V_{in}')$ are applied to the bases of Q_1 and Q_2. The difference in potential between V_{in} and V_{in}' determines which transistor I_{bias} is steered through, resulting in a voltage drop across either load resistance R_{L1} or R_{L2}. If $V_{in} = V_{OH}$ and $V_{in}' = V_{OL}$ $(V_{in,High} > V_{in,Low})$, Q_1 is on and Q_2 is off. Consequently, I_{bias} completely flows through R_{L1}, causing V_{out}' to drop for a logic low. With Q_2 off, V_{out} floats to ground for a logic high. If the terminal assignment of V_{out} and V_{out}' were reversed, this CML stage would be an inverter instead of a buffer.

The logic high V_{OH} of a CML gate is 0 V. The logic low output is determined by $V_{OL} = -R_{L1}I_{bias}$. With $R_{L1}/R_{L2} = 200$ Ωs, and $I_{bias} = 2$ mA, the traditional logic low of a CML gate is −400 mV. As CML gates are cascaded together, the outputs of one stage directly feed the inputs of another CML gate. As a result, the base-collector of the "on" transistor is slightly forward-biased (by 400 mV in this example). For high-speed operation, it is necessary to keep the switching transistors out of saturation. With a GaAs base-collector turn-on voltage near 1V, 500 to 600 mV forward-bias is typically tolerated without any saturation effects. In fact, this bias shortens the base-collector depletion region, resulting in the highest f_t vs. V_{ce} (f_{max} suffers due to increase in C_{bc}). As a result, maximum logic swing of a CML gate is constrained by the need to keep the transistors out of saturation. As the transistor is turned on, the logic high is actively pulled to a logic low; however, as the transistor is turned off, the logic low is pulled up by a RC time constant. With a large capacitive loading, it is possible that the risetime is slower than the falltime, and that may result in some complications with high-speed data.

A current mirror (Q_{cs} and R_{cs}) sets the bias (I_{bias}) of the differential pair. This is an essential parameter in determining the performance of CML logic. In HBTs, the f_t dependence on I_c is as follows:

$$1/2\pi f_t = \tau_{ec} = nkT \Big/ \Big(qI_c \big(C_{bej} + C_{bc} + C_{bed}\big)\Big) + R_c \big(C_{bej} + C_{bc}\big) + \tau_b + \tau_{cscl} \qquad (17.16)$$

where τ_{ec} is the total emitter-to-collector transit time, qI_c/nkT is the transconductance (g_m), C_{bc} is the base-collector capacitance, C_{bej} is the base–emitter junction capacitance, C_{bed} is the B–E diffusion capacitance, R_c is the collector resistance, τ_b is the base transit time, and τ_{cscl} is the collector space charge layer transit time. At low currents, the transit time is dominated by the device g_m and device capacitance. As the bias increases, τ_{ec} is eventually limited by τ_b and τ_{cscl}. As this limit approaches, Kirk effect typically starts to increase τ_b/τ_{cscl}, which decreases f_t. In some HBTs, the peak f_{max} occurs a bit after the peak f_t. With this in mind, optimal performance is typically achieved when I_{bias} is near $I_{c,maxft}$ or $I_{c,maxfmax}$. In some HBT technologies, the maximum bias may be constrained by thermal or reliability concerns. As a rule of thumb, the maximum current density of HBTs is typically on the order of 5×10^4 A/cm^2.

The bias of CML and ECL logic is typically set with a bias reference generator, where the simplest generator is shown in Fig. 17.10. Much effort has been invested in the design of the reference generator to maintain constant bias with power supply and temperature variation. In HBT, secondary effects of heterojunction design typically result in slightly varying ideality factor with bias. This makes the design of bandgap reference circuits quite difficult in most HBT technologies, which complicates the design of the reference generator. Nevertheless, the reference generators used today typically result in a 2% variation in bias current from –40 to 100 C with a 10% variation in power supply. In most applications, the voltage drop across R_{cs} is set to around 400 mV. With V_{ee} set at –5.2 V, V_{ref} is typically near –3.4 V. With constant bias, as V_{ee} moves by $\pm 10\%$, then the voltage drop across R_{cs} remains constant, so V_{ref} moves by the change in power supply (about ± 0.5 V). Since the logic levels are referenced to ground, the average value of V_{cm} (around –1.4 V) remains constant. This implies that changes in the power supply are absorbed by the base–collector junction of Q_{cs}, and it is important that this transistor is not deeply saturated.

Since the device goes from the cutoff mode to the forward active mode as it switches, the gate delay is difficult to predict analytically with small-signal analysis. Thus, large-signal models are typically used to numerically compute the delay in order to optimize the performance of a CML gate. Nevertheless, the small-signal model, frequency-domain approach described in Chapter 16.3 (Elmore) leads to the following approximation of a CML delay gate with unity fan-out:[19a]

$$\tau_{d,cml} = \big(1 + g_m R_L\big) R_b C_{bc} + R_b \big(C_{be} + C_d\big) + \big(2C_{bc} + 1/2 C_{be} + 1/2 C_d\big)\Big/ g_m \qquad (17.17)$$

where C_d is the diffusion capacitance of $g_m(\tau_b + \tau_{cscl})$. Furthermore, by considering the difference in charge storage at logic high and logic low, divided by the logic swing, the effective CML gate capacitance can be expressed[19b] as

$$C_{cml} = C_{be}/2 + 2C_{bc} + C_s + \big(\tau_b + \tau_{cscl}\big)\Big/ R_L \qquad (17.18)$$

where C_s is collector-substrate and interconnect capacitances. Both equations show that the load resistor and bias (which affects g_m and device capacitors) have a strong effect on performance.

For a rough estimate of the CML maximum speed without loading, one can assume that f_t is the gain bandwidth product. With the voltage gain set at $g_m R_L$, the maximum speed is $f_t/(g_m R_L)$. In the above example, at 1 mA average bias, $g_{m,int} = 1/26$ S at room temperature. Assuming the internal parasitic emitter resistance R_E is 10 ohms and using the fact that $g_{m,ext} = g_{m,int}/(1 + R_E g_{m,int})$, the effective extrinsic g_m is 1/36 mhos. With a 200-ohm load resistor, the voltage gain is approximately 5.5. With a 70-GHz f_t HBT process, the maximum switching rate is about 13 GHz. Although this estimate is quite rough, it does show that high-speed CML logic desires high device bias and low logic swing. In most differential circuits, only 3 to 4 kT/q is needed to switch the transistors and overcome the noise floor. With such low levels

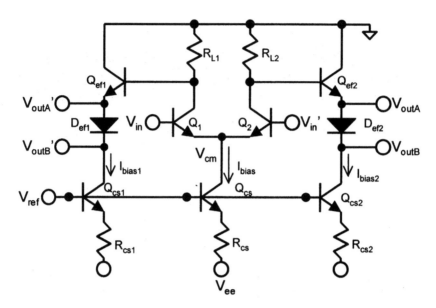

FIGURE 17.11 Standard differential ECL buffer with outputs taken at two different voltage levels.

and limited gain, the output may not saturate to the logic extremes, resulting in decreasing noise margin. In practice, the differential logic level should not be allowed to drop below 225 mV. With a 225-mV swing vs. 400 mV, the maximum gate bandwidth improves to 23 GHz from 13 GHz.

Emitter-Coupled Logic

By adding emitter followers to the basic HBT CML buffer, the HBT ECL buffer is formed in Fig. 17.11. From a dc perspective, the emitter followers (Q_{ef1} and Q_{ef2}) shift the CML logic level down by V_{be}. With the outputs at V_{outA}/V_{outA}' and 400 mV swing from the differential pair, the first level ECL logic high is –1.4 V and the ECL logic low is –1.8 V. For some ECL logic gates, a second level is created through a Schottky diode voltage shift (D_{ef1}/D_{ef2}). The typical Schottky diode turn-on voltage for GaAs is 0.7 V, so the output at V_{outB}/V_{outB}' is –2.1 V for a logic high and –2.5 V for a logic low. In general, the HBT ECL levels differ quite a bit from the standard Si ECL levels. Although resistors can be used to bias Q_{ef1}/Q_{ef2}, current mirrors (Q_{cs1}/Q_{cs2} and R_{cs1}/R_{cs2}) are typically used. Current mirrors offer stable bias with logic level at the expense of higher capacitance, while resistors offer lower capacitance but the bias varies more and may be physically quite large.

From an ac point of view, emitter followers have high input impedance (approximately β times larger than an unbuffered input) and low output impedance (approx. $1/g_m$), which makes it an ideal buffer. In comparison with CML gates, since the differential pair now drives a higher load impedance, the effect of loading is reduced, yielding increased bandwidth, faster edge rates, and higher fan-out. The cost of this improvement is the increase in power due to the bias current of the emitter followers. For example, in a 50 GHz HBT process, a CML buffer (fan-out = 1, I_{bias} = 2mA, R_{L1}/R_{L2} = 150 Ω), the propagation delay (t_D) is 14.8 ps with a risetime [20 to 80%] (t_r) of 31 ps and a falltime [20 to 80%] (t_f) of 21 ps. In comparison, an ECL buffer with level 1 outputs (fan-out = 1, I_{bias} = 2 mA, I_{bias1}/I_{bias2} = 2 mA, R_{L1}/R_{L2} = 150 Ω) has t_d = 14 ps, t_f = 9 ps, and t_r = 17 ps. With a threefold increase in P_{diss}, the impedance transformation of the EF stage results in slightly reduced gate delays and significant improvements in the rise/falltimes. With the above ECL buffer modified for level 2 (level shifted) outputs, the performance is only slightly lower with t_D = 14.2 ps, t_f = 11 ps, and t_r = 22 ps.

In general, emitter followers tend to have bandwidths approaching the f_t of the device, which is significantly higher than the differential pair. Consequently, it is possible to obtain high-speed operation with the EF biased lower than would be necessary to obtain the maximum device f_t. With the ECL level 1 buffer, if the I_{bias1}/I_{bias2} is lowered to 1 mA from 2 mA, the performance is still quite high, with $t_d = 15$ ps, $t_f = 13$ ps, and $t_r = 18$ ps. Although t_D approaches the CML case, the t_f and t_r are still significantly better.

As the EF bias is increased, its driving ability is also increased; however, at some point with high bias, the output impedance of the EF becomes increasingly inductive. When combined with large load capacitance (as in the case of high fan-out or long interconnect), it may result in severe ringing in the output that can result in excessive jitter on data edges. The addition of a series resistor between the EF output and the next stage can help to dampen the ringing by increasing the real part of the load. This change, however, increases the RC time constant, which usually results in a significant reduction in performance. In practice, changing the impedance of the EF bias source (high impedance current source or resistor bias) does not have a significant effect on the ringing. As a result, the primary method to control the ringing is through the EF bias, which places a very real constraint on bandwidth, fan-out, and jitter that needs to be considered in the topology of real designs. In some FET DCFL designs, several source followers are cascaded together to increase the input impedance and lower the output resistance between two differential pairs for high-bandwidth drive. Due to voltage headroom limits, it is very difficult to cascade two HBT emitter followers without causing the current source to enter deep saturation. In general, ECL gates are typically used for the high-speed sections due to significant improvement in rise/falltimes (bandwidth) and drive ability, although the power dissipation is higher.

ECL/CML Logic Examples

Typically, ECL and CML logic is mixed throughout high-speed GaAs/AlGaAs HBT designs. As a result, there are three available logic levels that can be used to interconnect various gates. The levels are CML (0/–400 mV), ECL1 (–1.4/–1/8 V), and ECL2 (–2.1/–2.5 V). To form more complex logic functions, typically two levels of transistors are used to steer the bias current. Figure 17.12 shows an example of an CML AND/NAND gate. For the ECL counterpart, it is only necessary to add the emitter followers. The top input is V_{inA}/V_{inA}'. The bottom input is V_{inB}/V_{inB}'. In general, the top can be driven with either the

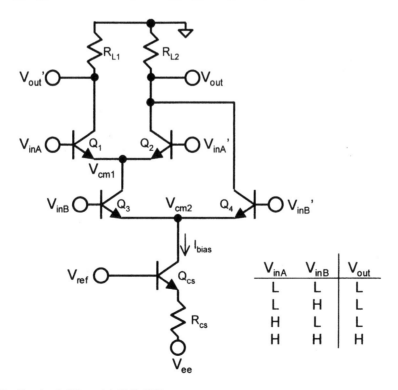

V_{inA}	V_{inB}	V_{out}
L	L	L
L	H	L
H	L	L
H	H	H

FIGURE 17.12 Two-level differential CML AND gate.

CML or the ECL1 inputs, and the bottom level can be driven by ECL1 and ECL2 levels. The choice of logic input levels is typically dictated by the design tradeoff between bandwidth, power dissipation, and fan-out. As seen in Fig. 17.12, only when V_{inA} and V_{inB} are high will I_{bias} current be steered into the load resistor that makes $V_{out} = V_{OH}$. All other combinations will make $V_{out} = V_{OL}$, as required by the AND function. Due to the differential nature, if the output terminal labels were reversed, this would be a NAND gate.

For the worst-case voltage headroom, $V_{inA}/V_{inA}{}'$ is driven with ECL1 levels, resulting in V_{cm1} of -2.8 V. With an ECL2 high on V_{inB} (-2.1 V), the lower stage (Q_3/Q_4) has a B-C forward-bias of 0.7 V, which may result in a slight saturation. This also implies that V_{cm2} is around -3.5 V, which results in an acceptable nominal 100 mV forward-bias on the current source transistor (Q_{cs}). As V_{ee} becomes less negative, the change in V_{ee} is absorbed across Q_{cs}, which places Q_{cs} closer into saturation. In saturation, the current source holds I_e in Q_{cs} constant; so if I_b increases (due to saturation), then I_c decreases. For some current source reference designs that cannot source the increased I_b, the increased loading due to saturated I_b may lower V_{ref}, which would have a global effect on the circuit bias. If the current source reference can support the increase in I_b, then the bias of only the local saturated differential pair starts to decrease leading to the potential of lower speed and lower logic swing. For HBTs, the worst-case Q_{cs} saturation occurs at low temperature, and the worst-case saturation for Q_3/Q_4 occurs at high temperature since V_{be} changes by -1.4 mV/C and $V_{diode} = -1.1$ mV/C (for constant-current bias). It is possible to decrease the forward-bias of the lower stage by using the base-emitter diode as the level shift to generate the second ECL levels; however, the power supply voltage needs to increase from -5.2 to possibly -6 V.

With the two-level issues in mind, Fig. 17.13 illustrates the topology for a two-level OR/NOR gate. This design is similar to an AND gate except that $V_{out} = V_{OL}$ if both V_{inA} and V_{inB} are low. Otherwise, Vout = V_{OH}. By using the bottom differential pair to select one of the two top differential pairs, many other prime logic functions can be implemented. In Fig. 17.14, the top pairs are wired such that $V_{out} = V_{OL}$ if $V_{inA} = V_{inB}$, forming the XOR/XNOR block. If the top differential pairs are thought of as selectable

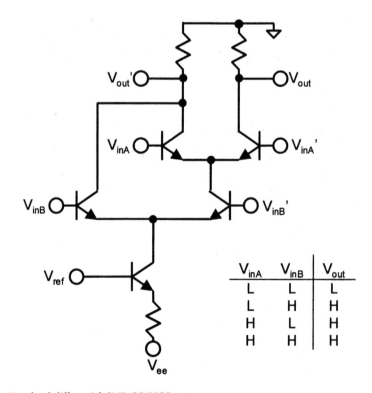

V_{inA}	V_{inB}	V_{out}
L	L	L
L	H	H
H	L	H
H	H	H

FIGURE 17.13 Two-level differential CML OR/NOR gate.

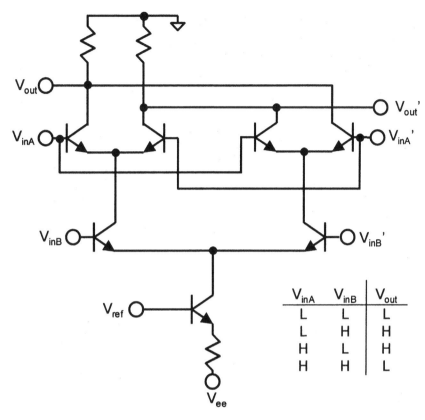

FIGURE 17.14 Two-level differential CML XOR gate.

V_{inA}	V_{inB}	V_{out}
L	L	L
L	H	H
H	L	H
H	H	L

buffers with a common output as shown in Fig. 17.15, then a basic 2:1 MUX cell is formed. Here, V_{inB}/V_{inB}' determines which input (V_{inA1}/V_{inA1}' or V_{inA2}/V_{inA2}') is selected to the output. This concept can be further extended to a 4:1 MUX if the top signals are CML and the control signals are ECL1 and ECL2, as shown in Fig. 17.16. Here, the MSB (ECL1) and LSB (ECL2) determine which of the four inputs are selected. With the 2:1 MUX in mind, if each top differential pair had separate output resistors with a common input, a 1:2 DEMUX is formed as shown in Fig. 17.17.

The last primary cell of importance is the latch. This is shown in Fig. 17.18. Here, the first differential pair is configured as a buffer. The second pair is configured as a buffer with positive feedback. The positive feedback causes any voltage difference between the input transistors to be amplified to full logic swing and that state is held as long as the bias is applied. With this in mind, as the first buffer is selected ($V_{inB} = V_{OH}$), the output is transparent to the input. As $V_{inB} = V_{OL}$, the last value stored in the buffer is held, forming a latch, which, in this case, is triggered on the falling edge of the ECL2 level. When two of these blocks are connected together in series, it forms the basic master-slave flip-flop.

Advanced ECL/CML Logic Examples

With small signal amplifiers, the cascode configuration (common base on top of a common emitter stage) typically reduces the Miller capacitance for higher bandwidth. With the top-level transistor on, the top transistor forms a cascode stage with the lower differential amplifier. This can lead to higher bandwidths and reduced rise/falltimes. However, in large-signal logic, the bottom transistor must first turn on the top cascode stage before the output can change. This added delay results in a larger propagation delay for the bottom pair vs. the top switching pair. In the case of the OR/NOR, as in Fig. 17.13, if Q_1 or Q_2 switches with Q_3 on or if Q_4 switches, the propagation delay is short. If Q_3 switches, it must first turn

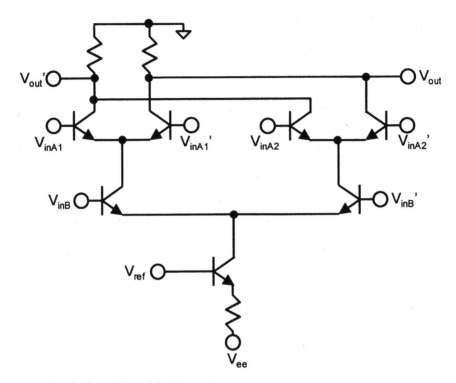

FIGURE 17.15   Two-level 2:1 differential CML MUX gate.

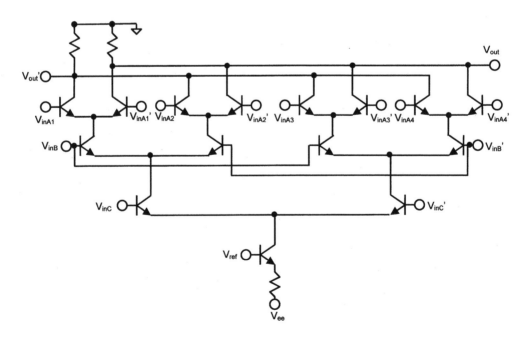

FIGURE 17.16   Three-level 4:1 differential CML MUX gate.

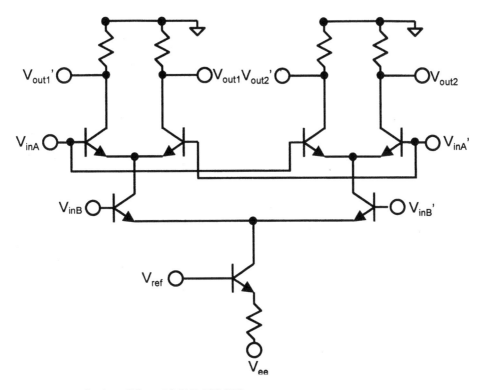

FIGURE 17.17 Two-level 1:2 differential CML DEMUX gate.

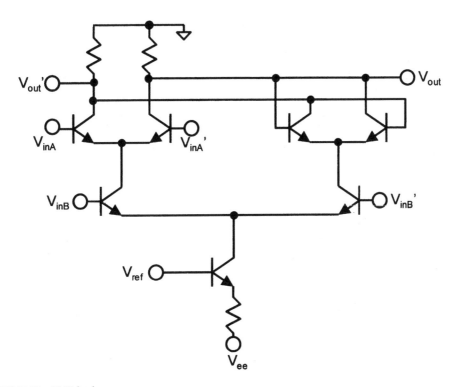

FIGURE 17.18 CML latch.

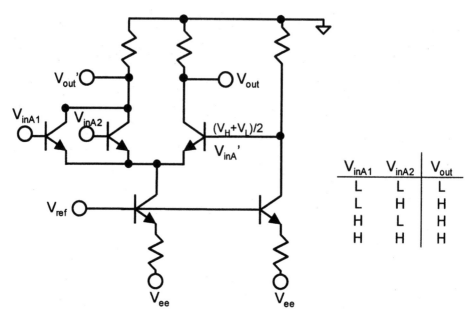

FIGURE 17.19 Single-level CML quasi-differential OR gate.

V_{inA1}	V_{inA2}	V_{out}
L	L	L
L	H	H
H	L	H
H	H	H

on either Q_1/Q_2, leading to the longest propagation delay. In this case, the longest delay limits the usable bandwidth of the AND gate. Likewise, in the XOR case, the delay of the top input is shorter than the lower input, which results in asymmetric behavior and reduced bandwidth. For a 10-GHz flip-flop, this can result in as much as a 10-ps delay from the rising edge of the clock to the sample point of the data. This issue must be taken into account in determining the optimal input data phase for lowest bit errors when dealing with digital data.

One solution to the delay issue is to use a quasi-differential signal. In Fig. 17.19, a single-ended single-level OR/NOR gate is shown. Here, a reference generator of $(V_H + V_L)/2$ is applied to V_{inA}'. If either of the V_{inA1} or V_{inA2} is high, then $V_{out} = V_{OH}$. This design has more bandwidth than the two-level topology shown in Fig. 17.12, but some of the noise margin may be sacrificed.

Figure 17.20 shows an example of a single-level XOR gate with a similar input level reference. In this case, $I_{bias1} = I_{bias2} = I_{bias3}$. The additional I_{bias3} is used to make the output symmetric. Ignoring I_{bias3}, when V_{inA} is not equal to V_{inB}, I_{bias1} and I_{bias2} are used to force $V_{out}' = V_{OL}$. When $V_{inA} = V_{inB}$, $V_{out} = V_{out}'$ since both are lowered by I_{bias}, resulting in an indeterminate state. To remedy this, I_{bias3} is added to V_{out} to make the outputs symmetric. This design results in higher speed due to the single-level design; however, the noise margin is somewhat reduced due to the quasi-differential approach and the outputs have a common-mode voltage offset of $R_L I_{bias}$.

In a standard differential pair, the output load capacitance can be broken into three parts. The base-collector capacitance of the driving pair, the interconnect capacitance, and the input capacitance of the next stage. The interconnect capacitance is on the order of 5 to 25 fF for adjacent to nearby gates. The base-collector depletion capacitance is on the order of 25 fF. Assuming that the voltage gain is 5.5, the effective C_{bc} or Miller capacitance is about 140 fF. $C_{be,j}$, when the transistor is off, is typically less than 6 fF. The $C_{be,d}$ capacitance when the transistor is on is of the order of 50 to 200 fF. These rough numbers show that the Miller effect has a significant effect on the effective load capacitance. For the switching transistor, the Miller effect increases both the effective internal C_{bc} as well as the external load. In these situations, a cascode stage may result in higher bandwidth and sharper rise/falltimes with a slight increase in propagation delay. Figure 17.21 shows a CML gate with an added cascode stage. Due to the 400-mV logic swing, the cascode bases are connected to ground. For higher swings, the cascode bases can be biased to a more negative voltage to avoid saturation. The cascode requires that the input level be either

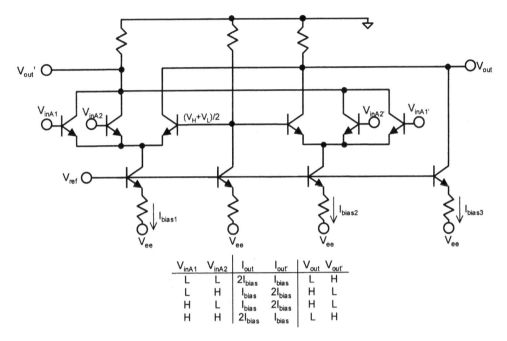

FIGURE 17.20 Single-level CML quasi-differential XOR gate.

V_{inA1}	V_{inA2}	I_{out}	$I_{out'}$	V_{out}	$V_{out'}$
L	L	$2I_{bias}$	I_{bias}	L	H
L	H	I_{bias}	$2I_{bias}$	H	L
H	L	I_{bias}	$2I_{bias}$	H	L
H	H	$2I_{bias}$	I_{bias}	L	H

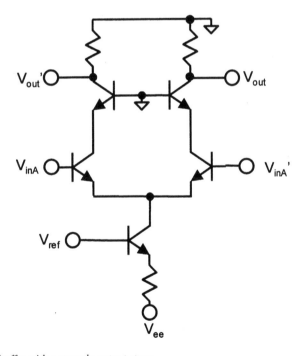

FIGURE 17.21 CML buffer with a cascode output stage.

ECL1 or ECL2 to account for the V_{be} drop of the cascode. Since the base of the cascode is held at ac ground, the Miller effect is not seen at the input of the common-base stage as the output voltage swings. From the common-emitter point of view, the collector swings only about 60 mV per decade change in I_c; thus, the Miller effect is greatly reduced. The reduction of the Miller effect through cascoding reduces

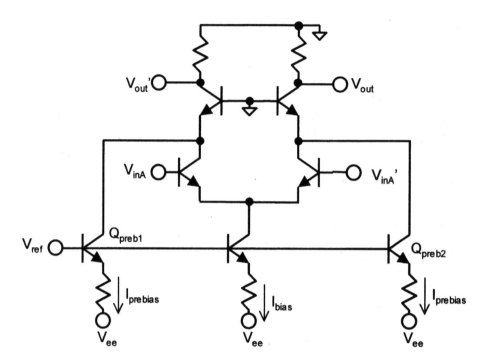

FIGURE 17.22 CML buffer with a cascode output stage and bleed current to keep the cascode "on."

the effect of both the internal transistor C_{bc} and load capacitance due to C_{bc} of the other stages, which results in the reduced rise/falltimes, especially at the logic transition region.

As both of the transistors in a cascode turn off, the increased charge stored in both transistors that has to discharge through an RC time constant may result in a slower edge rate near the logic high of a rising edge. In poorly designed cascode stages, the corner point between the fast-rising edge to the slower-rising edge may occur near the 20/80% point, canceling out some of the desired gains. Furthermore, with the emitter node of the off common-base stage floating in a high-impedance state, its actual voltage varies with time (large RC discharge compared to the switching time). This can result in some "memory" effects where the actual node voltage depends on the previous bit patterns. In this case, as the transistor turns on, the initial voltage may vary, which can result in increased jitter with digital data. With these effects in mind, the cascoded CML design can be employed with performance advantages in carefully considered situations.

One way to remedy the off cascode issues is to use prebias circuits as shown in Fig. 17.22. Here, the current sources formed with Q_{preb1} and Q_{preb2} ($I_{prebias} \ll I_{bias}$) ensures that the cascode is always slightly on by bleeding a small bias current. This results in improvements in the overall rise- and falltimes, since the cascode does not completely turn off. This circuit does, however, introduce a common-mode offset in the output that may reduce the headroom in a two-level ECL gate that it must drive. Furthermore, a series resistor can be introduced between the bleed point and the current source to decouple the current source capacitance into the high-speed node. This design requires careful consideration to the design tradeoffs involving the ratio of $I_{bias}/I_{prebias}$ as well as the potential size of the cascode transistor vs. the switch transistors for optimal performance. When properly designed, the bleed cascode can lead to significant performance advantages.

In general, high-speed HBT circuits require careful consideration and design of each high-speed node with respect to the required level of performance, allowable power dissipation, and fan-out. The primary tools the designer has to work with are device bias, device size, ECL/CML gate topology, and logic level to optimize the design. Once the tradeoff is understood, CML/ECL HBT-based circuits have formed some of the faster circuits to date. The performance and capability of HBT technology in circuit applications are summarized below.

HBT Circuit Design Examples

A traditional method to benchmark the high-speed capability of a technology is to determine the maximum switching rate of a static frequency divider. This basic building block is employed in a variety of high-speed circuits, which include frequency synthesizers, demultiplexers, and ADCs. The basic static frequency divider consists of a master/slave flip-flop where the output data of the slave flip-flop is fed back to the input data of the master flip-flop. The clock of the master and the clock of the slave flip-flop are connected together, as shown in Fig. 17.23. Due to the low transistor count and importance in many larger high-speed circuits, the frequency divider has emerged as the primary circuit used to demonstrate the high-speed potential of new technologies. As HBT started to achieve SSI capability in 1984, a frequency divider with a toggle rate of 8.5 GHz[19c] was demonstrated. During the transition from research to pilot production in 1992, a research-based AlInAs/GaInAs HBT (f_t of 130 GHz and f_{max} of 90 GHz) was able to demonstrate a 39.5 GHz divide-by 4.[20] Recently, an advanced AlInAs/GaInAs HBT technology (f_t of 164 GHz and f_{max} of 800 GHz) demonstrated a static frequency divider operating at 60 GHz.[21] This HBT ECL-based design, to date, reports the fastest results for any semiconductor technology, which illustrates the potential of HBTs and ECL/CML circuit topology for high-speed circuits.

Besides high-speed operation, production GaAs HBTs have also achieved a high degree of integration for LSI circuits. For ADCs and DACs, the turn-on voltage of the transistor (V_{be}) is determined by material constants; thus, there is significantly less threshold variation when compared to FET-based technologies. This enables the design of high-speed and accurate comparators. Furthermore, the high linearity characteristics of HBTs enable the design of wide dynamic range and high linearity sample-and-hold circuits. These paramount characteristics result in the dominance of GaAs HBTs in the super-high performance/high-speed ADCs. An 8-bit 2 gigasamples/s ADC has been fabricated with 2500 transistors. The input bandwidth is from dc to 1.5 GHz, with a spur-free dynamic range of about 48 dB.[22]

Another lucrative area for digital HBTs is in the area of high-speed circuits that are employed in fiber-optic based telecommunications systems. The essential circuit blocks (such as a 40-Gb/s 4:1 multiplexers[23]

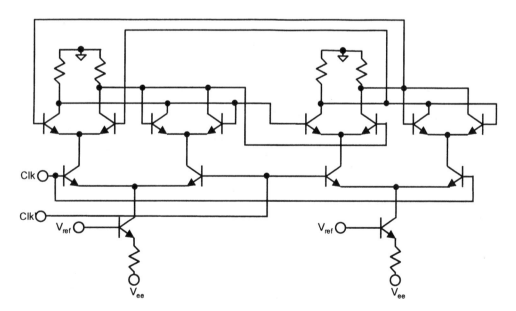

FIGURE 17.23   2:1 Frequency divider based on two CML latches (master/slave flip-flop).

and 26-GHz variable gain-limiting amplifiers[24]) have been demonstrated with HBTs in the research lab. In general, the system-level specifications (SONET) for telecommunication systems are typically very stringent compared with data communication applications at the same bit rate. The tighter specifications in telecom applications are due to the long-haul nature and the need to regenerate the data several times before the destination is reached. Today, there are many ICs that claim to be SONET-compliant at OC-48 (2.5 Gb/s) and some at OC-192 (10 Gb/s) bit rates. Since the SONET specifications apply on a system level, in truth, there are very few ICs having the performance margin over the SONET specification for use in real systems. Due to the integration level, high-speed performance, and reliability of HBTs, some of the first OC-48 (2.5 Gb/s) and OC-192 (10 Gb/s) chip sets (e.g., preamplifiers, limiting amplifiers, clock and data recovery circuits, multiplexers, demultiplexers, and laser/modulator drivers) deployed are based on GaAs HBTs. A 16×16 OC-192 crosspoint switch has been fabricated with a production 50 GHz f_t and f_{max} process.[25] The LSI capability of HBT technology is showcased with this 9000 transistor switch on a 6730×6130 μm^2 chip. The high-speed performance is illustrated with a 10 Gb/s eye diagram shown in Fig. 17.24. With less than 3.1 ps of RMS jitter (with four channels running), this is the lowest jitter 10-Gb/s switch to date. At this time, only two 16×16 OC-192 switches have been demonstrated[25,26] and both were achieved with HBTs. With a throughput of 160,000 Mb/s, these HBT parts have the largest amount of aggregate data running through it of any IC technology.

In summary, III-V HBT technology is a viable high-speed circuit technology with mature levels of integration and reliability for real-world applications. Repeatedly, research labs have demonstrated the world's fastest benchmark circuits with HBTs with ECL/CML-based circuit topology. The production line has shown that current HBTs can achieve both the integration and performance level required for high-performance analog, digital circuits, and hybrid circuits that operate in the high gigahertz range. Today, the commercial success of HBTs can be exemplified by that fact that HBT production lines ship several million HBT ICs every month and that several new HBT production lines are in the works. In the future, it is expected that advances in Si based technology will start to compete in the markets currently held by III-V technology; however, it is also expected that III-V technology will move on to address ever higher speed and performance issues to satisfy our insatiable demand for bandwidth.

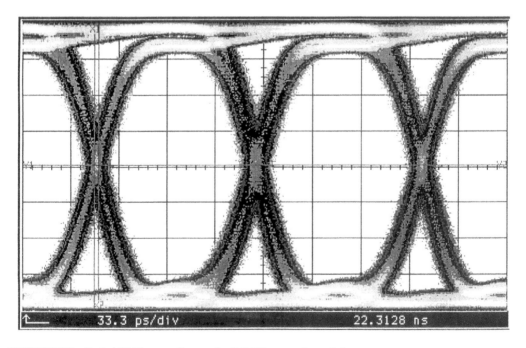

FIGURE 17.24   Typical 10 Gbps eye diagram for OC-192 crosspoint switch.

References

1. Long, S. and Butner, S., *Gallium Arsenide Digital Integrated Circuit Design,* McGraw-Hill, New York, 1990, 210.
2. Vitesse Semiconductor, *1998 Product Selection Guide,* 164, 1998.
3. Troutman, R. R., Subthreshold Design Considerations for Insulated Gate Field-Effect Transistors, *IEEE J. Solid-State Circuits,* SC-9, 55, 1974.
4. Lee, S. J. et al., Ultra-low Power, High-Speed GaAs 256 bit Static RAM, presented at *IEEE GaAs IC Symp.,* Phoenix, AZ, 1983, 74.
5. Long, S. and Butner, S., *Gallium Arsenide Digital Integrated Circuit Design,* McGraw-Hill, New York, 1990, Chap. 2.
5a. Long, S. and Butner, S., *Gallium Arsenide Digital Integrated Circuit Design,* McGraw-Hill, New York, 1990, Chap. 3.
6. Lassen, P. S., High-Speed GaAs Digital Integrated Circuits for Optical Communication Systems, Ph.D Dissertation, Tech. U. Denmark, Lyngby, Denmark, 1993.
7. Miyamoto, Y., Yoneyama, M., and Otsuji, T., 40-Gbit/s TDM Transmission Technologies Based on High-Speed ICs, presented at *IEEE GaAs IC Symp.,* Atlanta, GA, 51, 1998.
8. Otsuji, T. et al., 40 Gb/s IC's for Future Lightwave Communications Systems, *IEEE J. Solid State Circuits,* 32, 1363, 1997.
9. Otsuji, T. et al., A Super-Dynamic Flip-Flop Circuit for Broadband Applications up to 24 Gb/s Utilizing Production-Level 0.2 μm GaAs MESFETs, *IEEE J. Solid State Circuits,* 32, 1357, 1997.
10. Lang, M., Wang, Z. G., Thiede, A., Lienhart, H., Jakobus, T. et al., A Complete GaAs HEMT Single Chip Data Receiver for 40 Gbit/s Data Rates, presented at *IEEE GaAs IC Symposium,* Atlanta, GA, 55, 1998.
11. Ichioka, T., Tanaka, K., Saito, T., Nishi, S., and Akiyama, M., An Ultra-High Speed DCFL Dynamic Frequency Divider, presented at *IEEE 1989 Microwave and Millimeter-Wave Monolithic Circuits Symposium,* 61, 1989.
12. Thiede, A. et al., Digital Dynamic Frequency Dividers for Broad Band Application up to 60 GHz, presented at *IEEE GaAs IC Symposium,* San Jose, CA, 91, 1993.
13. Rocchi, M. and Gabillard, B., GaAs Digital Dynamic IC's for Applications up to 10 GHz, *IEEE J. Solid-State Circuits,* SC-18, 369, 1983.
14. Shikata, M., Tanaka, K., Inokuchi, K., Sano, Y., and Akiyama, M., An Ultra-High Speed GaAs DCFL Flip Flop – MCFF (Memory Cell type Flip Flop), presented at *IEEE GaAs IC Symp.,* Nashville, TN, 27, 1988.
15. Maeda, T., Numata, K. et al., A Novel High-Speed Low-Power Tri-state Driver Flip Flop (TD-FF) for Ultra-low Supply Voltage GaAs Heterojunction FET LSIs, presented at *IEEE GaAs IC Symp.,* San Jose, CA, 75, 1993.
16. Madden, C. J., Snook, D. R., Van Tuyl, R. L., Le, M. V., and Nguyen, L. D., A Novel 75 GHz InP HEMT Dynamic Divider, presented at *IEEE GaAs IC Symposium,* Orlando, FL, 137, 1996.
17. Maeda, T. et al., An Ultra-Low-Power Consumption High-Speed GaAs Quasi-Differential Switch Flip-Flop (QD-FF), *IEEE J. Solid-State Circuits,* 31, 1361, 1996.
17a. Ware, R., Higgins, W., O'Hearn, K., and Tiernan, M., Growth and Properties of Very Large Crystals of Semi-Insulating Gallium Arsenide, presented at *18th IEEE GaAs IC Symp.,* Orlando, FL, 54, 1996.
18. Yuan, H. T., Shih, H. D., Delaney, J., and Fuller, C., The Development of Heterojunction Integrated Injection Logic, *IEEE Trans. Elect. Dev.,* 36, 2083, 1989.
19. Johnson, R. A. et al., Comparison of Microwave Inductors Fabricated on Silicon-on-Sapphire and Bulk Silicon, *IEEE Microwave and Guided Wave Letters,* 6, 323, 1996.
19a. Ashar, K. G., The Method of Estimating Delay in Switching Circuits and the Fig. of Merit of a Switching Transistor, *IEEE Trans. Elect. Dev.,* ED-11, 497, 1964.

19b. Asbeck, P. M., Bipolar Transistors, *High-Speed Semiconductor Devices*, S. M. Sze, Ed., John Wiley & Sons, New York, 1990, Chap. 6.

19c. Matthews, J. W. and Blakeslee, A. E., Coherent Strain in Epitaxially Grown Films, *J. Crystal Growth*, 27, 118, 1974.

20. Jensen, J., Hafizi, M., Stanchina, W., Metzger, R., and Rensch, D., 39.5 GHz Static Frequency Divider Implemented in AlInAs/GaInAs HBT Technology, presented at *IEEE GaAs IC Symposium*, Miami, FL, 103, 1992.

21. Lee, Q., Mensa, D., Guthrie, J., Jaganathan, S., Mathew, T. et al., 60 GHz Static Frequency Divider in Transferred-substrate HBT Technology, presented at *IEEE International Microwave Symposium*, Anaheim, CA, 1999.

22. Nary, K. R., Nubling, R., Beccue, S., Colleran, W. T. et al., An 8-bit, 2 Gigasample per Second Analog to Digital Converter, presented at *17th Annual IEEE GaAs IC Symposium*, San Diego, CA, 303, 1995.

23. Runge, K., Pierson, R. L., Zampardi, P. J., Thomas, P. B., Yu, J. et al., 40 Gbit/s AlGaAs/GaAs HBT 4:1 Multiplexer IC, *Electronics Letters*, 31, 876, 1995.

24. Yu, R., Beccue, S., Zampardi, P., Pierson, R., Petersen, A. et al., A Packaged Broadband Monolithic Variable Gain Amplifier Implemented in AlGaAs/GaAs HBT Technology, presented at *17th Annual IEEE GaAs IC Symposium*, San Diego, CA, 197, 1995.

25. Metzger, A. G., Chang, C. E., Campana, A. D., Pedrotti, K. D., Price, A. et al., A 10 Gb/s High Isolation 16×16 Crosspoint Switch, Implemented with AlGaAs/GaAs HBT's, to be published, 1999.

26. Lowe, K., A GaAs HBT 16×16 10 Gb/s/channel Cross-Point Switch, *IEEE J. Solid-State Circuits*, 32, 1263, 1997.

Index